Wildlife
Responses
to
Climate Change

Wildlife Responses *to* Climate Change

NORTH AMERICAN CASE STUDIES

Edited by
Stephen H. Schneider
and Terry L. Root

ISLAND PRESS
Washington • Covelo • London

ISBN 1-55963-924-5 (hardcover)
ISBN 1-55963-925-3 (paperback)

Cataloging-in-Publication Data is available from the Library of Congress
and the British Library.

Printed on recycled, acid-free paper

Manufactured in the United States of America
10 9 8 7 6 5 4 3 2 1

Contents

Foreword

It was not long ago that the big question about climate change was whether or not it actually was taking place. Now, there is broad scientific consensus that it is, in fact, happening, and that human activities are largely to blame.

The most recent body of scientific evidence sounds a warning that global climate change could occur more quickly than previously thought, and take an even greater toll on people and wildlife. We are beginning to see convincing evidence that climate trends are already affecting the Earth's natural systems: butterflies are moving north, birds are breeding earlier, and coral reefs are bleaching across the globe.

While the challenge of global climate change is daunting, the first step toward creating the solution is understanding the problem. The National Wildlife Federation is committed to providing and supporting the sound science necessary to build that understanding and to shape the solutions that will mitigate an environmental problem that is global in scale.

This book is the culmination of a three-year project to research and study the impacts of global climate change on U.S. ecosystems and individual wildlife species. In 1997, we provided fellowships to eight outstanding graduate students to conduct research on global climate change. NWF engaged Steven Schneider, Ph.D., and Terry Root, Ph.D., two of the world's leading scientists studying the effects of climate change on wildlife, to shepherd the students through their studies. This work, combined with the generous support of a Dorothy Chancellor Estate grant, enables NWF to add to the body of knowledge critical to scientists, resource managers, and policymakers in shaping a solution to the climate change crisis.

The book has another important function too. By detailing the potentially devastating effects of climate change on wildlife and their habitat, we hope to stir the concern and the action of countless people who care deeply about the fate of our natural world. Such people have previously rallied America behind historic efforts to clean up our air and water and to save imperiled species. Addressing the crisis of global climate change will require an even greater commitment to action and, in keeping with its long and distinguished history, the National Wildlife Federation will help to provide the knowledge, the understanding, and the tools to make it possible. Nothing less than the future of both people and wildlife depends on it.

—MARK VAN PUTTEN
President and CEO
National Wildlife Federation

Introduction: The Rationale for the National Wildlife Federation Cohort of Young Scientists Studying Wildlife Responses to Climate Change

STEPHEN H. SCHNEIDER AND TERRY L. ROOT

The project that has culminated in this book has two main purposes. The first is to establish a credible scientific link between the health of natural systems (wildlife in particular) and human-induced climate change. The second is to help create a community of young scholars who can demonstrate that interdisciplinary science and outreach activities connecting wildlife and climate change disciplines can be accomplished with a high level of scientific quality. Often, academia discourages scientists from engaging in research outside traditional disciplinary boundaries or research with strong policy components. Implicitly, and occasionally explicitly, scientists are also discouraged from providing information to the public, particularly when the subject is plagued by many inherent uncertainties, as is the case of wildlife and climate change. The disincentives are often overwhelming. We expect this group of young scientists will join the nucleus of a growing community of scholars that help provide a positive example of the rewards of pursuing interdisciplinary research and outreach.

In 1997 the National Wildlife Federation (NWF) asked us to

comment on their idea that the time was ripe to create a cohort of young scholars interested in branching out from traditional ecological inquiries (concerning the structure or functioning of particular parts of ecosystems) to address fundamental scientific issues concerning the threats to wildlife conservation imposed by the prospect of human-induced climate change. Furthermore, the NWF identified a need to train young scholars to produce scientifically credible writings that are understandable to a broad audience and to help sponsor the publication of their respective works as a collection. Moreover, we strongly agree that science should not be isolated from important questions in biological conservation, in general, and about wildlife conservation, in particular. We responded enthusiastically to their inquiry, but warned that traditional research paradigms in ecology (or climatology) tend to encourage narrowly defined laboratory, field, or theoretical problems, especially for graduate student dissertations or postdoctoral projects, which was the level of scholars the NWF had targeted.

We fundamentally agree that public acceptance of policies to promote conservation activities requires understanding of the reasons for actions by the public and its elected officials, which in turn requires literacy in the issues and knowledge of the trade-offs involved. Thus, we share with NWF a belief in the value of encouraging and training young scientists in the art of clear, jargon-free communication of scientific research· and the assessment of its implications, and that such communication is an important skill.

Not all practicing scientists, however, agree that public outreach—let alone advocacy—is an appropriate activity for scientists (e.g., see the reported debate in *Science* [Kaiser 2000]). Young scientists are often discouraged from such pursuits during formative stages in their careers. Nevertheless, we believed that a cohesive group of young scientists could show by example that it is possible to do high quality scientific research on multiscale and multidisciplinary questions that attend ecological impact analyses of climate change. In particular, a single, peer-reviewed volume that publishes the activities of such a cohort could serve as an example and help reduce skepticism. Such a volume could demonstrate to hesitant, more senior colleagues that working on important conservation issues across disciplinary boundaries and across scales of analysis can produce first-rate science while sending a message to other young scientists that such work is both rewarding and respectable.

To help achieve these goals, the NWF solicited proposals from junior scientists for funding. From a competitive applicant pool, we and a few other scientists selected the initial cohort of "NWF Fellows." This volume summarizes the research efforts of the fellows, along with a chapter from another young scientist who joined the cohort at a later date, although she did not receive funding. In addition, the NWF anticipates that the fellows will work with NWF staff to produce more accessible accounts of their works that will be available to government officials and the public. These accounts will provide in lay terms scientifically grounded knowledge that has policy implications for wildlife conservation.

To establish an esprit de corps among the cohort and to discuss common concerns about career aspects of interdisciplinary research, conservation focus, and outreach activities, we held a symposium at the Ecological Society of America's annual meeting in Spokane, Washington, in August 1999. We were highly gratified at the large turnout of ESA members to hear the preliminary contributions of this cohort. Moreover, we led extensive dialogues with the young scientists on issues of weighing fundamental scientific curiosity with problem-driven activities and balancing scholarly work with public outreach. We also discussed the responses of the first round of peer reviews on each participant's chapter and discussed strategies for revisions.

The success of creating a cohort of young scientists to work on the interdisciplinary issues of studying the effects of climate change on wildlife (or wildlife-dependent vegetation systems) is quite promising. This volume represents the completed chapters of the NWF cohort after two rounds of external, scientific peer review as well as exchanges among the fellows themselves on each other's work. We believe the volume represents high quality science, as well as a positive model for younger scientists also wanting to work on tough problems of conservation in the face of global change disturbances. In addition, this cohort and this volume provide an example to more senior scientists that such exemplary activity can be done by our best young scientists. In fact, our observation from working with this group is that many young scientists insist that they not be distanced from pursuit of important scientific problems inherent in the conservation of wildlife, nor that they be discouraged from the public dissemination of such work. Yet, such work must and is being done within the traditional scientific culture of peer review—and with appropriate care to present a balanced account of the science and implication of their works.

The fellows were not chosen deliberately to have their work span both terrestrial and marine locations, nor to have representative examples of climatic impacts on insects, mammals, and vegetation important to wildlife conservation; nevertheless, their research spans such broad topics. For example, the chapters by Sagarin and Sanford study marine responses to recent and projected climatic changes and demonstrate that a discernible impact of recent regional climatic changes is already being felt by marine animals. In addition, chapters by Koteen and Saavedra address climatic effects on plants that are critical to animals—bears in Yellowstone Park in one case and alpine meadow communities subjected to a deliberate warming experiment in another. Hellmann and Crozier's chapters on butterflies demonstrate that ecological theory suggests these animals should respond to climatic changes; one of these chapters demonstrates that recent data indicate that range shifts are already occurring for recent climatic variations. Zavaleta and Royval argue that conservation efforts must be cognizant of the multiple stresses imposed by human disturbances and show that exotic invasive species of plants in the United States, combined with climatic changes, will likely pose a synergistic combination of disturbances that will confront native species—especially endangered ones— in the twenty-first century. Finally, Shevliakova addresses the potential impacts of climate change on the geographic distribution of vegetation in the United States—by using a probabilistic model of the likelihood various vegetation types would occur as climate changes. Such alterations in the geographic patterns and likelihood of various vegetation types could reverberate significantly through animal communities.

Reflecting this point, our opening chapter sketches out the nature of the climate change debate, the need for cross-scale and cross-disciplinary research, and addresses some ecological implications of climate change. We also report briefly on our recent work in which a meta-analysis of more than 2000 studies shows highly significant associations between recent observed regional climate changes and response by environmental systems like lake ice, flowering dates, and wildlife behavior. The most consistent explanation for this set of associations is that there is indeed already a discernible impact of global climate change on ecosystems.

We are pleased with the breadth and depth of this set of chapters, and we believe this volume helps to launch this cohort of young scholars into careers that will be both scientifically credible and influential on the conservation efforts so critically required to protect ecosystems in the twenty-first century.

We are very gratified to have been asked by the NWF to guide this effort and to edit this volume. We also provide, as mentioned, an opening chapter that summarizes some of our own work in the area of climate change and its potential impacts on ecosystems, both to lay out the context for the specific contributions of the cohort that follow and to lend our weight to the growing movement of environmental scientists who believe that excellent science and work on important conservation projects and their communication to a wider public are all fundamental components of a scientist's job. Moreover, we believe that coupling submodels of climate to those of ecology (and even social systems) could well give rise to emergent properties that would not be discovered by disciplinary analyses alone, and that the search for emergent behavior of coupled physical, biological, and social systems will define a major thrust of basic environmental science research in the twenty-first century. That such systems science is also needed for conservation efforts is an added benefit in our view. We look forward to working with the young scientists in this cohort many times in the future, as we are certain they will become leaders in both science and conservation as they progress. We also eagerly anticipate getting to know other young scholars who, thanks to the efforts of these NWF Fellows, will feel a bit freer to follow in the fellows' footsteps in the years ahead.

Finally, we thank Patty Glick of the NWF and Camille Parmesan of the University of Texas for their extra efforts to work with these young scientists in their development as scholars, communicators, and concerned conservationists.

Literature Cited

Kaiser, J. 2000. Ecologists on a mission to save the world. *Science* 287: 1188–1192.

Wildlife
Responses
to
Climate Change

Climate Change: Overview and Implications for Wildlife

TERRY L. ROOT AND STEPHEN H. SCHNEIDER

Synergisms of Climate Change and Ecology

The Earth's climate is vastly different now from what it was 100 million years ago when dinosaurs roamed the planet and tropical plants thrived closer to the poles. It is different from what it was only 20,000 years ago when ice sheets covered much of the Northern Hemisphere. Although the Earth's climate will surely continue to change, climatic changes in the distant past were driven by natural causes, such as variations in the Earth's orbit or the carbon dioxide (CO_2) content of the atmosphere. Future climatic changes, however, will probably have another source as well—human activities. Humans cannot directly rival the power of natural forces driving the climate—for example, the immense energy input to the Earth from the sun that powers the climate. We can, however, indirectly alter the natural flows of energy enough to create significant climatic changes.

The best-known way people could inadvertently modify climate is by enhancing the natural capacity of the atmosphere to trap radiant heat near the Earth's surface—the so-called greenhouse effect. This natural phenomenon allows solar energy that reaches the Earth's surface to warm the climate. Gases in the atmosphere such

1

as water vapor and CO_2, however, trap a large fraction of long-wavelength radiant energy, called terrestrial infrared radiation, near the Earth's surface. This causes the natural greenhouse effect to be responsible for some 33°C (60°F) of surface warming. Thus, seemingly small, human-induced changes to the natural greenhouse gases are typically projected to result in a global warming of about 1.5°C to 6°C in the 21st century (IPCC 1990, 1996a, 2001a—the latter reference suggesting the upper range limit of nearly 6°C warming by 2100). This range, especially if beyond 2–3°C (IPCC 2001b), would likely result in ecologically significant changes, which are why climatic considerations are fundamental in the discussion of possible ecological consequences that may involve wildlife.

We may already be feeling the climatic effects of having disturbed the atmosphere with gases such as CO_2. Many activities associated with human economic development have changed our physical and chemical environment in ways that modify natural resources. When these changes—such as burning fossil fuels that release CO_2 or using land for agriculture or urbanization and thereby causing deforestation—become large enough, significant global (worldwide) changes are expected. Such modifications can disturb the natural flows of energy in Earth systems and thus can force climatic changes. These disturbances are also known as global change forcings (e.g., melted sea ice). We need quantitative evaluations of the potential for human activities to effect physical ecological changes around the globe. Such evaluations are central to potential policy responses to mitigate such global changes (Schneider 1990; IPCC 1996b,c; IPCC 2001b,c).

Synergisms

The synergistic, or combined, effects of habitat fragmentation and climate change represent one of the most potentially serious global change problems. People destroy or divide natural habitats for farmland, settlements, mines, or other developmental activities. Changes in climate will force individual species of plants and animals to adjust, if they can, as they have in the past. During the Ice Age many species survived by migrating to appropriate habitats. Today such migrations would be much more difficult because they would entail migration across freeways, agricultural zones, industrial parks, military bases, and cities of the 21st century. An even further complication arises with the imposition of the direct effects of changes in CO_2, which can change terrestrial and marine primary productivity as well as alter the competitive relations among photosynthesizing organisms.

The Kirtland's warbler in northern Michigan provides one example of syngergism. The species is restricted to a narrow area of jack pines that grow in sandy soil (Botkin et al. 1991). Forest gap models of growth and decline of jack pines indicate that this tree will move north with warming, but the warbler is not likely to survive the transition. This bird nests on the ground under relatively young pines, and the soil to the north is not generally sandy enough to allow sufficient drainage for successful fledging of young (Cohn 1989). Consequently, global warming could well doom the warbler to extinction in 30 to 60 years. This potential for extinction indicates how the already high rate of extinctions around the world could be exacerbated by climatic changes occurring more rapidly than species can adapt (see Pimm 1991, Peters and Lovejoy 1992, Wilson 1992).

The synergism question raises management problems of anticipating and responding to global change risk. For example, one controversial management plan would be to set up interconnected nature reserves that run north to south or from lower to higher elevation, which could reduce the likelihood of some species being driven to extinction in the event of climate changes. Alternatively, we could simply let the remnants of relatively immobile wildlife and natural plant communities remain in existing isolated reserves and parks, which could lead to some extirpations. If we do opt for more environmental safeguards by interconnecting our parks, the question then becomes how we interconnect the nature reserves. Priorities must be set and money made available for constructing natural corridors through which species can travel. For example, elevated sections of highways may be needed to allow for migration routes, similar to what was done for the caribou in the Arctic when the Alaskan pipeline was built. To examine such questions in scientific and economic detail, a multidisciplinary examination of various aspects of climatology, economics, and ecology is needed. Here we begin with a background discussion of climatic history, processes, modeling, and validation as a prelude to focusing on ecological processes, which need to be examined in order to project possible synergisms among ecology and climate change.

Climate History: What Has Happened

Scientists can reconstruct the cyclical expansion and contraction of polar caps and other ice masses from ice core samples taken from Greenland and Antarctica. When snow falls on high, cold glaciers the air trapped between snow grains is eventually transformed into air bubbles as the snow is compressed into ice from the weight of subsequent accumulations. The ratio of two oxygen molecules with different molecular weights (O^{16} and O^{18} isotopes) is a proxy record for the temperature conditions that existed when the snow was deposited. By studying these air bubbles and ice, scientists have been able to determine that the ice buildup from 90,000 years ago to 20,000 years ago was quite variable and was followed by a (geologically speaking) fairly rapid 10,000-year transition to the (current) climatically very stable Holocene period. The Holocene is the 10,000-year interglacial period in which human civilization developed and modern plant and animal distributions shifted into their current states (Eddy and Oeschger 1993). These ice cores also provide information on the presence of CO_2, an important greenhouse-effect gas. Carbon dioxide was in much lower concentrations during cold periods than in interglacials (which is similar for the greenhouse gas methane, CH_4). This implies an amplifying effect, or a positive feedback, because less of these gases during glacials means less infrared radiative heat trapped in the atmosphere—thus amplifying the cooling—and vice versa during interglacials. The ice cores also show that concentrations of CO_2 and CH_4 and temperature were remarkably constant for about the past 10,000 years (before A.D. 1700), particularly when compared with the longer record. That relative constancy in chemical composition of the greenhouse gases held until the industrial age during the last few centuries.

The transition from extensive glaciation of the Ice Age to the more hospitable landscapes of the Holocene took from 5000 to 10,000 years, during which time the average global temperature increased 5–7°C and the sea level rose some 100 meters. Thus we estimate that natural rates of warming on a sustained global basis are about 0.5°C to 1°C per thousand years. In addition to the sustained rates, there is growing evidence of rapid so-called abrupt nonlinear changes as well (Severinghaus and Brook 1999). Both the slower and more rapid changes were large enough to have radically influenced where species live and to have potentially contributed to the well-known extinctions of large animals (e.g., woolly mammoths, sabertooth cats, and enormous salamanders).

A large interdisciplinary team of scientists, including ecologists, palynologists (scientists who study pollen), paleontologists (scientists who study prehistoric life, especially fossils), climatologists, and geologists, formed a research consortium (Cooperative Holocene Mapping Project 1988, Wright et al. 1993) to study the dramatic ecological changes accompanying the transition from Ice Age to the recent interglacial period. One group of these researchers used a variety of proxy indicators to reconstruct vegetation patterns over the past 18,000 years for a significant fraction of the Earth's land areas. In particular, cores of fossil pollen from dozens of sites around North America clearly showed how boreal, coniferous tree pollen, now the dominant pollen type in the boreal zone in central Canada, was a prime pollen type during the last Ice Age (15,000–20,000 years ago) in what are now the mixed hardwood and Corn Belt regions of the United States. During the last Ice Age, most of Canada was under ice; pollen cores indicate that as the ice receded, boreal trees moved northward, "chasing" the ice cap. One interpretation of this information was that biological communities moved intact with a changing climate. In fact, Darwin (1859) asserted as much:

> As the arctic forms moved first southward and afterward backward to the north, in unison with the changing climate, they will not have been exposed during their long migrations to any great diversity of temperature; and as they all migrated in a body together, their mutual relations will not have been much disturbed. Hence, in accordance with the principles inculcated in this volume, these forms will not have been liable to much modification.

If this were true, the principal ecological concern over the prospect of future climate change would be that human land-use patterns might block what had previously been the free-ranging movement of natural communities in response to climate change. The Cooperative Holocene Mapping Project, however, investigated multiple pollen types, including not only boreal species but also herbs and more arid (xeric) species, as well as oaks and other mesic species.

They discovered that during the transition from the last Ice Age to the present interglacial, nearly all species moved north, as expected. During a significant portion of the transition period, however, the distribution and combinations of pollen types provided no analogous associations to today's vegetation communities (Overpeck et al. 1992). That is, whereas all species moved, they moved at different rates and directions, not as groups. Consequently, the groupings of species during the transition period were often dissimilar to those present today. The relevance of this is that, in the future, ecosystems will not necessarily move as a unit as climate changes (even if there were time and space enough for such a migration).

Past vegetation responses to climate change at a sustained average rate of 1°C per millennium indicates that credible predictions of future vegetation changes cannot neglect transient (i.e., time-evolving) dynamics of the ecological system. Furthermore, because the forecasted global average rate of temperature increase over the next century or two exceeds those typical of the sustained average rates experienced during the last 120,000 years, it is unlikely that paleoclimatic conditions reconstructed from millennial time scale conditions would be near analogs for a rapidly changing anthropogenically warmed world (e.g., see Schneider and Root 1998, from which some of this chapter is adapted). Future climates may not only be quite different from more recent past climates, they may also be quite different from those inferred from paleoclimatic data and from those to which some existing species are evolutionarily adapted. Therefore, past changes do provide a backdrop or context to gauge future changes, but not primarily as a spatial analog; rather as means to verify the behavior of models of climate or ecosystem dynamics that are then used to project the future conditions given the rapid time-evolving patterns of anthropogenic forcing (Crowley 1993, Schneider 1993a).

Forces of Climate Change

Causes of climate change are broadly categorized as external and internal. These terms, however, are defined relative to the focus of study; stating which components are external or internal to the climatic system depends on the time period and spatial scale being examined, as well as on the phenomena being considered. External causes of climate change do not have to be physically external to the Earth (such as the sun), but do occur outside of the climate system under examination. If our focus is on atmospheric change on a one-week time scale (i.e., the weather), the oceans, land surfaces, biota, and human activities that produce CO_2 are all external (i.e., they are not influenced much by the atmosphere in such a short time). If our focus is on 100,000-year ice age interglacial cycles, however, the oceans, ice sheets, and biota are all part of the internal climatic system and vary as an integral part of the Earth's environmental systems. On this longer scale we must also include as part of our internal system the "solid" Earth, which really is not solid but viscous and elastic.

Fluctuations in heat radiated by the sun—perhaps related to varying sunspots—are external to the climate system. Influences of the gravitational tugs of other planets on the Earth's orbit are also external. Human-caused changes in the Earth's climate could not perceptibly alter either one of these cycles. Carbon dioxide and methane levels rise and fall with ice age cycles, meaning they are certainly internal on a 10,000-year time scale. But on a 20-year scale these greenhouse gases become largely an external cause of climate change because small changes in climate have little feedback effect on, for example, humans burning fossil fuels or clearing land.

Changes in the character of the land surface, such as those caused by human activities, are largely external. If vegetation cover changes because of climate change, however, land surface change then becomes internal because changes in plant cover can influence the climate by changing greenhouse gas concentrations, albedo (reflectivity to sunlight), evapotranspiration, surface roughness, and relative humidity (Henderson-Sellers et al. 1993).

Snow and ice are important factors in climate change because they have higher albedo than warmer surfaces and, in the instance of sea ice, can inhibit transfer of heat and moisture between air and wet surfaces. Salinity, which affects changes in both sea ice and the density of seawater (which helps control where ocean waters sink), may also be an internal cause of climatic variation. The sinking and upwelling of ocean waters are biologically significant because the upwelling waters are often nutrient-rich.

Unusual patterns of ocean surface temperature—such as the El Niño—demonstrate the importance of internally caused climatic fluctuations because the atmospheric circulation can change simultaneously with ocean surface temperatures. When the atmosphere rubs on the ocean, the ocean responds by modifying its motions and temperature pattern, which forces the atmosphere to adjust, which changes the winds, which changes the way the atmosphere rubs on the ocean, and so forth (Trenberth 1993). As a result, air and water interact internally in this coupled system like blobs of gelatin of different size and stiffness, connected by elastic bands or springs, all interacting with one another while also being pushed from the outside (by solar, volcanic, or human-caused change).

Climate Change Projections: What May Happen?

To predict the ecologically significant ways the climate might change, one must specify what people do that modifies how energy is exchanged among the atmosphere, land surface, and space, because such energy flows are the driving forces behind climate. Air pollution is an example of such a so-called societal forcing of the climate system. Estimating societal forcing involves forecasting a plausible set of human activities affecting pollution or land use over the next century. The next step is to estimate the response of the various components of the Earth system to such forcings.

The Earth system itself consists of the following interacting subcomponents: atmosphere, oceans, cryosphere (snow, seasonal ice, and glaciers), and land-surface (biota and soils) systems.

Research in the field and in laboratories provides an understanding about various processes affecting the subcomponents of the Earth system. This understanding allows a modeling of the behavior of particular components of the Earth system. In practice, models of the atmosphere are connected to models of the oceans, ice, biota, and land surfaces to simulate the consequences of some scenario of societal forcing on climate and ecosystems. Controversy arises because both the societal forcing that will actually occur and the scientific knowledge of each subsystem are still incomplete (e.g., Schneider 2001). Because models are not perfect replicas of the actual natural system, they must be tested against the expanding base of field and laboratory data. This not only allows assessment of the credibility of current simulations, it also reveals improvements for the next generation of models.

Elements of Global Warming Forecasts

The societal driving forces behind global-warming scenarios are projections of population, consumption, land use, and technology. Typical 21st century projections both for human population size and affluence show drastic increases when aggregated over less highly developed countries and more highly developed countries. When these factors are multiplied by the amount of energy used to produce a unit of economic product (the so-called energy intensity) and the amount of CO_2 emitted per unit of energy (the technology factor called carbon intensity), carbon emissions are typically predicted to rise several-fold over the next 100 years. Making such projections credibly is difficult. Therefore analysts disagree by as much as a factor of 10 about how much CO_2 will be emitted by 2100 (Johansson et al. 1993, IPCC 1996c, Nakicenovic and Swart 2000). Specific scenarios are debatable because the amount of carbon emitted through human activities will significantly depend on social-structural projections, such as what kinds of energy systems will be developed and deployed globally and what the standards of living will be over the next several decades, not to mention population growth.

To turn estimates of CO_2 emissions into estimates of CO_2 concentrations in the atmosphere, which is the variable needed to calculate potential climate changes, one must estimate what fraction of CO_2 emitted will remain in the atmosphere. This airborne fraction was estimated during the last few decades of the 20th century at about 50%, because the amount of CO_2 buildup in the atmosphere each year (about 3 billion tons of carbon as CO_2) was about half the fossil fuel–injected CO_2. The atmospheric concentration of CO_2 should, however, be computed by using carbon cycle models, which account for the time-evolving amounts of carbon in vegetation, soils, and oceanic and atmospheric subcomponents (IPCC 1996a,b and 2001a). The estimated CO_2 concentration can then be fed into computerized climatic models to estimate its effects on climate (e.g., Wigley and Schimel 2000).

Climate prediction, like most other forecasts involving complex systems, generally involves subjective judgments (e.g., see Moss and Schneider 2000). Those attempting to determine the future behavior of the climate system from knowledge of its past behavior and present state basically can take two approaches. One approach, the empirical-statistical, uses statistical methods such as regression equations that connect past and present observations statistically to obtain the most probable extrapolation. The second approach, usually called climate modeling, focuses on first principles, which are equations representing laws believed to describe the physical, chemical, and biological processes governing climate. Because the statistical approach depends on historical data, it is obviously limited to predicting climates that have been observed or are caused by processes appropriately represented in the past conditions. The statistical method cannot reliably answer questions such as what would happen if atmospheric CO_2 increased at rates much faster than in the known past. Thus the more promising approach to climate prediction for conditions or forcings different from the historic or ancient past is climate modeling. A significant component of empirical-statistical information, though, is often embedded in these models (Washington and Parkinson 1986, Root and Schneider 1995). This often makes modelers uncomfortable about the validity of predictions of such models on unusual or unprecedented situations unless a great deal of effort is expended to test the models against present and paleoclimatic baseline data.

Climate models vary in their spatial resolution, that is, the number of dimensions they simulate and the spatial detail they include. The simplest model calculates only the average temperature of the Earth, independent of the average greenhouse properties of the atmosphere. Such a model is called zero-dimensional—it reduces the real temperature distribution on the Earth to a single point, a global average. In contrast, three-dimensional climate models produce the variation of temperature with latitude, longitude, and altitude. The most complex atmospheric models, the general circulation models (GCMs), predict the time evolution of temperature plus humidity, wind, soil moisture, sea ice, and other variables through three dimensions in space (Washington and Parkinson 1986).

Verifying Climate Forecasts

The most perplexing question about climate models is whether they can be trusted as a reliable basis for altering social policies, such as those governing CO_2 emissions or the shape and location of wildlife reserves. Even though these models are fraught with uncertainties, several methods are available for verification tests. Although no method is sufficient by itself, several methods together can provide significant, albeit circumstantial, evidence of a forecast's credibility.

The first validation-testing method involves checking the model's ability to simulate the present climate. The seasonal cycle is one good test because temperature changes in a seasonal cycle are larger on a hemispheric average than the change from an ice age to an interglacial period (that is, 15°C seasonal range in the Northern Hemisphere versus 5–7°C glacial/interglacial cycle). General circulation models map the seasonal cycle well. This supports the scientific consensus about the plausibility of global warming of several degrees in the 21st century. The seasonal test, however, does not indicate how well a model simulates slow processes such as changes in deep ocean circulation, ice cover, forests, or soil carbon storage, which may have significant effects on the decade- to century-long time scales over which atmospheric CO_2 is expected to double.

A second verification technique involves isolating individual physical components of the model and testing them against actual data. A model should reproduce the flow of thermal energy among the atmosphere, the surface, and space with no more than about a 10 to 20% error. Together, these energy flows make up the well-established natural greenhouse effect on Earth and constitute a formidable and necessary test for all models. A model's performance in simulating these energy flows is an example of physical validation of model components.

A third validation method involves a model's ability to reproduce the diverse climates of the past. This method is aided by recording instrumental observations made during the past few centuries and paleorecords that serve as a proxy for climatic conditions of the ancient Earth. This method may even include testing the model's ability to simulate climates of other planets (Kasting et al. 1988). Paleoclimatic simulations of the Mesozoic (age of the dinosaurs), glacial–interglacial cycles, or other extreme past climates help scientists understand the coevolution of the Earth's climate and living things (Schneider and Londer 1984). As verification tests of climate models, they are also crucial to predicting future climates and changes in biological systems.

Using these techniques, much has been learned from examining the global climatic trends of the past century. The years 1997 and 1998 were the warmest years on record for the Earth's surface in the past century; at the same time the stratosphere was at its coldest (IPCC 1996a and 2001a). These data are consistent with an enhanced greenhouse-effect signal that might be anticipated from the greenhouse-gas injections over the past 150 years, which saw a 30% increase in CO_2, a 150% increase in CH_4, and the introduction of human-generated heat-trapping chemicals such as chlorofluorocarbons and halons. Industrial activities since the 1950s have contributed to the increase of sulfur dioxide and other aerosol particles into the atmosphere, the net effect of which is likely to reduce surface temperature by reflecting sunlight back to space. This is complicated by the fact that light hazes like sulfur oxides reflect much more radiation than they absorb but dark particles like soot produced in fires or diesel engines can warm the climate by absorbing more energy than they reflect. The IPCC (2001a) estimates the net effect to be a cooling, with a large range of uncertainty. Although such cooling effects may have counteracted global warming by only several tenths of a degree, the hazes occur regionally and could be producing ecologically significant, unexpected regional changes in climate patterns (Schneider 1994).

Although the 0.6°C ± 0.2°C surface warming in the 20th century is consistent with the human-induced greenhouse gas buildup, some have argued that this warming could possibly be largely natural—either a natural internal fluctuation of the system or driven by natural forcing like a change in the energy output of the sun. However, the IPCC (2001a) assessment states on page 11 of the Summary for Policymakers that there is too much consistent evidence of a human influence to assign all the 20th century warming to natural causes: "There is new and stronger evidence that most of the warming observed over the last 50 years is attributable to human activities" (for estimates of the subjective probability of human induced global warming amounts given a doubling of CO_2, see Morgan and Keith, 1995). However, if one argues that there could have been a natural warming trend in the 20th century, then by symmetry it is also not possible to rule out the converse that, independent of the enhanced greenhouse effect due to human activity, there was a natural cooling fluctuation taking place during the 20th century. If so, the world would then have warmed up much more than observed had we not had such a fortuitous natural cooling trend. One could even speculate that the dramatic temperature rise since the 1970s with global high temperature records reflects the termination of a natural cooling trend combined with the rapid establishment of the expected enhanced greenhouse effect. We are certainly not suggesting this to be the most probable case, simply showing why a fairly wide range of possibilities is plausible given the remaining uncertainties.

Santer et al. (1996) suggest that when aerosols and greenhouse gas forcings are combined, climate models more closely match 30 years of observations. Nevertheless, wide ranges of climate sensitivities—from as low as a 0.5°C warming to well above a 5.0°C warming (e.g., Wigley and Raper 2001, Andronova and Schlesinger 2000)—are still consistent with current observations. Several reasons exist for such a wide range of uncertainty: difficulty in knowing how to model delays in global warming because of the large heat capacity of the oceans; not knowing what other global-change forcings may have opposed warming (e.g., sulfate aerosols from burning high-sulfur coal and oil or undetectable changes in the sun's light output before 1980); and large, unknown, internal, natural climatic fluctuations. As mentioned previously, though, the ecologically important forecasts of time-evolving regional climatic changes are much less credible than global-average projections and require that ecologists use many alternative scenarios of possible climatic changes. IPCC (2001a) suggests that by the end of the 21st century, there will likely be overall warming between about 1.5 and 6°C; land will warm more than oceans, and higher latitudes more than tropical latitudes; precipitation will increase on average; midlatitude, midcontinental drying in summer is likely in some areas, and the hydrological cycle will intensify thus raising the possibility of enhanced extremes like droughts and floods, as well as more extreme heat waves and fewer cold snaps. Increased intensity of tropical cyclones is considered likely, but frequency changes remain speculative. In short, the future climate could be very altered from that to which modern ecosystems have become adapted.

In summary, no clear physical objection or direct empirical evidence has contradicted the consensus of scientists (IPCC 1990, 1996a, 2001a; NAS 2001) that the world is warming, nor has credible evidence emerged to contradict the substantial probability that temperatures will rise because of increases in greenhouse gases (Morgan and Keith 1995). Even in the mid-1990s many scientists thought the evidence sufficient to believe that recently observed climatic variations and human activities are probably connected (Karl et al. 1995). The IPCC (1996a, page 5) carefully weighed the uncertainties and concluded that "[n]evertheless, the balance of evidence suggests that there is a discernable human influence on global climate." The IPCC (2001a) reinforces and strengthens that judgment, particularly since recent studies of the past 1000 years (Mann et al. 1999) show that for the Northern Hemisphere the last half of the 20th century is likely to be warmer than at any time during the past 1000 years.

Relevance of Climate Modeling to Regional Climate Change and Ecosystem Studies

Scientists who estimate the future climatic changes that are relevant to ecosystems have focused on the GCMs that attempt to represent mathematically the complex physical and chemical interactions among the atmosphere, oceans, ice, biota, and land. As these models have evolved, more and more information has become available, and more comprehensive simulations have been performed. Nevertheless, the complexities of the real climate system still vastly exceed the GCMs and the capabilities of even the most advanced computers (IPCC 1990, 1996a, 2001a). Simulating 1 year of weather in 30-minute time steps with the crude resolution of 40 latitudinal lines × 48 longitudinal lines and 10 vertical layers—nearly 20,000 grid cells around the globe—takes many minutes on a supercomputer. This level of resolution, however, cannot resolve the Sierra Nevada of California and the Rocky Mountains as separate mountain chains. Refining the resolution to 50-square-kilometer grid squares or less would so dramatically increase the number of computations that it could take roughly months of computer time to simulate weather statistics for one year.

Even the highest-resolution, three-dimensional GCMs will not have a grid with nodes much less than 10 kilometers apart within the foreseeable future; individual clouds and most ecological research (to say nothing of cloud droplets) occur on scales far smaller than that. Therefore, GCMs will not be able to resolve the local or regional details of weather affecting most local biological communities or the importance of regional effects of hills, coastlines, lakes, vegetation boundaries, and heterogeneous soil. It is, nonetheless, important to have climatic forecasts and ecological-response analyses on the same physical scales (Root and Schneider 1993).

What is most needed to evaluate potential biological effects of temperature change is a regional projection of climatic changes that can be applied to ecosystems at a regional or local scale. Analyses of large, prehistoric climatic changes (Barron and Hecht 1985, Budyko et al. 1987, Schneider 1987, Cooperative Holocene Mapping Project 1988) and historical weather analogs (Pittock and Salinger 1982, Jager and Kellogg 1983, Lough et al. 1983, Shabalova and Können 1995) provide some insights into such changes. Historical weather analogs, however, because they are empirically and statistically based, rely on climatic cause-and-effect processes that probably differ from those that will be driven by future greenhouse gas effects (Schneider 1984, Mearns et al. 1990, Crowley 1993). Consequently, ecologists turn to climatic models to produce forecasts of regional climatic changes for the decades ahead.

Regional Changes

Although the consensus among researchers about the plausibility of significant human-induced global climate change is growing, no assessment (e.g., IPCC 1996c, 2001a) has suggested the existence of a strong consensus about how that global climate change might be distributed regionally. For example, the world is not actually undergoing a dramatic and instantaneous doubling of CO_2, which is the hypothesis that has been used in most standard computer model experiments applied to ecological assessments. Instead, the world is undergoing a steady increase in greenhouse gas forcing. Because that increase is heating the Earth in a reasonably uniform way, one might expect a uniform global response, though this is far from likely. For example, the centers of continents have relatively low heat-retaining capacity, and the temperatures there would move relatively rapidly toward whatever their new equilibrium climate would be compared with the centers of oceans, which have high heat-retaining capacity. Tropical oceans, though, have a thin (about 50 meters) mixed layer that interacts primarily with the atmosphere. It takes about 10 years for that mixed layer to substantially change its temperature, which is still much slower than the response time of the middle of the continents, but is much faster than that of the oceans closer to the poles. At high latitudes, in places like the Weddell or Norwegian seas, waters can mix down to the bottom of the ocean, thereby continuously bringing up cold water and creating a deepwater column that takes a century or more to substantially change its temperature.

During the transient phase of climate change over the next century, therefore, one would expect the middle of continents, the middle of oceans, and the polar and subpolar oceans all to change toward their new equilibrium temperatures at different rates. Thus the temperature differences from land to sea and equator to pole will evolve over time, which, in turn, implies that the transient character of regional climatic changes could be very different from the expected long-term equilibrium (Schneider and Thompson 1981, Stouffer et al. 1989, Washington and Meehl 1989). This does not imply that transient regional changes are inherently unpredictable, only that at present they are very difficult to predict confidently.

Even more uncertain than regional averages, but perhaps more important to long-term ecosystem responses (Parmesan et al. 2000), are estimates of climatic variability during the transition to a new equilibrium, particularly at the regional scale. These include estimates of such events as the frequency and magnitude of severe storms, enhanced heat waves, temperature extremes, sea-level rises (Titus and Narayanan 1995), and reduced frost probabilities (Mearns et al. 1984, 1990; Parry and Carter 1985; Wigley 1985; Rind et al. 1989). For example, evaporation increases dramatically as surface-water temperature increases. Because hurricanes are powered by evaporation and condensation of water, if all other factors are unchanged, the intensity of hurricanes and the length of the hurricane season could increase with warming of the oceans (Emanuel 1987, Knutson 1998). Such changes would significantly affect susceptible terrestrial and marine ecosystems (Doyle 1981, O'Brien et al. 1992).

Downscaling Climate Predictions to Regional Effects

Empirical Mapping Techniques

Techniques exist that can translate the output of climate models so that it is closer to most ecological scales. One method that uses actual climatic data at both large and small scales can help provide maps that may allow small-scale analysis of large-scale climate change scenarios. As mentioned above, the Sierra Nevada of California or the Cascades in the northwestern United States are north–south mountain chains whose east–west dimensions are smaller than the grid size of a typical general circulation model (GCM). In the actual climate system, onshore winds on the Pacific coast would produce cool upslope and rainy conditions on the western slope and a high probability of warmer and drier conditions associated with that flow pattern on the downslope or eastern slope.

A regional map has been generated for Oregon (Gates 1985) in which a high-resolution network of meteorological stations was used to plot temperature and precipitation isopleths based on observed climatic fluctuations at large (e.g., state-sized) scales. This map shows that the dominant mode of variation for this area is warm and dry on one side of the mountains, cold and wet on the other. Although this empirical mapping technique seems appropriate for translating low-resolution, grid-scale climate model forecasts to local applications, a strong caveat must be provided. That is, the processes in the climate system that give rise to internal variability or natural fluctuations are not necessarily the same processes that would give rise to local deviations from large-scale patterns if the climate change were driven by external forces rather than an internal variation of the system. For example, the Oregon map would indicate that if the grid-box average temperature were warmer on the eastern slope, then it should be cooler and wetter on the western slope. That condition is the most probable regional situation for today's naturally fluctuating climate. However, if 50 years from now the warming on the eastern slope were, say, a result of doubled atmospheric CO_2 causing an enhanced downward infrared radiative (greenhouse) heating, then both eastern and western slopes would probably experience warming. Although the degree of warming and associated precipitation changes would not necessarily be uniform, an entirely different climate change pattern would probably occur as opposed to that obtained from the empirical mapping technique if one used the naturally varying weather conditions existing today rather than the anthropogenically forced conditions of the 21st century (Schneider 1993b).

Therefore, techniques to shrink climate forecasts that use current distributions of environmental variables at local scales and correlate them with current large-scale regional patterns will not necessarily provide a good guideline about how large-scale patterns would be distributed regionally. The reason is that the causes of the future change may be physically or biologically different from the causes of the historical fluctuations that led to the empirical maps in the first place. This caveat is so important that it requires scientists to use extreme caution before adopting such empirical techniques for global change applications.

Regional-Scale Models with GCM Inputs

Other techniques can still translate large-scale patterns to smaller scales, but these techniques are based on known processes rather than on empirical maps of today's conditions. One such technique is to drive a high-resolution, process-based model for a limited region with the large-scale patterns produced by a GCM. In essence, this approach uses a mesoscale model (i.e., 10–50-kilometer grid cells) based on physical laws to solve the problem of translating GCM grid-scale averages into a finer-scale mesh much closer to the dimensions of most ecological applications. Of course, even this mesoscale grid will still be too coarse to assess many impacts, necessitating further downscaling techniques. Neither are the problems of GCMs entirely eliminated by mesoscale grids, because such grids are bigger than individual clouds or trees. But such methods do bring climate-model scales and ecological-response scales much closer.

Examples of Ecological Responses
to Climate Changes

Bringing climatic forecasts down to ecological applications at local and regional scales is one way to bridge the scale gap across ecological and climatological studies. Ecologists, however, have also analyzed data and constructed models that apply over large scales, including the size of climatic model grids. A long tradition in ecology has associated the occurrence of vegetation types or the range limits of different plant species with physical factors such as temperature, soil moisture, or elevation. Biogeography is the field that deals with such associations, and its results have been applied to estimate the large-scale ecological response to climate change.

Predicting Vegetation Responses to Climate Change

The Holdridge (1967) life-zone classification assigns biomes (for example, tundra, grassland, desert, or tropical moist forest) according to two measurable variables, temperature and precipitation. Other more complicated large-scale formulas have been developed to predict vegetation patterns from a combination of large-scale predictors (for example, temperature, soil moisture, or solar radiation); the vegetation modeled includes individual species (Davis and Zabinski 1992), limited groups of vegetation types (Box 1981), or biomes (Prentice 1992, Melillo et al. 1993, Neilson 1993). These kinds of models predict vegetation patterns that represent the gross features of actual vegetation patterns, which is an incentive to use them to predict vegetation change with changing climate, but they have some serious drawbacks as well. That is, they are typically static, not time-evolving dynamic simulations, and thus cannot capture the transient sequence of changes that would take place in reality. In addition, such static biome models occasionally make "commission errors"—they predict vegetation types to occur in certain zones where climate would indeed permit such vegetation, but other factors like soils, topography, or disturbances like fire actually preclude it. Furthermore, local patterns may influence vegetation dynamics at scales not captured in some simulations, and seed germination and dispersal mechanisms are also either not explicitly simulated or simulated only crudely with such models. Remarkably they are still able to produce generalized maps of vegetation types that do indeed resemble current or even paleoclimatic patterns in a broad sense. Their details, however, do not provide confident projections for future vegetation states. Fortunately, progress is being made to include some of the deficiencies mentioned above, and so-called dynamical global vegetation models are being developed to treat the transient nature of vegetation change that would likely accompany climate change.

Predicting Animal Responses to Climate Change

Scientists of the U.S. Geological Survey (USGS), in cooperation with Canadian scientists, conduct the annual North American Breeding Bird Survey, which provides distribution and abundance information for birds across the United States and Canada. From these data, collected by volunteers under strict guidance from the USGS, shifts in bird ranges and densities can be examined. Because these censuses were begun in the 1960s, these data can provide a wealth of baseline information. Price (1995) has used these data to examine the birds that breed in the Great Plains. By using the present-day ranges and abundances for each of the species, Price derived large-scale, empirical-statistical models based on various climate variables (for example, maximum temperature in the hottest month and total precipitation in the wettest month) that provided estimates of the current bird ranges. Then, by using a GCM to forecast how doubling of CO_2 would affect the climate variables in the models, he applied statistical models to predict the possible shape and location of the birds' ranges.

Significant changes were found for nearly all birds examined. The ranges of most species moved north, up mountain slopes, or both. The empirical models assume that these species are capable of moving into these new areas, provided habitat is available and no major barriers exist. Such shifting of ranges and abundances could cause local extirpations in the more southern portions of the birds' ranges, and, if movement to the north is impossible, extinctions of entire species could occur. We must bear in mind, however, that this empirical-statistical technique, which associates large-scale patterns of bird ranges with large-scale patterns of climate, does not explicitly represent the physical and biological mechanisms that could lead to changes in birds' ranges. Therefore, such detailed maps should be viewed only as illustrative of the potential for very significant shifts with doubled CO_2 climate change scenarios. More refined techniques that also attempt to include actual mechanisms for ecological changes are discussed later.

Reptiles and amphibians, which together are called herpetofauna (or herps), are different from birds in many ways that are important to our discussion. First, because herps are ectotherms—meaning their body temperatures adjust to the ambient temperature and radiation of the environment—they must avoid environments where temperatures are too cold or too hot. Second, amphibians must live near water, not only because the reproductive part of their life cycle is dependent on water, but also because they must keep their skin moist to allow them to respire through their skin. Third, herps are not able to disperse as easily as birds, and the habitat through which they crawl must not be too dry or otherwise impassible (for example, high mountains or busy superhighways).

As the climate changes, the character of extreme weather events, such as cold snaps and droughts, will also change (Karl et al. 1995), necessitating relatively rapid habitat changes for most animals (Parmesan et al. 2000). Rapid movements by birds are possible because they can fly, but for herps such movements are much more difficult. For example, R. L. Burke (then at University of Michigan, Ann Arbor, pers. comm. 1995) noted that during the 1988 drought in Michigan, many more turtles than usual were found dead on the roads. He assumed they were trying to move from their usual water holes to others that had not yet dried up or that were cooler. For such species, moving across roads usually means high mortality. In the long term, most birds can readily colonize new habitat as climatic regimes shift, but herp dispersal (colonization) rates are slow. Indeed, some reptile and amphibian species may still be expanding their ranges north even now, thousands of years after the last glacial retreat.

R. L. Burke and T. L. Root (pers. comm. 1995) performed a preliminary analysis of North American herp ranges in an attempt to determine which, if any, are associated with climatic factors such as temperature, vegetation-greening duration, and solar radiation. Their evidence suggests that northern boundaries of some species ranges are associated with these factors, implying that climate change could have a dramatic impact on the occurrence of these species (e.g., see Schneider and Root 1998 for details and a literature survey). Furthermore, most North American turtles and several other reptile species could exhibit vulnerability to climate change because the temperature experienced as they develop inside the egg determines their sex. Such temperature-dependent sex determination makes these animals uniquely sensitive to temperature change, meaning that climate change could potentially cause severely skewed sex ratios, which could result in dramatic range contractions. Many more extinctions are possible in herps than in birds because the forecasted human-induced climatic changes could occur rapidly when compared with the rate of natural climatic changes, and because the dispersal ability of most herps is painfully slow, even without considering the additional difficulties associated with human land-use changes disturbing their migration paths.

In general, animals most likely to be affected earliest by climate change are those in which populations are fairly small and limited to isolated habitat islands. There are estimates that a number of small mammals living near isolated mountaintops (which are essentially habitat islands) in the Great Basin would become extinct given typical global change scenarios (MacDonald and Brown 1992). Recent studies of small mammals in Yellowstone National Park show that statistically significant changes in both abundances and physical sizes of some species occurred with historical climate variations (which were much smaller than most projected climate changes for the next century), but there appear to have been no simultaneous genetic changes (Hadley 1997). Therefore, climate change in the 21st century could likely cause substantial alteration to biotic communities, even in protected habitats such as Yellowstone National Park.

Current Animal Responses to Climate Change

Animals are showing many different types of changes related to climate. These include changes in ranges; abundances; phenology (timing of an event); morphology and physiology; and community composition, biotic interactions, and behavior. Changes are being seen in all different types of taxa, from insects to mammals, and on many of the continents (Price et al. 2000). For example, the ranges of butterflies in Europe and North America have been found to shift poleward and upward in elevation as temperatures have increased (Pollard 1979, Parmesan 1996, Ellis et al. 1997, Parmesan et al. 1999). From 1979 to 1989, population densities of the Puerto Rican coqui (*Eleutherodactylus coqui*) showed a negative correlation with the longest dry period during the previous year (Stewart 1995). Similarly, the disappearance of the golden toad (*Bufo periglenes*) and the harlequin frog (*Atelopus varius*) from Costa Rica's Monteverde Cloud Forest Reserve seemed to be linked to the extremely dry weather associated with the 1986–87 El Niño–Southern Oscillation (Pounds and Crump 1994). Birds' ranges reportedly have extended poleward in Antarctica (Fraser et al. 1992, Emslie et al. 1998), and Europe (Thomas and Lennon 1999). For instance, the northern movement of the spring range of barnacle geese (*Branta leucopsis*) along the Norwegian coast correlates significantly with an increase in the number of April and May days with temperatures above 6°C (Prop et al. 1998). Reproductive success of the California quail (*Calipepla californica*) is positively correlated with the previous winter's precipitation (Botsford et al. 1988). Rainfall affects the chemistry of plants eaten by quail, with the plants producing phytoestrogens, compounds similar to hormones that regulate reproduction in birds and mammals. Drought-stunted plants tend to have higher concentrations of these compounds (Leopold et al. 1976). The northern extension of the porcupine's (*Erethizon dorsatum*) range in central Canada has been associated with a warming-associated poleward shift in the location of tree line (Payette 1987). In the United Kingdom, the dormouse (*Muscardinus avellanarius*) has disappeared from approximately half of its range over the last 100 years (Bright and Morris 1996). This disappearance appears to be linked to a complex set of factors including climatic changes, fragmentation, and the deterioration and loss of specialized habitat.

Warmer conditions during autumn–spring adversely affect the phenology of some cold-hardy insects. Experimental work on spittlebugs (*Philaenus spumarius*) found that they hatched earlier in winter-warmed (3°C above ambient) grassland plots (Masters et al. 1998). Chorusing behavior in frogs, an indication of breeding activities, appears to be triggered by rain and temperature (Busby and Brecheisen 1997). Two frog species, at their northern range limit in the United Kingdom, spawned 2 to 3 weeks earlier in 1994 than in 1978 (Beebee 1995). Three newt species also showed highly significant trends toward earlier breeding, with the first individuals arriving 5 to 7 weeks earlier over the course of the same study period. This study also examined temperature data, finding strong correlations with average minimum temperature in March and April (negative) and maximum temperature in March (positive) for the two frogs with significant trends, and a strong negative correlation between lateness of pond arrival and average maximum temperature in the month before arrival for the newts. Using less precise methods, a family of naturalists in England recorded the timing of first frog and toad croaks for the period from 1736 to 1947 (Sparks and Carey 1995). The date of spring calling for these amphibians became earlier over time, and was positively correlated with spring temperature, which was positively correlated with year. Changes in phenology or links between phenology and climate have been noted for earlier breeding of some birds in the United Kingdom (Thompson et al. 1986), Germany (Winkel and Hudde 1996, Ludwichowski

1997) and the United States (Brown et al. 1999). Changes in bird migration have also been noted with earlier arrival dates of spring migrants in the United States (Ball 1983), later autumn departure dates in Europe (Bezzel and Jetz 1995), and changes in migratory patterns in Africa (Gatter 1992).

The effect of temperature on the metabolism of dormant horned toads in Brazil was found to be stronger than the effect on resting toads at most temperatures (Bastos and Abe 1998). Reptile physiology is temperature sensitive also. Painted turtles grew larger in warmer years, and during warm sets of years turtles reached sexual maturity faster (Frazer et al. 1993). Physiological effects of temperature can also occur while reptiles are still within their eggs. Leopard geckos (*Eublepharis macularius*) produced from eggs incubated at a high temperature of 32°C showed reproductive behavioral changes and possible female sterility (Gutzke and Crews 1988). Spring and summer temperatures have been linked to variations in the size of the eggs of the pied flycatcher (*Ficedula hypoleuca*) (Järvinen 1996). The early summer mean temperatures explaining approximately 34% of the annual variation in egg size between the years 1975 and 1994. Body mass, which correlates with many life-history traits including reproduction, diet, and size of home ranges of the North American wood rat (*Neotoma* spp.) has shown a significant decline inversely correlated with a significant increase in temperature over the last 8 years in one arid region of North America (Smith et al. 1998). In studies of spring temperature effects on red deer (*Cervus elaphus*) in Scotland, juvenile deer grew faster in warm springs leading to increases in adult body size, a trait positively correlated with adult reproductive success. In Norway, red deer born following warm winters (that have more snow) were smaller than those born after cold winters—a difference persisting into adulthood (Post et al. 1997).

Differential responses by species could cause existing animal communities to undergo a reformulation (Root and Schneider 1993). Peach–potato aphids grown on plants kept in elevated CO_2 (700 ppm) showed a reduced response to alarm pheromones in comparison to those grown on plants in ambient CO_2 (350 ppm) (Awmack et al. 1997a). The aphids were more likely to remain on leaves, rather than move away, in response to the pheromones, possibly making them more susceptible to predators and parasitoids. Temperature and dissolved-oxygen concentrations can alter the behavior of amphibian larvae, and changes in thermal environments can alter the outcome of predator–prey interactions (Moore and Townsend 1998). Climate change may be causing mismatching in the timing of breeding of great tits (*Parus major*) in the United Kingdom and other species in their communities (Visser et al. 1998). Post et al. (1999) documented a positive correlation between gray wolf (*Canis lupus*) pack size in winter and snow depth on Isle Royale (U.S.). In years with deeper snow, wolves formed larger packs, which led to more than three times as many moose kills.

Top-Down Approaches

The biogeographic approach just summarized is an example of a top-down technique (like that of the Holdridge life-zone classification), in which data on abundances or range limits of species, vegetation types, or biomes are overlain on data of large-scale environmental factors such as temperature or precipitation. When associations among large-scale biological and climatic patterns are revealed, biogeographic rules expressing these correlations graphically or mathematically can be used to forecast changes driven by given climate changes.

Bottom-Up Approaches

The next traditional analysis and forecasting technique is often referred to as bottom-up. Small-scale ecological studies have been undertaken at the scale of a plant (Idso and Kimball 1993) or even a single leaf to understand how, for example, increased atmospheric CO_2 concentrations might directly enhance photosynthesis, net primary production, or water-use efficiency. Most studies such as these indicate increases in all these factors, increases that some researchers have extrapolated to ecosystems (Idso and Brazel 1984, Ellsaesser 1990).

However, at the scale of a forest, the relative humidity within the canopy, which significantly influences the evapotranspiration rate, is itself regulated by the forest. In other words, if an increase in water-use efficiency decreased the transpiration from each tree, the aggregate forest effect would be to lower relative humidity. This, in turn, would increase transpiration, thereby offsetting some of the direct

CO_2/water-use efficiency improvements observed experimentally at the scale of a single leaf or plant. Regardless of the extent to which this forest-scale negative feedback effect (or "emergent property" of the coupled forest atmosphere system) will offset inferences made from bottom-up studies of isolated plants, the following general conclusion emerges: the bottom-up methods may be appropriate for some processes at some scales in environmental science, but they cannot be considered to produce highly confident conclusions without some sort of validation testing at the scale of the system under study.

Combined Top-Down and Bottom-Up Approaches

To help resolve the deficiencies of the top-down models mentioned previously, more process-based, bottom-up approaches such as forest gap models have been developed (Botkin et al. 1972, Pastor and Post 1988, Smith et al. 1992). These models include individual species and can calculate vegetation dynamics driven by time-changing climate change scenarios. But the actual growth rate calculated in the model for each species has usually been determined by multiplying the ideal growth-rate curve by a series of growth-modifying functions that attempt to account for the limiting effects of nutrient availability, temperature stress, and so forth. These growth-modifying functions for temperature are usually determined empirically at a large scale by fitting an upside-down U-shaped curve, whose maximum is at the temperature midway between the average temperature of the species' northern range limit and the average temperature of its southern range limit. Growing degree-days (related to average temperature but not average temperature per se) are used in this scenario.

In essence, this technique combines large-scale, top-down empirical pattern correlations into an otherwise mechanistic bottom-up modeling approach. Although this combined technique refines both approaches, it has been criticized because such large-scale, top-down inclusions are not based on the physiology of individual species and lead to confusion about the fundamental and realized ranges (Pacala and Hurtt 1993). (The fundamental range is the geographic space in which a given species could theoretically survive—for example, if its competitors were absent—and the realized range is where it actually exists.) The question then is, What limits the realized range, particularly at the southern boundary? Further, more refined models should include factors such as seed dispersal, so that plant recruitment is related to the preexisting population and is not simply the result of a random number generator in the computer code.

Studies of More Refined Approaches

As noted, problems with the singular use of either top-down or bottom-up methods have led to well-known criticisms. For bottom-up models, the primary problem is that some of the most conspicuous processes observable at the smaller scales may not be the dominant processes that generate large-scale patterns. Top-down approaches suffer because of the possibility that the discovered associations at large scales are statistical artifacts that do not, even implicitly, reflect the causal mechanisms needed for reliable forecasting. As Järvis (1993:121) states, "A major disadvantage of a top-down model is that predictions cannot be made safely outside the range of the variables encountered in the derivation of the lumped parameter function."

A search of the literature (Wright et al. 1993, Root 1994, Harte et al. 1995) provides examples of a refined approach to analyzing across large and small scales, which Root and Schneider (1995) labeled strategic cyclical scaling (SCS). This method builds upon the combined techniques in which top-down and bottom-up approaches are applied cyclically in a strategic design that addresses a practical problem: in our context, the ecological consequences of global climate change. Large-scale associations are used to focus small-scale investigations; this helps ensure that tested causal mechanisms are generating the large-scale relations. Such mechanisms become the laws that allow more credible forecasts of the consequences of global change disturbances. According to Levin (1993:14), "Although it is well understood that correlations are no substitute for mechanistic understanding of relationships, correlations can play an invaluable role in suggesting candidate mechanisms for (small-scale) investigation." SCS, however, is not only intended as a two-step process, but also as a continuous cycling process between large- and small-scale studies, with each successive investigation building on previous insights from all scales. This approach is designed to enhance the credibility of the overall assessment process (see also Vitousek 1993, Harte and Shaw 1995), which is why *strategic* is the first word in SCS.

Bir∂ Ca∂e Stu∂y

If the rate at which humans are injecting greenhouse gases into the atmosphere is not greatly decreased, there is a significant chance that the Earth's climate will warm by several degrees Celsius by the year 2050 (Titus and Narayanan 1995, IPCC 2001a). With that in mind, Root (1988a) examined the biogeographic patterns of all wintering North American birds. She chose this group of species because birds are important parts of ecosystems and because of the availability of the necessary data. The National Audubon Society has volunteer forces amassed to aid in the collection of Christmas Bird Count data. By using these data, Root determined that for a large proportion of species, average distribution and abundance patterns are associated with various environmental factors (e.g., northern range limits of some species are apparently limited by average minimum January temperature [Root 1988a,b,c, 1989; Repasky 1991]).

The scaling question is, What mechanisms (such as competition or thermal stress) at small scales may have given rise to the large-scale associations? Root first tested the hypothesis that local physiological constraints may be causing most of the particular large-scale, temperature-range boundary associations. She used published, small-scale studies on the wintering physiology of key species to determine that about half of the songbird species wintering in North America extend their ranges no farther north than the regions where, to avoid hypothermia during winter nights, they need not increase their metabolic rates more than roughly 2.5 times their basal metabolic rate (Root 1988b). Root embarked on a larger, regional study to determine whether the longer nights, hence fewer hours of daylight available for foraging, or the colder temperatures in the more northerly locations are relatively more important. Preliminary results indicate that temperatures are more likely than day length to explain this effect (Root 2000). Thus global temperature changes would probably cause a rapid range and abundance shift, at least by selected bird species. Indeed, Root found significant year-to-year shifts in ranges and abundances; these shifts are apparently associated with year-to-year changes in winter temperatures (Root

1994). No claim is made at this point in the research for the generality of the preliminary results indicating strong and quantitative links between bird distributions and climate change. This example does permit, however, a clear demonstration of refined methods for cycling across scales to estimate ecological responses to climate change.

Three-Way Linkages and Community Ecology

The anticipated changes in plant ranges will probably have dramatic effects on animals, both on the large biogeographic scale and on the local scale. The ranges of many animals are strongly linked to vegetation. For example, red-cockaded woodpeckers are endemic to mature longleaf pine and pine–oak forests (Mengel and Jackson 1977), and the winter range of Sprague's pipit is coincident with bluestem, a grass (Root 1988a). Consequently, the ranges of various animals that rely on specific vegetation will change as the ranges of these plants shift, assuming that some other factor is not limiting these animals. If the climate changes more rapidly than the dispersal rates of the plants, it could result in extensive plant die-offs in the south or downslope before individuals can disperse and become established in the north or upslope. Thus the ranges of animals relying on these plants could become compressed, and in some instances, both the plants and the animals could become extinct. For instance, the red-cockaded woodpecker needs mature, living trees for nesting sites (Jackson 1974), and if rising temperature causes most large trees to die before the newly established dispersing trees grow large enough, then this woodpecker, federally listed as endangered, could become extinct.

Many animal species have ranges that are not directly limited by vegetation but are instead restricted by temperature (Root 1988c). This is true for most ectotherms (insects and related arthropods, amphibians, reptiles) as well as some endotherms (mammals and birds). For example, the eastern phoebe, a North American songbird, winters in the United States in areas with average minimum temperatures warmer than 4°C (Root 1988a). As the Earth warms, those species directly limited by temperature will be able to expand northward as rapidly as their dispersal mechanisms will allow, again assuming other factors are not limiting them. The animals limited by vegetation will be able to expand their ranges only as rapidly as the vegetation changes. Consequently, the potential for significant disruption among communities is high (Root and Schneider 1993). For instance, some animals may no longer be able to coexist because an invading species disrupts the balance between competing species or between predator and prey. Therefore, to understand the ecological consequences of global climate change on animals, the three-way linkages among animals, plants, and climate need to be investigated. Animals and plants affect each other and are affected by climate. At the same time, altered surface vegetation can affect climate because mid-continental summer precipitation is significantly influenced by water vapor from evapotranspiration (Ye 1989, Salati and Nobre 1991).

A Discernible Impact of Climate on Wildlife in the 20th Century?

Attributing observed changes in populations of plants and animals to climate change, specifically temperature increases, is possible because the patterns created by this large-scale pressure (global warming) are broad, often predictable, and generally continuous, rather than spotty. Additionally, these changes are expected to be concentrated in areas where the temperature change is largest, and less evident elsewhere. Certainly, climate change is only one of a long list of pressures influencing population distributions, health, morphology, and traits such as timing of activities. Other key pressures include: conversion of natural and semi-natural habitats, human persecution (e.g., legal and illegal by-catch, harassment), wildlife trade, war and other civil conflict, pollution and other biochemical poisoning, introduction of exotic species, and physical obstructions (e.g., roads, farm fields, tall towers, and buildings). Changes caused by these localized pressures would create a pattern of response that is irregular and patchy, and often centered around rapidly developing areas. Therefore, to document a strong role for climate in explaining many of the observed changes in animal and plant populations, increased confidence is obtained using examples that, when observed over a decade or longer and over large spatial scales, show changes in the direction predicted by the physiological tolerances of the particular species. The result—a "fingerprint"—of a coherent large-scale pattern exhibited in many species around the globe, which is consistent with the understanding of the causal mechanisms, provides the greatest confidence in the attribution of observed changes in wildlife or plants to global climate change.

Using information from the literature, Root et al. (2001) examined if animals and plants are already exhibiting a discernible change consistent with changing temperatures and predicted by our understanding of the physiological constraints of species. They amassed over 1300 research papers with keywords concerning climate and animals and used the literature assessed by Chapter 5 of IPCC (2001b) for plants (over 1000 articles). Of these, over 500 address changes over time, with around 100 studies analyzing at least 10 years of data. A priori, they developed three criteria to determine which of these studies would be included in their analysis. Each article met at least two of the following three criteria: (1) a trait of at least one animal or plant must show a statistically significant positive or negative change over time, (2) changes in this trait need to correlate significantly with changes in local temperature, and (3) local temperature must significantly change with time. The 45 studies meeting these criteria report significant and nonsignificant findings on more than 450 animal species in Europe, North America, Central America, the Southern Ocean islands, Antarctica, along the North American shoreline of the Pacific Ocean, in the North Pacific Ocean, and in the Antarctic Ocean, and approximately 50 plant species (in Europe, North America, and Antarctica). Of these 500 species, around 440 (88%) exhibit a significant change over time, and 84% (about 370) of these 440 species exhibit change in a manner predicted, based on the physiology of the species (an application of the SCS technique of Root and Schneider 1995).

Documented Responses of Animals and Plants Used in IPCC Third Assessment Report

Forty-five studies were included in these analyses for the IPCC Third Assessment Report (TAR). The number would be higher if Root et al. (2001) had examined other climatic variables in addition to temperature, which is the variable predicted with most confidence to change with increasing greenhouse gases (IPCC 1996a). Precipitation (e.g., drought) was not considered in the IPCC effort primarily for three reasons: (1) It is more difficult to use than temperature to determine mechanisms of how changing precipitation might influence many animals. (2) It is more difficult to use than temperature to determine how precipitation changed in the past at local scales because of its high degree of regional heterogeneity. (3) How precipitation may change in the future is more difficult to model confidently than temperature. Additionally, the effects of temperature on the physiology of species are fairly well understood and reliably demonstrated in the literature to cause changes in traits of species (Root 1988c). These 45 studies indicate significant changes are occurring in Europe and northern Africa (32 studies), North America (6 studies), Central America (1 study), Antarctica (2 studies), Southern Oceanic islands (1 study), the North American Pacific Ocean shoreline (1 study), the North Pacific Ocean (1 study), and the Antarctic Ocean (1 study) (Table OV.1). If drought had been included, then studies from many other regions, such as Australia and Asia, would have been included. Of those species exhibiting change (about 90%), the vast majority (about 80%) show changes consistent with a hypothesized response to climate warming.

Quantifying such a wide array of changes is problematic. Meta-analyses, however, provide a statistical method of summarizing results from many studies, even though such studies may not use common methods or databases (Hedges and Olkin 1985). If warming causes changes in phenology, then it would be reasonable to expect that phenological changes observed regionally might be associated with regional temperature changes. To test this hypothesis, Root et al. (2001) performed a meta-analysis on all 18 animal studies (195 species), and a meta-analysis on 4 plant studies (49 species) reporting spring phenological changes (Table OV.2).

Eight animal studies lacked information needed to determine correlation coefficients (r). Consequently, Root et al. (2001) performed two types of meta-analyses on these data: one taking advantage of the r values that were reported (12 studies, 34 species),[1] and a second including all 18 studies and 193 species, but only taking into account the sign of the phenological change (i.e., negative for earlier in the year and positive for later). A meta-analysis using 49 plant species was also performed. The meta-analysis of the correlation coefficients between animal traits and time of year allows one to make an estimate of a common fingerprint—a common correlation coefficient underlying the several studies (see Hedges and Olkin [1985] for more detail on the method). The estimated under-

Table OV.I. Information from 45 Studies Used by the IPCC (2001b).

| Location | Taxa | Number of Species | | | Type of Change | Citation |
		Exp	No Chg	Opp		
Antarctic Ocean	Invertebrates	2	0	0	Density	Loeb et al. 1997
Antarctica	Birds	2	0	0	Shift Range and Density	Fraser et al. 1992, Smith et al. 1999
Antarctica	Vascular Plants	2	0	0	Density	Smith 1994
Central America	Amphibians	20	30	0	Shift Range and Density	Pounds et al. 1999
Central America	Reptiles	2	1	0	Shift Range and Density	Pounds et al. 1999
Central America	Birds	15	0	0	Shift Range and Density	Pounds et al. 1999
Europe	Invertebrates	5	0	0	Spring Phenology	Fleming and Tatchell 1995
Europe	Invertebrates	4	0	0	Spring Phenology	Zhou et al. 1995
Europe	Invertebrate	1	0	0	Spring Phenology	Visser et al. 1998
Europe and Northern Africa	Invertebrates	34+	17+	1+	Shift Range and Density	Parmesan et al. 1999
Europe	Invertebrates	1	0	0	Morphology	De Jong and Brakefield 1998
Europe	Invertebrate	1	0	0	Genetics	Rodriguez-Trelles et al. 1998
Europe	Amphibian	1	0	0	Spring Phenology	Forchhammer et al. 1998
Europe	Amphibians	5	0	0	Spring Phenology	Beebee 1995
Europe	Amphibian	1	0	0	Morphology	Reading 1998, Reading and Clarke 1995

continues

Table OVI. *Continued*

Location	Taxa	Number of Species			Type of Change	Citation
		Species	*No*	*Opp*		
Europe	Bird	1	0	0	Morphology and Spring Phenology	Järvinen 1989, 1994
Europe	Birds	0	0	1	Spring Phenology	Forchhammer et al. 1998
Europe	Birds	51	0	14	Spring Phenology	Crick et al. 1997, Crick and Sparks 1999
Europe	Bird	1	0	0	Spring Phenology	Ludwichowski 1997
Europe	Bird	1	0	0	Spring Phenology	McCleery and Perrins 1998
Europe	Bird	1	0	0	Spring Phenology	Slater 1999
Europe	Bird	1	0	0	Spring Phenology	Winkel and Hudde 1996
Europe	Bird	1	0	0	Spring Phenology	Visser et al. 1998
Europe	Birds	3	0	0	Spring Phenology	Winkel and Hudde 1997
Europe	Birds	122	0	24	Spring Phenology	Sparks 1999, Mason 1995
Europe	Bird	1	0	0	Fall Phenology	Bezzel and Jetz 1995
Europe	Birds	27	6	13	Fall Phenology	Gatter 1992
Europe	Bird	1	0	0	Shift Range and Density	Bergmann 1999
Europe	Bird	1	0	0	Shift Range and Density	Prop et al. 1998
Europe	Birds	56	7	38	Shift Range and Density	Thomas and Lennon 1999
Europe	Birds	0	0	2	Density	Forchhammer et al. 1998

Location	Taxa				Type of Change	Citation
Europe	Mammals	7	0	0	Morphology	Post and Stenseth 1999
Europe	Forbs	11	0	1	Spring Phenology	Post and Stenseth 1999
Europe	Tree	1	0	0	Spring Phenology	Walkovszky 1998
Europe	Trees and Shrubs	?	?	?	Spring and Fall Phenology	Menzel and Fabian 1999
Europe	Mt. Plants	?	?	?	Shift Range	Grabherr et al. 1994, Pauli et al. 1996
Europe	Tree	1	0	0	Morphology	Hasenauer et al. 1999
North America	Bird	1	0	0	Spring Phenology	Brown et al. 1999
North America	Bird	1	0	0	Spring Phenology	Dunn and Winkler 1999
North America	Birds	15	0	4	Spring Phenology	Bradley et al. 1999
North America	Mammals	3	0	0	Morphology	Post and Stenseth 1999
North America	Grasses	6	0	0	Density	Alward et al. 1999
North America	Forbs	24	0	11	Spring Phenology	Bradley et al. 1999
North America	Tree	1	0	0	Morphology	Barber et al. 2000
North American	Tree	1	0	0	Spring Phenology	Bradley et al. 1999
S. Ocean Islands	Bird	1	0	0	Shift Range and Density	Cunningham and Moors 1994
N. Am. Shoreline	Invertebrates	15	?	3	Density	Sagarin et al. 1999
N. Pacific Ocean	Fish	1	0	0	Morphology	Ishida et al. 1995

Includes location of the study, taxa of the species examined, number of species changing, either significantly or non-significantly, in the direction expected with temperature change based on physiological studies (Exp), the number of species exhibiting no change (No chg), the number significantly or non-significantly changing opposite to that expected (Opp), the type of change observed, and the citation. When the number of species is not specified in the citation, then "?" is used. (After Root et al., 2001.)

lying common correlation is -0.38, which is statistically significantly different from zero ($P < 0.05$) with a 95% confidence interval of $-0.45 \leq r \leq -0.31$. Consequently, a strong pattern of consistent change—a shift toward earlier spring activities—is occurring among those species with negative changes in some measure of their spring phenology.

The "vote counting" meta-analysis of all animals included in spring phenology studies incorporated data for species for which either a correlation coefficient or slope of the relationship between the changing species trait and time was reported. In total, Root et al. (2001) analyzed data with this method for 195 species from 17 studies (Table OV.2). This vote-counting statistic is based on the

Table OV.2. Information from 21 Studies Addressing Changes in Spring Phenology Used by the IPCC (2001b).

Taxa	Species	No. of Years	Corr Coef	P	Sign	Type of Change	Citation
Invertebrate	*Drepanosiphum platanoidis*	25+	?	0.0001	—	Peak Abundance	Fleming and Tatchell 1995
Invertebrate	*Elatobium abietinum*	25+	?	0.0001	—	Peak Abundance	Fleming and Tatchell 1995
Invertebrate	*Microlophium carnosum*	25+	?	0.0001	—	Peak Abundance	Fleming and Tatchell 1995
Invertebrate	*Periphyllus testudinaceus*	25+	?	0.0001	—	Peak Abundance	Fleming and Tatchell 1995
Invertebrate	*Phorodon humuli*	25+	?	0.0001	—	Peak Abundance	Fleming and Tatchell 1995
Invertebrate	caterpillar	23	?	‹0.05	—	Peak Abundance	Visser et al. 1998
Invertebrate	*Brachycaudus helichrysi*	28	?	‹0.05	—	Peak Abundance	Zhou et al. 1995
Invertebrate	*Sitobion avenae*	28	?	‹0.05	—	Peak Abundance	Zhou et al. 1995
Invertebrate	*Metopolophium dirhodum*	28	?	‹0.05	—	Peak Abundance	Zhou et al. 1995
Invertebrate	*Myzus persicae*	28	?	‹0.05	—	Peak Abundance	Zhou et al. 1995
Amphibian	*Triturus vulgaris*	17	-0.78	‹0.001	—	Breeding	Beebee 1995
Amphibian	*T. cristatus*	17	-0.59	‹0.02	—	Breeding	Beebee 1995
Amphibian	*T. helveticus*	17	-0.6	‹0.02	—	Breeding	Beebee 1995
Amphibian	*Bufo calamita*	17	-0.71	‹0.01	—	Breeding	Beebee 1995
Amphibian	*Rana kl. esculenta*	17	-0.58	‹0.05	—	Breeding	Beebee 1995
Amphibian	*Rana temporaria*	18	?	‹0.05	+	Breeding	Forchhammer et al. 1998
Bird	*Ardea herodias*	13	-0.24	0.09	—	Spring Arrival	Bradley et al. 1999
Bird	*Monothrus ater*	14	0.27	0.05	+	Spring Arrival	Bradley et al. 1999
Bird	*Caprimulgus vociferus*	17	-0.3	0.02	—	Spring Arrival	Bradley et al. 1999
Bird	*Sialia sialis*	18	0.004	0.79	+	Spring Arrival	Bradley et al. 1999
Bird	*Passarella iliaca*	18	0	0	0	Spring Arrival	Bradley et al. 1999
Bird	*Hylocicia mustelina*	20	-0.08	0.22	—	Spring Arrival	Bradley et al. 1999
Bird	*Pipilio erythrophithalamus*	22	0.04	0.38	+	Spring Arrival	Bradley et al. 1999
Bird	*Ficedula hypoleuca*	22	-0.39	0.07	—	Breeding	Jarvinen 1989, 1994, 1996
Bird	*Ceryls alcon*	23	-0.09	0.17	—	Spring Arrival	Bradley et al. 1999
Bird	*Ficedula hypoleuca*	23	-0.58	‹0.01	-	Breeding	Slater 1999
Bird	*Parus major*	23	?	0.33	-	Breeding	Visser et al. 1998
Bird	*Troglodytes aedon*	24	-0.34	0	-	Spring Arrival	Bradley et al. 1999
Bird	*Turdus migratorius*	25	-0.24	0.01	-	Spring Arrival	Bradley et al. 1999
Bird	*Sturnella magna*	25	-0.09	0.15	-	Spring Arrival	Bradley et al. 1999
Bird	*Bucephala clangula*	25	-0.64	0.006	-	Breeding	Ludwichowski 1997
Bird	*Parus major*	25	-0.49	0.01	-	Hatch date	Winkel and Hudde 1997
Bird	*Parus caeruleus*	25	-0.46	‹0.05	-	Hatch date	Winkel and Hudde 1997
Bird	*Ficedula hypoleuca*	25	-0.53	‹0.01	-	Hatch date	Winkel and Hudde 1997
Bird	*Pheucticus ludovicianus*	26	-0.4	0	-	Spring Arrival	Bradley et al. 1999
Bird	*Sitta europaea*	26	-0.56	‹0.01	-	Hatch date	Winkel and Hudde 1996
Bird	*Branta canadensis*	27	-0.54	0	-	Spring Arrival	Bradley et al. 1999

	Species	Years	r	P	Sign	Type of Change	Citation
:d	*Scolopax minor*	27	-0.14	0.05	-	Breeding	Bradley et al. 1999
:d	*Aphelocoma ultramarina*	27	-0.15	0.05	-	Breeding	Brown et al. 1999
:d	*Parus major*	27	-0.3	0.003	-	Breeding	McCleery and Perrins 1998
:d	*Sayornis phoebe*	28	-0.38	0.09	-	Spring Arrival	Bradley et al. 1999
:d	*Toxostomum rufum*	28	-0.02	0.46	-	Spring Arrival	Bradley et al. 1999
:d	*Cardinalis cardinalis*	29	-0.22	0.01	-	Breeding	Bradley et al. 1999
:d	*Agelaius phoeniceus*	30	-0.15	0.04	-	Spring Arrival	Bradley et al. 1999
:d	*Icterus galbula*	30	-0.09	0.11	-	Spring Arrival	Bradley et al. 1999
·d	*Tachycineta bicolor*	30	-0.64	0.001	-	Breeding	Dunn and Winkler 1999
:ds	Unspecified	30	?	<0.05	55 -	Spring Arrival	Sparks 1999, Mason 1995
:ds	Unspecified	30	?	>0.05	91 -	Spring Arrival	Sparks 1999, Mason 1995
:ds	*Miliaria calandra*	50	?	?	?	Spring Arrival	Forchhammer et al. 1998
:ds	*Pica pica*	50	?	?	?	Spring Arrival	Forchhammer et al. 1998
:ds	*Phylloscopus collybita*	50	?	?	?	Spring Arrival	Forchhammer et al. 1998
:d	*Alauda arvensis*	51	?	?	+	Spring Arrival	Forchhammer et al. 1998
rb	*Apocynum androsaemifolium*	10	-0.02	0.69	-	Flowering Date	Bradley et al. 1999
rb	*Anemone quinquefolia*	10	-0.01	0.74	-	Flowering Date	Bradley et al. 1999
rb	*Oxalis violacea*	11	0.01	0.82	+	Flowering Date	Bradley et al. 1999
rb	*Erigeron strilgosus*	12	0.01	0.77	+	Flowering Date	Bradley et al. 1999
rb	*Linaria vulgaris*	12	0.01	0.23	+	Flowering Date	Bradley et al. 1999
rb	*Pentstemon gracilis*	13	0.00	0.39	+	Flowering Date	Bradley et al. 1999
rb	*Asclepias incarnata*	15	-0.51	0.00	-	Flowering Date	Bradley et al. 1999
rb	*Achillea millefolium*	15	0.02	0.16	+	Flowering Date	Bradley et al. 1999
rb	*Viola pedata*	15	0.03	0.52	+	Flowering Date	Bradley et al. 1999
rb	*Hypericum perforatum*	16	-0.00	0.89	-	Flowering Date	Bradley et al. 1999
rb	*Lithospermum canescens*	16	0.24	0.06	+	Flowering Date	Bradley et al. 1999
rb	*Asclepias syriaca*	17	-0.29	0.02	-	Flowering Date	Bradley et al. 1999
rb	*Anemone patens*	19	-0.15	0.11	-	Flowering Date	Bradley et al. 1999
rb	*Uvularia grandiflora*	19	-0.09	0.22	-	Flowering Date	Bradley et al. 1999
rb	*Campanula rotundifolia*	19	0.01	0.64	+	Flowering Date	Bradley et al. 1999
rb	*Euphorbia corollata*	20	0.03	0.47	+	Flowering Date	Bradley et al. 1999
rb	*Phlos philosa*	21	-0.13	0.10	-	Flowering Date	Bradley et al. 1999
rb	*Phlox divaricata*	22	-0.32	0.01	-	Flowering Date	Bradley et al. 1999
rb	*Dodecatheon media*	22	-0.23	0.02	-	Flowering Date	Bradley et al. 1999
rb	*Caltha palustris*	22	-0.13	0.10	-	Flowering Date	Bradley et al. 1999
rb	*Anemone canadensis*	23	-0.17	0.05	-	Flowering Date	Bradley et al. 1999
rb	*Lupinus perennis*	23	-0.09	0.16	-	Flowering Date	Bradley et al. 1999
rb	*Rosa carolina*	23	0.01	0.63	+	Flowering Date	Bradley et al. 1999
rb	*Sisryinchium campestre*	23	0.04	0.36	+	Flowering Date	Bradley et al. 1999
rb	*Rudbeckia hirta*	24	-0.17	0.04	-	Flowering Date	Bradley et al. 1999
rb	*Geranium maculatum*	24	-0.02	0.51	-	Flowering Date	Bradley et al. 1999
rb	*Antennaria neglecta*	24	-0.01	0.67	-	Flowering Date	Bradley et al. 1999
rb	*Trillium grandiflorum*	25	-0.70	0.20	-	Flowering Date	Bradley et al. 1999
rb	*Baptisia leucantha*	25	-0.41	0.00	-	Flowering Date	Bradley et al. 1999
rb	*Asclepias tuberosa*	25	-0.41	0.00	-	Flowering Date	Bradley et al. 1999
rb	*Aquilegia canadensis*	25	-0.26	0.01	-	Flowering Date	Bradley et al. 1999
rb	*Amelanchier laevis*	25	-0.04	0.33	-	Flowering Date	Bradley et al. 1999
rb	*Tradescantia ohiensis*	26	-0.12	0.08	-	Flowering Date	Bradley et al. 1999
rb	*Dicentra cucullaria*	28	-0.07	0.19	-	Flowering Date	Bradley et al. 1999
rb	*Hepatica acutiloba*	31	-0.14	0.04	-	Flowering Date	Bradley et al. 1999
rb	*Linnaea borealis*	44	-0.45	<0.05	-	Flowering Date	Post and Stenseth 1999
rb	*Primula officinalis*	45	-0.34	<0.05	-	Flowering Date	Post and Stenseth 1999
rb	*Oxalis acetosella*	47	-0.38	<0.05	-	Flowering Date	Post and Stenseth 1999
rb	*Trientalis europaea*	47	-0.17	<0.05	-	Flowering Date	Post and Stenseth 1999
rb	*Epilobium angustifolium*	49	-0.44	<0.05	-	Flowering Date	Post and Stenseth 1999
rb	*Caltha palustris*	49	-0.40	<0.05	-	Flowering Date	Post and Stenseth 1999
rb	*Calluna vulgaris*	49	0.05	<0.05	+	Flowering Date	Post and Stenseth 1999
rb	*Anemone hapatica*	50	-0.60	<0.05	-	Flowering Date	Post and Stenseth 1999
rb	*Anemone nemorosa*	50	-0.60	<0.05	-	Flowering Date	Post and Stenseth 1999
rb	*Convallaria majalis*	50	-0.52	<0.05	-	Flowering Date	Post and Stenseth 1999
rb	*Tussilago farfara*	50	-0.43	<0.05	-	Flowering Date	Post and Stenseth 1999
rb	*Vaccinium myrtillus*	50	-0.43	<0.05	-	Flowering Date	Post and Stenseth 1999
ee	*Robinia pseudoacacia*	12	?	?	?	Flowering Date	Walkovsky 1998
ee	*Prunus virginiana*	18	0.15	0.11	+	Flowering Date	Bradley et al. 1999
ee	*Betula odorata*	50	-0.24	<0.05	-	Flowering Date	Post and Stenseth 1999
cular plants	Unspecified	?	?	?	?	Flowering Date	Menzel and Fabian 1999

cludes taxa of the species examined, scientific name of the species, number of years spanned by the data, correlation coefficients hen provided or information available to calculate the coefficients in the citation), P value of the correlation coefficients (when)vided in the citation), sign of the relationship between species trait and year, type of change observed, and citations. An known value is indicated by "?". (Root et al., 2001.)

number of these associations indicating an earlier phenological shift compared to all reported associations. The phenologies of 94 out of 105 species were recorded as shifting earlier in time. The probability that an estimated phenological shift is earlier, based on a sample size of 13 years, was found to be 0.895 (with a population-wide correlation coefficient of -0.36).

For the meta-analysis of plants showing a change in their blooming or budding dates, the common correlational "fingerprint" for the 48 species from North America and Europe is -0.26, which is statistically different from zero ($P < 0.05$, 95% confidence interval $-0.31 \leq r \leq -0.20$). Again, a strong pattern of consistent shifting toward earlier spring activities is occurring in species of plants investigated.

Numerous studies examined shifts in density, which can be created by a change in abundance within the range of a species, a shift in the range boundary, or both. To test for an underlying pattern using the data available the "vote counting" method was used. For animals and plants, about 200 species show a change in density, with approximately 160 of these changes in the expected direction. The meta-analysis indicates that there is statistically significant movement in the direction expected for these species. While "vote-counting" is often insensitive to detecting underlying effects, the strength of this result indicates that there is most likely a fingerprint in the shifts of densities in both plants and animals.

Results from most studies using long-term data sets provide circumstantial (e.g., correlational) evidence about the association between changes in climate-related environmental factors and animal traits. Circumstantial evidence, insufficient for "proving" causation by itself, is highly suggestive when numerous studies, examining many different taxa from several different locations, are found to be consistent with one phenological fingerprint. Unfortunately, other changes seen in species that are apparently associated with climate change, such as morphological shifts, do not lend themselves as easily to quantification. This does not mean that changes in traits are not shifting in concert. Indeed, about 80% of the species showing change are changing in the manner expected based on species' physiological tolerances (Table OV.1). The Working Group II IPCC TAR authors examining the results of the Root et al. (2001) study concluded that many animals and plants are already responding in concert with the increase in global average temperature of 0.6°C (Summary for Policymakers and Chapter 5, IPCC 2001b).

Meta-analyses provide a way to combine results, whether significant or not, from various studies and find an underlying consistent shift, or fingerprint, among species from different taxa examined at disparate locations. Hence, for studies meeting the strict criteria used by Root et al. (2001), the balance of evidence suggests that a significant impact from climatic warming is discernible in the form of long-term, large-scale alteration of several animal and plant populations. (Note that no claim is made that most plants and animals have been or will be affected by warming. Although that is possible, data are insufficient to show that. Rather, the high confidence conclusion is that many systems are already affected, and the number and scale of the changes will increase with further warming.) The "balance of evidence" conclusion is extended by IPCC (2001b) to include "environmental systems"—sea and lake ice cover and mountain glaciers in addition to the plant and animal taxa examined by Root et al. (2001). Taken together, the IPCC concluded with "high confidence" that the consistent broad-scale pattern of changes observed strongly suggests that a warming of the globe is the most likely explanation of these observed phenomena. Thus, the "discernible statement" of IPCC (1996a) for detection of an anthropogenic climate signal is broadened in IPCC (2001b) to include a "discernible statement" about observed global climatic changes affecting environmental systems. Clearly, if such climatic and ecological signals are now being detected above the background of climatic and ecological noise for a 20th century warming of only 0.6°C, it is likely that the expected impacts on ecosystems of changes up to an order of magnitude larger by A.D. 2100 could be dramatic.

Climate Forecasts, Ecosystem Responses, and Synergistic Effects

Improve Regional Analysis, Study Transients, and Include Many Variables

The most reliable projections from climatic models are for global-scale temperature changes. Ecological impact assessments, however, need time-evolving (transient) scenarios of regional-to-local-scale climate changes. Included are changes in precipitation; severe storm intensity, frequency, and duration; drought frequency, intensity, and duration; soil moisture; frost-free days; intense heat waves; ocean currents; upwelling zones; near-ground ozone; forest canopy humidity; and ultraviolet radiation and total solar radiation reaching the surface, where photosynthesis may be affected. Data gathered at many scales and by coordinated volunteer and professional sources are needed for archives of these regional and local variables, which, in turn, can be used to develop and test models or other techniques for climatic forecasting.

Abrupt Climatic Changes

We have argued that sustained globally averaged rates of surface temperature change from the past Ice Age to the present were about 1°C per 1000 years (and the large changes in ice masses also occurred on time scales of thousands of years). Alarmingly, a change of a few degrees Celsius per millennia is a factor of 10 or so slower than the expected changes of several degrees Celsius per 100 years typically projected for the 21st century due to human effects. We emphasize the words sustained globally averaged because comparably rapid variations have occurred, at least regionally. For example, about 13,000 years ago, after warm-weather fauna had returned to northern Europe and the North Atlantic, there was a dramatic return to ice age–like conditions in less than 100 years. This Younger Dryas miniglacial lasted hundreds of years before the stable recent period was established (Berger and Labeyrie 1987). The Younger Dryas was also accompanied by dramatic disturbances to plants and animals in the North Atlantic and Europe (Coope 1977; Ruddiman and McIntyre 1981). During the same period, dramatic shifts occurred outside of the North Atlantic Region (e.g., Severinghaus and Brook 1999), but no comparable climate change is evident in Antarctic ice cores. Even so, studies of fossils in the North Atlantic show that the warm Gulf Stream current deviated many degrees of latitude to the south and that the overall structure of deep ocean circulation may have returned to near ice age form in only decades— a weakening of the vertical circulation known as the conveyor-belt current (Broecker et al. 1985).

Plausible speculations about the cause of the Younger Dryas center on the injection of fresh meltwater into the North Atlantic, presumably associated with the breakdown of the North American ice sheet (Boyle and Weaver 1994, Paillard and Labeyrie 1994).

Could such a rapid change to the conveyor-belt current be induced today by pushing the present climatic system with human disturbances such as greenhouse gases? The potential for this is speculative, of course, but its possibility has concerned many scientists (Broecker 1994, Rahmstorf 1999). The prospect of climatic surprises in general is concerning enough to lend considerable urgency to the need to speed up the rate of our understanding, slow down the rates at which we are forcing nature to change, or both.

If the complexity of the coupled climate–ecological system is daunting, then recognition of what we actually will experience is even more so. That is, the actual system to be simulated is the coupled physical, biological, and social systems in which human behavior causes disturbances that propagate through natural systems and create responses that, in turn, feed back on human behavior in the form of policies for adaptation or mitigation to the human-induced disturbances. In fact, some very recent studies (e.g., Mastrandrea and Schneider 2001) show that when the socio-natural system is integrated over several hundred years—the time scale of overturning in the oceans—emergent properties can be uncovered. These properties would be difficult if not impossible to find by studying any of the sub-systems in disciplinary isolation. The search for emergent behaviors of complex coupled multicomponent systems will be a primary challenge for the next generation of scientists interested in climate and wildlife connections and their management implications (Kinzig et al. 2000).

Adaptability

Our current inability to credibly predict time-evolving regional climatic changes has many implications, one of which concerns the adaptability of agricultural ecosystems. That is, any experience farmers might have with anomalous weather in, say, the 2020s, may not help them adapt to the evolving climate change in the 2030s, because a transient climate change could differ dramatically over time. This would inhibit learning by doing, creating a potential lack of adaptability associated with the difficulty of reliably predicting regional climatic consequences (Schneider, Easterling, and Mearns 2000). Such rapid climate changes would be especially difficult for natural ecosystems to adapt to because habitats do not have the luxury of "choosing" to plant new seeds or change irrigation systems, soil tillage practices, or other agricultural practices.

The capacity of a system to adapt is a function of the exposure and sensitivity to climate change, as well as the knowledge, infrastructure, and resources available to each adaptive agent. This in turn depends on the social and economic status of each sector or region. Thus the very development conditions that determine how much greenhouse gas is emitted by future society also helps to determine how vulnerable each sector might be to the climate changes so created. Ecosystems, of course, have no capacity to anticipate anthropogenic climate changes nor to deploy a host of technological defenses. That is why most assessments (e.g., IPCC 2001b) continue to claim that human systems are likely to be more adaptable than most natural systems to climate change, at least if there is some foresight as to what changes might occur.

Ecological Applications-Driven Climatic Research

Regional projections of climatic changes arising from a variety of greenhouse gas and sulfur oxide emissions scenarios are essential for ecological applications. Such studies must stress the climatic variables most likely to have significant effects on biological resources. For example, extreme variability measures such as high temperature and low relative humidity are important for evaluating the risk of forest fires (Torn and Fried 1992). Identifying such variables of ecological importance and communicating this information to climate scientists require close interdisciplinary, multi-institutional, and cross-scale research efforts to ensure that combinations of variables relevant to ecological applications receive research priority by climatologists. A focus of climate research toward changing climatic variability (Mearns et al. 1984, 1990; Rind et al. 1989) might be more useful for ecological impact assessments than the current focus among climatic modelers on climatic means. Recent assessments, fortunately, are beginning to grapple with the difficult problem of variability changes and consequent impacts (e.g., IPCC 2001a, b).

Interactive, Multiscale, Ecological Studies Needed

Most ecological studies project the response of one species at small scales or shifts in biomes at large scales to an equilibrium, CO_2-doubled climate model (e.g., the Vegetation/Ecosystem Modeling and Analysis Project 1995). What is needed for more realistic and useful ecological impact assessments is a multiscale, multispecies, multi-taxa analysis driven by regionally specific, transient climate change forecasts. The construction of ecological forecast models first requires large-scale data sets gathered locally by professional (e.g., U.S. Geological Survey land-cover data sets) and volunteer (e.g., National Audubon Society Christmas Bird Count) workers. Without such data sets, virtually no credible progress is possible in determining large-scale patterns of associations among ecological and climatic variables. Small-scale studies informed by large-scale patterns are then needed to refine causal mechanisms underlying such large-scale associations, thereby testing the formulas used to make projections of various species or biome responses to hypothesized global changes. For example, Pacala and Hurtt (1993) suggested small- to medium-scale experiments to improve forest gap models. Their criticisms suggest that largely first principles, bottom-up models may still be unrealistic if some top-down parameters (that is, growth-modifying functions in the instance of gap models) are not appropriately derived from data at the scale at which the model is being applied (Root and Schneider 1995).

One obvious truism emerges: credible modeling required for forecasting across many scales and for complex interacting systems is a formidable task requiring repeated testing of many approaches. Nevertheless, tractable improvements in refining combined top-down and bottom-up techniques can be made. It will, however, take more than one cycle of interactions and testing with both large- and small-scale data sets to more credibly address the cross-scale and multicomponent problems of ecological assessment—another instance of what we (Root and Schneider 1995) have labeled SCS. The SCS paradigm has two motivations: (1) better explanatory capabilities for multiscale, multicomponent interlinked environmental (e.g., climate–ecosystem interactions or behavior of adaptive agents in responding to the advent or prospect of climatic changes) and (2) more reliable impact assessments and problem-solving capabilities—predictive capacity—as has been called for by the policy community.

Finally, to mobilize action to mitigate potential risks to environment or society, it is often necessary to establish that a discernible trend has been detected in some variable of importance (e.g., the first arrival of a spring migrant or the latitudinal extent of the sea ice boundary) and that the trend can be attributed to some causal mechanism (e.g., a warming of the globe from increases in anthropogenic greenhouse gases). Pure association of trends in some variables of interest are not, by themselves, sufficient to attribute any detectable change above background noise levels to any particular cause. Explanatory mechanistic models are needed, and the predictions from such models should be consistent with the observed trend before a high confidence can be assessed that a particular impact can be pinned on any suspected causal agent. We have argued that conventional scaling paradigms—top-down associations among variables believed to be cause and effect; bottom-up mechanistic models run to predict associations but for which there is no large-scale data time series to confirm—are not by themselves sufficient to provide high confidence in cause and effect relationships embedded in integrated assessments. Rather, we suggested that a cycling between top-down associations and bottom-up mechanistic models is needed. Moreover, assessors cannot assign high confidence to cause and effect claims until repeated cycles of testing are done in which mechanistic models predict outcomes and large scale data "verifies" that these mechanistic models are at least partially explanatory. Very high confidence usually requires a considerable degree of convergence in the cycling between top-down and bottom-up components, as well as sufficient baseline data for testing over long periods. We believe there are a number of taxa that have already demonstrated likely responses to regional-scale climatic changes over the past century, and that these responses are fairly coherent over widely separated geographical regions (see the Summary for Policymakers and Chapter 5 of IPCC 2001b, Parmesan et al. 2000, Sagarin this volume, Crozier this volume, and Root et al. 2001). Taken together, the consistent broad-scale pattern of changes observed strongly suggests a warming of the globe to be the most likely explanation of these observed phenomena, although for the reasons given in the above sentences, we would not yet assign a very high confidence to that inference. The efforts of the NWF authors in this volume are examples of the kinds of studies that will be needed to further increase our confidence in the recognition that climate change and other human disturbances not only have the potential to alter wildlife patterns significantly and create serious stresses for vulnerable species, but that this process has already begun and can be demonstrated with a disturbingly high level of confidence.

Notes

1. The numerical results for the meta-analyses (generally in parenthesis) were the values used for the IPCC, Working Group II, Third Assessment Report (IPCC 2001b). The actual numbers in Root et al. (2001) may differ owing to the ongoing review process.

Literature Cited

Alward, R. D., J. K. Detling, and D. G. Milchunas. 1999. Grassland vegetation changes and nocturnal global warming. *Science* 283: 229–231.

Andronova, N. G., and M. E. Schlesinger. 2000. Causes of global temperature changes during the 19th and 20th centuries. *Geophysical Research Letter* 27(14): 2137.

Awmack, C. S., C. M. Woodcock, and R. Harrington. 1997. Climate change may increase vulnerability of aphids to natural enemies. *Ecological Entomology* 22: 366–368.

Ball, T. 1983. The migration of geese as an indicator of climate change in the southern Hudson Bay region between 1715 and 1851. *Climatic Change* 5: 85–93.

Barber, V. A., G. P. Juday, and B. P. Finney. 2000. Reduced growth of Alaskan white spruce in the twentieth century from temperature-induced drought stress. *Nature* 405: 668–673.

Barron, J., and A. D. Hecht, eds. 1985. *Historical and Paleoclimatic Analysis and Modeling.* New York: John Wiley & Sons.

Bastos, R. P., and A. Abe. 1998. Dormancy in the Brazilian horned toad *Ceratophrys aurita* (Anusr, Lepdodactylidae). *Ciencia e Cultura* (Sao Paulo) 50: 68–70.

Beebee, T. J. C. 1995. Amphibian breeding and climate. *Nature* 374: 219–220.

Berger, W. H., and L. D. Labeyrie, eds. 1987. *Abrupt Climate Change.* Dordrecht, the Netherlands: D. Reidel.

Bergmann, F. 1999. Long-term increase in numbers of early-fledged reed warblers (*Acrocephalus scirpaceus*) at Lake Constance (Southern Germany) [German]. *Journal Fuer Ornithologie* 140: 81–86.

Bezzel, E., and W. Jetz. 1995. Delay of the autumn migratory period in the blackcap (*Sylvia atricappila*) 1966–1993: A reaction to global warming? [German] *Journal Fuer Ornithologie* 136: 83–87.

Botkin, D. B., J. R. Janak, and J. R. Wallis. 1972. Some ecological consequences of a computer model of forest growth. *Journal of Ecology* 60:849–872.

Botkin, D. P., D. A. Woodby, and R. A. Nisbet. 1991. Kirtland's warbler habitat: A possible early indicator of climate warming. *Biological Conservation* 56: 63–78.

Botsford, L. W., T. C. Wainwright, and J. T. Smith. 1988. Population dynamics of California quail related to meteorological conditions. *Journal of Wildlife Management* 52: 469–477.

Box, E. O. 1981. *Macroclimate and Plant Forms: An Introduction to Predictive Modeling in Phytogeography.* The Hague: Junk.

Boyle, E., and A. Weaver. 1994. Conveying past climates. *Nature* 372: 41–42.

Bradley, N. L., A. C. Leopold, J. Ross, and W. Huffaker. 1999. Phenological changes reflect climate change in Wisconsin. *Proceedings of the National Academy of Sciences of the United States of America* 96: 9701–9704.

Bright, P. W., and P. A. Morris. 1996. Why are dormice rare? A case study in conservation biology. *Mammal Review* 26: 157–187.

Broecker, W S. 1994. Massive iceberg discharges as triggers for global climate change. *Nature* 372: 421–424.

Broecker, W. S., D. M. Peteet, and D. Rind. 1985. Does the ocean–atmosphere system have more than one stable mode of operation? *Nature* 315: 21–25.

Brown, J. L., S.-H. Li, and N. Bhagabati. 1999. Long-term trend toward earlier breeding in an American bird: A response to global warming? *Proceedings of the National Academy of Sciences of the United States of America* 96: 5565–5569.

Budyko, M. I., A. B. Ronov, and A. L. Yanshin. 1987. *History of the Earth's Atmosphere.* New York: Springer-Verlag.

Busby, W. H., and W. R. Brecheisen. 1997. Chorusing phenology and habitat associations of the crawfish frog, *Rana areolata* (Anura: Ranidae), in Kansas. *Southwestern Naturalist* 42: 210–217.

Cohn, J. P. 1989. Gauging the biological impacts of the greenhouse effect. *BioScience* 39: 142–146.

Coope, G. R. 1977. Fossil coleopteran assemblages as sensitive indicators of climate changes during the Devensian (last) cold state. *Proceedings of the Philosophical Transactions of the Royal Society of London* B 280: 313–340.

Cooperative Holocene Mapping Project. 1988. Climatic changes of the last 18,000 years: Observations and model simulations. *Science* 241: 1043–1052.

Crick, H. Q. P., and T. H. Sparks. 1999. Climate change related to egg-laying trends. *Nature* 399: 423–424.

Crick, H. Q., C. Dudley, D. E. Glue, and D. L. Thomson. 1997. UK birds are laying eggs earlier. *Nature* 388: 526.

Crowley, T. 1993. Use and misuse of the geologic "analogs" concept. Pages 17–27 in J. A. Eddy and H. Oeschger, eds. *Global Changes in the Perspective of the Past.* Dahlem Workshop Report ES12. Chichester, England: John Wiley & Sons.

Cunningham, D. M., and P. J. Moors. 1994. The decline of Rockhopper penguins (*Eudyptes chrysocome*) at Campbell Island, Southern Ocean and the influence of rising sea temperatures. *Emu* 94: 27–36.

Darwin, C. 1859. *On the Origin of Species by Means of Natural Selection.* London: John Murray.

Davis, M. B., and C. Zabinski. 1992. Changes in geographical range resulting from greenhouse warming effects on biodiversity in forests. Pages 297–308 in R. L. Peters and T. E. Lovejoy, eds., *Global Warming and Biological Diversity.* New Haven: Yale University Press.

De Jong, P. W., and P. M. Brakefield. 1998. Climate and change in clines for melanism in the two-spot ladybird, *Adalia bipunctata. Proceedings of the Royal Society of London* B 25: 39–43.

Doyle, T.W. 1981. The role of disturbance in the gap dynamics of a montane rain forest: An application of a tropical forest succession model. Pages 56–73 in D. C. West, H. H. Shugart, and D. B. Botkin, eds., *Forest Succession: Concepts and Applications.* New York: Springer-Verlag.

Dunn, P. O., and D. W. Winkler. 1999. Climate change has affected the breeding date of tree swallows throughout North America. *Proceedings of the Royal Society of London* B 266: 2487–2490.

Eddy, J. A., and H. Oeschger, eds. 1993. *Global changes in the Perspective of the Past.* New York: John Wiley & Sons.

Ellis, W. N., J. H. Donner, and J. H. Kuchlein. 1997. Recent shifts in distribution of microlepidoptera in the Netherlands. *Entomologische Berichten* (Amsterdam) 57: 119–125.

Ellsaesser, H. W. 1990. A different view of the climatic effect of CO_2-updated. *Atmósfera* 3: 3–29.

Emanuel, K. A. 1987. The dependence of hurricane intensity on climate. *Nature* 326: 483–485.

Emslie, S. D., W. Fraser, R. C. Smith, and W. Walker. 1998. Abandoned penguin colonies and environmental change in the Palmer Station area, Anvers Island, Antarctic Peninsula. *Antarctic Science* 10: 257–268.

Fleming, R. A., and G. M. Tatchell. 1995. Shifts in the flight periods of British aphids: A response to climate warming? Pages 505–508 in R. Harrington and N. E. Stork, eds., *Insects in a Changing Environment.* San Diego: Academic Press.

Forchhammer, M. C., E. Post, and N. C. Stenseth. 1998. Breeding phenology and climate. *Nature* 391: 29–30.

Fraser, W. R., W. Z. Trivelpiece, D. C. Ainley, and S. G. Trivelpiece. 1992. Increases in Antarctic penguin populations: Reduced competition with whales or a loss of sea ice due to environmental warming? *Polar Biology* 11: 525–531.

Frazer, N. B., J. L. Greene, and J. W. Gibbons. 1993. Temporal variation in growth rate and age at maturity of male painted turtles, *Chrysemys picta. American Midland Naturalist* 130: 314–324.

Gates, W. L. 1985. The use of general circulation models in the analysis of the ecosystem impacts of climatic change. *Climatic Change* 7: 267–284.

Gatter, W. 1992. Timing and patterns of visible autumn migration: Can effects of global warming be detected? *Journal Fuer Ornithologie* 133: 427–436.

Grabherr, G., M. Gottfried, and H. Pauli. 1994. Climate effects on mountain plants. *Nature* 369: 448.

Gutzke, W. H. N., and D. Crews. 1988. Embryonic temperature determines adult sexuality in a reptile. *Nature* 332: 832–834.

Hadley, E. A. 1997. Evolutionary and ecological response of pocket gophers (*Thomomvs talpoides*) to late-Holocene climate change. *Biological Journal of the Linnean Society* 60:277–296.

Harte, J., and R. Shaw. 1995. Shifting dominance within a montane vegetation community: results of a climate-warming experiment. *Science* 267: 876–880.

Harte, J., M. Torn, F. R. Chang, B. Feifarek, A. Kinzig, R. Shaw, and K. Shen. 1995. Global warming and soil microclimate: results from a meadow-warming experiment. *Ecological Applications* 5: 132–150.

Hasenauer, H., R. R. Nemani, K. Schadauer, and S. W. Running. 1999. Forest growth response to changing climate between 1961 and 1990 in Austria. *Forest Ecology and Management* 122: 209–219.

Hedges, L. V., and I. Olkin. 1985. *Statistical Methods for Meta-Analysis.* New York: Academic Press.

Henderson-Sellers, A., Z. L. Yang, and R. E. Dickinson. 1993. The project for intercomparison of land-surface parameterization schemes. *Bulletin of the American Meteorological Society* 74: 1335–1349.

Holdridge, L. R. 1967. *Life Zone Ecology.* San Jose, Costa Rica: Tropical Science Center. 206 pp.

Idso, S. B., and A. J. Brazel. 1984. Rising atmospheric carbon dioxide concentrations may increase streamflow. *Nature* 312: 51–53.

Idso, S. B., and B. A. Kimball. 1993. Tree growth in carbon dioxide enriched air and its implications for global carbon cycling and maximum levels of atmospheric CO_2. *Global Biogeochemical Cycle* 7: 537–555.

Intergovernmental Panel on Climate Change (IPCC). 1990. *Climate Change: The IPCC Scientific Assessment,* eds., J. T. Houghton, G. J. Jenkins, and J. J. Ephraums. Contribution of working group I to the first assessment report of the IPCC. Cambridge: Cambridge University Press.

———. 1996a. *Climate Change 1995: The Science of Climate Change,* eds., J. T. Houghton, L. G. Meira Filho, B. A. Callander, N. Harris, A. Kattenberg, and K. Maskell. Contribution of working group I to the second assessment report of the IPCC. Cambridge: Cambridge University Press.

———. 1996b. *Climate Change 1995: Impacts, Adaptations and Mitigation of Climate Change: Scientific Technical Analysis,* eds., R. T. Watson, M. C. Zinyowera, and R. H. Moss. Contribution of working group II to the second assessment report of the IPCC. Cambridge: Cambridge University Press.

————. 1996c. *Climate Change 1995: Economic and Social Dimensions of Climate Change,* eds., J. Bruce, H. Lee, and E. Haites. Contribution of working group III to the second assessment report of the IPCC. Cambridge: Cambridge University Press.

————. 2001a. *Climate Change 2001: The Scientific Basis,* eds., J. T. Houghton, Y. Ding, D. J. Griggs, M. Noguer, P. J. van der Linden, and D. Xiaosu. Contribution of working group I to the third assessment report of the IPCC. Cambridge: Cambridge University Press.

————.2001b. *Climate Change 2001: Impacts, Adaptation, and Vulnerability,* eds., J. J. McCarthy, O. F. Canziani, N. A. Leary, D. J. Dokken, and K. S. White. Contribution of working group II to the third assessment report of the IPCC. Cambridge: Cambridge University Press.

————. 2001c. *Climate Change 2001: Mitigation,* eds., B. Metz, O. Davidson, R. Swart, and J. Pan. Contribution of working group III to the third assessment report of the IPCC. Cambridge: Cambridge University Press.

Ishida, Y., D. W. Welch, and M. Ogura. 1995. Potential influence of North Pacific sea–surface temperature on increased production of chum salmon (*Oncorhynchus keta*) from Japan. Pages 271–275 in R. J. Beamish, ed., *Climatic Change and Northern Fish Populations.* Canadian Special Publication of Fisheries and Aquatic Science, vol. 121.

Jackson, J. A. 1974. Gray rat snakes versus red-cockaded woodpeckers: Predator–prey adaptation. *Auk* 91: 342–347.

Jager, J., and W. W. Kellogg. 1983. Anomalies in temperature and rainfall during warm arctic seasons. *Climatic Change* 5: 39–60.

Järvinen, A. 1989. Patterns and causes of long-term variation in reproductive traits of the Pied Flycatcher *Ficedula hypoleuca* in Finnish Lapland. *Ornis Fennica* 66: 24–31.

————. 1994. Global warming and egg size of birds. *Ecography* 17: 108–110.

————. 1996. Correlation between egg size and clutch size in the pied flycatcher *Ficedula hypoleuca* in cold and warm summers. *Ibis* 138: 620–623.

Järvis, P. G. 1993. Prospects for bottom-up models. Pages 117–126 in J. R. Ehleringer and C. B. Field, eds., *Scaling Physiological Processes: Leaf to Globe.* New York: Academic Press.

Johansson, T. B., H. Kelly, A. K. N. Reddy, and R. H. Williams, eds. 1993. *Renewable Energy: Sources for Fuels and Electricity.* Washington, D.C.: Island Press.

Karl, T. R., R. W. Knight, D. R. Easterling, and R. G. Quayle. 1995. Trends in U.S. climate during the twentieth century. *Consequences* 1: 3–12.

Kasting, J. F., O. B. Toon, and J. B. Pollack. 1988. How climate evolved on the terrestrial planets. *Scientific American* 258: 90.

Kinzig, A. P., J. Antle, W. Ascher, W. Brock. S. Carpenter, F. S. Chapin III, R. Costanza, K. Cottingham, M. Dove, H. Dowlatabadi, E. Elliot, K.

Ewel, A. Fisher, P. Gober, N. Grimm, T. Groves, S. Hanna, G. Heal, K. Lee, S. Levin, J. Lubchenco, D. Ludwig, J. Martinez-Alier, W. Murdoch, R. Naylor, R. Norgaard, M. Oppenheimer, A. Pfaff, S. Picket, S. Polasky, H. R. Pulliam. C. Redman, J. P. Rodriguez, T. Root, S. Schneider, R. Schuler, T. Scudder, K. Segersen, R. Shaw, D. Simpson, A. Small, D. Starrett, P. Taylor, S. van der Leeuw, D. Wall, and M. Wilson. *Nature and Society: An Imperative for Integrated Environmental Research.* A report of a workshop, presented to the National Science Foundation, November 2000. Available online at http://lsweb.la.asu.edu/akinzig/report.htm.

Knutson T R. 1998. Simulated increase of hurricane intensities in a CO_2-warmed climate. *Science* 279: 1018–1020.

Leopold, A. S., M. Erwin, J. Oh, and B. Browning. 1976. Phytoestrogens: Adverse effects on reproduction in California quail. *Science* 191: 98–100.

Levin, S. A. 1993. Concepts of scale at the local level. Pages 7–19 in J. R. Ehleringer and C. B. Field, eds., *Scaling Physiological Processes: Leaf to Globe.* New York: Academic Press.

Loeb, V., V. Siegel, O. Holm-Hansen, R. Hewitt, W. Fraser, W. Trivelpiece, and S. Trivelpiece. 1997. Effects of sea-ice extent and krill or salp dominance on the Antarctic food web. *Nature* 387: 897–900.

Lough, J. M., T. M. L. Wigley, and J. P. Palutikof. 1983. Climate and climate impact scenarios for Europe in a warmer world. *Journal of Climate and Applied Meteorology* 22: 1673.

Ludwichowski, I. 1997. Long-term changes of wing-length, body mass and breeding parameters in first-time breeding females of goldeneyes (*Bucephala clangula clangula*) in northern Germany. *Vogelwarte* 39: 103–116.

MacDonald, K. A., and J. H. Brown. 1992. Using montane mammals to model extinctions due to global change. *Conservation Biology* 6: 409–425.

Mann, M. E., Bradley, R. S., and Hughes, M. K. 1999. Northern Hemisphere Temperatures during the Past Millennium: Inferences, Uncertainties, and Limitations. *Geophysical Research Letter* 26(6): 759.

Mason, C. F. 1995. Long-term trends in the arrival dates of spring migrants. *Bird Study* 42: 182–189.

Masters, G. J., V. K. Brown, I. P. Clarke, and J. B. Whittaker. 1998. Direct and indirect effects of climate change on insect herbivores: Auchenorrhyncha (Homoptera). *Ecological Entomology* 23: 45–52.

Mastrandrea, M. and Schneider, S. H. 2001. Integrated assessment of abrupt climatic changes. *Climate Policy* 53: 1–17.

McCleery, R. H., and C. M. Perrins. 1998. . . . temperature and egg-laying trends. *Nature* 391: 30–31.

Mearns, L. O., R. W Katz, and S. H. Schneider. 1984. Extreme high temperature events: Changes in their probabilities and changes in mean temperature. *Journal of Climate and Applied Meteorology* 23: 1601–1613.

Mearns, L. O., S. H. Schneider, S. L. Thompson, and L. McDaniel. 1990.

Analysis of climate variability in general circulation models: Compared with observations and changes in variability in 2 × CO_2 experiments. *Geophysical Research* 95: 20,469–20,490.

Melillo, J. M., A. D. McGuire, D. W. Kicklighter, B. Moore III, C. J. Vorosmarty, and A. L. Schloss. 1993. Global climate change and terrestrial net primary production. *Nature* 363: 234–240.

Mengel, R. M., and J. A. Jackson. 1977. Geographic variation of the redcockaded woodpecker. *Condor* 79: 349–355.

Menzel, A., and P. Fabian. 1999. Growing season extended in Europe. *Nature* 397: 659.

Moore, M. K., and V. R. Townsend. 1998. The interaction of temperature, dissolved oxygen and predation pressure in an aquatic predator–prey system. *Oikos* 81: 329–336.

Morgan, M. G., and D. W. Keith. 1995. Subjective judgments by climate experts. *Environmental Science and Technology* 29: 468A–477A.

Moss, R. H., and S. H. Schneider, 2000. Uncertainties in the IPCC TAR: Recommendations to lead authors for more consistent assessment and reporting. Pages 33–51 in R. Pachauri, T. Taniguchi, and K. Tanaka, eds., Guidance Papers on the Cross Cutting Issues of the Third Assessment Report of the IPCC, Intergovernmental Panel on Climate Change, Geneva; available from the Global Industrial and Social Progress Research Institute: http://www.gispri.or.jp.

Nakicenovic, N., and R. Swart. 2000. Special Report of the Intergovernmental Panel on Climatic Change (IPCC) on Emissions Scenarios (SRES). Cambridge: Cambridge University Press. Summary for Policymakers available online at http://www.ipcc.ch/.

National Academy of Sciences (NAS). 2001. *Climate Change Science: An Analysis of Some Key Questions.* Washington, D.C.: National Academy Press.

Neilson, R. P. 1993. Transient ecotone response to climatic change: Some conceptual and modelling approaches. *Ecological Applications* 3: 385–395.

O'Brien, S. T., B. P. Hayden, and H. H. Shugart. 1992. Global climate change, hurricanes, and a tropical rain forest. *Climatic Change* 22: 175–190.

Overpeck, J. T., R. S. Webb, and T. Webb III. 1992. Mapping eastern North American vegetation change over the past 18,000 years: No analogs and the future. *Geology* 20: 1071–1074.

Pacala, S. W., and G. C. Hurtt. 1993. Terrestrial vegetation and climate change: Integrating models and experiments. Pages 57–74 in P. Kareiva, J. Kingsolver, and R. Huey, eds., *Biotic Interactions and Global Change.* Sunderland, Mass.: Sinauer Associates.

Paillard, D., and L. Labeyrie. 1994. Role of the thermohaline circulation in the abrupt warming after Heinrich events. *Nature* 372: 162–164.

Parmesan, C. 1996. Climate and species' range. *Nature* 382: 765–766.

Parmesan, C., T. L. Root, and M. R. Willig. 2000. Impacts of extreme weather and climate on terrestrial biota. *Bulletin of the American Meteorological Society* 40: 443–450.

Parmesan, C., N. Ryrholm, C. Stefanescu, J. K. Hill, C. D. Thomas, H. Descimon, B. Huntley, L. Kaila, J. Kullberg, T. Tammaru, W. J. Tennent, J. A. Thomas, and M. Warren. 1999. Poleward shifts in geographical ranges of butterfly species associated with regional warming. *Nature* 399: 579–583.

Parry, M. L., and T. R. Carter. 1985. The effect of climatic variations on agricultural risk. *Climatic Change* 7: 95–110.

Pastor, J., and W. M. Post. 1988. Response of northern forests to CO_2-induced climate change. *Nature* 334: 55–58.

Pauli, H., M. Gottfried, and G. Grabherr. 1996. Effects of climate change on mountain ecosystems: Upward shifting of mountain plants. *World Resources Review* 8: 382–390.

Payette, S. 1987. Recent porcupine expansion at tree line: A dendroecological analysis. *Canadian Journal of Zoology* 65: 551–557.

Peters, R. L., and T. E. Lovejoy, eds. 1992. *Global Warming and Biological Diversity*. New Haven: Yale University Press.

Pimm, S. 1991. *The Balance of Nature*. Chicago: University of Chicago Press.

Pittock, A. B., and J. Salinger. 1982. Towards regional scenarios for a CO_2-warmed Earth. *Climatic Change* 4: 23–40.

Pollard, E. 1979. Population ecology and change in range of the white admiral butterfly *Ladoga camilla* L. in England. *Ecological Entomology* 4: 61–74.

Post, E., and N. C. Stenseth. 1999. Climatic variability, plant phenology, and northern ungulates. *Ecology* 80: 1322–1339.

Post, E., N. C. Stenseth, R. Langvatn, and J. M. Fromentin. 1997. Global climate change and phenotypic variation among red deer cohorts. *Proceedings of the Royal Society of London* B 264: 1317–1324.

Post, E., R. O. Peterson, N. C. Stenseth, and B. E. McLaren. 1999. Ecosystem consequences of wolf behavioural response to climate. *Nature* 401: 905–907.

Pounds, J. A., and M. L. Crump. 1994. Amphibian decline and climate disturbance: The case of the golden toad and the harlequin frog. *Conservation Biology* 8: 72–85.

Pounds, J. A., M. P. L. Fogden, and J. H. Campbell. 1999. Biological response to climate change on a tropical mountain. *Nature* 398: 611–615.

Prentice, I. C. 1992. Climate change and long-term vegetation dynamics. Pages 293–339 in D. C. Glenn-Lewin, R. A. Peet, and T. Veblen, eds., *Plant Succession: Theory and Prediction*. New York: Chapman & Hall.

Price, J. 1995. Potential Impacts of Global Climate Change on the Summer Distribution of Some North American Grasslands Birds. Ph.D. dissertation, Wayne State University, Detroit, Michigan.

Price, J. T., T. L. Root, K. R. Hall, G. Masters, L. Curran. W. Fraser, M. Hutchins, and N. Myers. 2000. Climate Change, Wildlife and Ecosystems. Available online at http://www.usgcrp.gov/ipcc/html/ecosystem.pdf.

Prop, J., J. M. Black, P. Shimmings, and M. Owen. 1998. The spring range of barnacle geese *Branta leucopsis* in relation to changes in land management and climate. *Biological Conservation* 86: 339–346.

Rahmstorf, S. 1999. Shifting seas in the greenhouse? *Nature* 399: 523–524.

Reading, C. J. 1998. The effect of winter temperatures on the timing of breeding activity in the common toad *Bufo bufo. Oecologia* 117: 469–475.

Reading, C. J., and R. T. Clarke. 1995. The effects of density, rainfall and environmental temperature on body condition and fecundity in the common toad, *Bufo bufo. Oecologia* 102: 453–459.

Repasky, R. R. 1991. Temperature and the northern distributions of wintering birds. *Ecology* 72: 2274–2285.

Rind, D., R. Goldberg, and R. Ruedy. 1989. Change in climate variability in the twenty-first century. *Climatic Change* 14: 5–37.

Rodriguez-Trelles, F., M. A. Rodriguez, and S. M. Sheiner. 1998. Tracking the genetic effects of global warming: *Drosophila* and other model systems. Conservation Ecology [Online] 2: 2. Available online at http://www.consecol.org/vol2/iss2/art2.

Root, T. L. 1988a. *Atlas of Wintering North American Birds.* Chicago: University of Chicago Press.

———. 1988b. Environmental factors associated with avian distributional boundaries. *Journal of Biogeography* 15: 489–505.

———. 1988c. Energy constraints on avian distributions and abundances. *Ecology* 69: 330–339.

———. 1989. Energy constraints on avian distributions: A reply to Castro. *Ecology* 70: 1183–1185.

———. 1994. Scientific/philosophical challenges of global change research: A case study of climatic changes on birds. *Proceedings of the American Philosophical Society* 138: 377–384.

———. 2000. Ecology: Possible consequences of rapid global change. Pages 315–324 in G. Ernst, ed., *Earth Systems.* Cambridge: Cambridge University Press.

Root, T. L., and S. H. Schneider. 1993. Can large-scale climatic models be linked with multi-scale ecological studies? *Conservation Biology* 7: 256–270.

———. 1995. Ecology and climate: Research strategies and implications. *Science* 269: 334–341.

Root, T. L., J. T. Price, K. R. Hall, C. Rosenzweig, and S. H. Schneider. 2001. The impact of climatic change on animals and plants: A meta-analysis. *Nature* (submitted).

Ruddiman, W. F., and A. McIntyre. 1981. The North Atlantic Ocean during the last deglaciation. *Paleogeography, Paleoclimatology, Paleoecology* 35: 145–214.

Sagarin, R. D., J. P. Barry, S. E. Gilman, and C. H. Baxter. 1999. Climate-related change in an intertidal community over short and long time scales. *Ecological Monographs* 69: 465–490.

Salati, E., and C. A. Nobre. 1991. Possible climatic impacts of tropical deforestation. *Climatic Change* 19: 177–196.

Santer, B. D., K. E. Taylor, T. M. L. Wigley, P. D. Jones, D. J. Karoly, J. F. B. Mitchell, A. H. Oort, J. E. Penner, V. Ramaswamy, M. D. Schwarzkopf, R. J. Stouffer, and S. F. B. Tett. 1996. A search for human influences on the thermal structure of the atmosphere. *Nature* 382: 39–46.

Schneider, S. H. 1984. On the empirical verification of model-predicted CO_2-induced climatic effects. Pages 187–201 in J. Hansen and T. Takahashi, eds., *Climate Processes and Climate Sensitivity*. Washington, D.C.: American Geophysical Union.

———. 1987. Climate modeling. *Scientific American* 256: 72–80.

———. 1990. *Global warming: Are We Entering the Greenhouse Century?* New York: Vintage Books.

———. 1993a. Can paleoclimatic and paleoecological analyses validate future global climate and ecological change projections? Pages 317–340 in J. A. Eddy and H. Oeschger, eds., *Global Changes in the Perspective of the Past*. Chichester: John Wiley & Sons.

———. 1993b. Scenarios of global warming. Pages 9–23 in P. Kareiva, J. Kingsolver, and R. Huey, eds., *Biotic Interactions and Global Change*. Sunderland, Mass.: Sinauer Associates.

———. 1994. Detecting climatic change signals: Are there any "fingerprints"? *Science* 263: 341–347.

———. 2001. What is "dangerous" climate change?. Commentary, *Nature* 411: 17–19.

Schneider, S. H., and R. Londer. 1984. *The Coevolution of Climate and Life*. San Francisco: Sierra Club Books.

Schneider, S. H. and T. L. Root. 1998. Impacts of climate changes on biological resources. Pages 89–116 in M. J. Mac, P. A. Opler, C. E. Puckett Haecker, and P. D. Doran, eds., *Status and Trends of the Nations Biological Resources*, vol. 1. Reston, Va.: U.S. Department of the Interior, U.S. Geological Survey.

Schneider, S. H., and S. L. Thompson. 1981. Atmospheric CO_2 and climate: Importance of the transient response. *Journal of Geophysical Research* 86: 3135–3147.

Schneider S. H., W. E. Easterling, L. O. Mearns. 2000. Adaptation: Sensitivity to natural variability, agent assumptions and dynamic climate changes. *Climatic Change* 45: 203–221.

Severinghaus, J. P., and E. J. Brook. 1999. Abrupt climate change at the end of the last glacial period inferred from trapped air in polar ice. *Science* 286: 930–934.

Shabalova, M. V., and G. P. Können. 1995. Climate change scenarios: Comparisons of paleoreconstructions with recent temperature changes. *Climate Change* 29: 409–428.

Slater, F. M. 1999. First-egg date fluctuations for the pied flycatcher *Ficedula hypoleuca* in the woodlands of mid-Wales in the twentieth century. *Ibis* 141: 497–499.

Smith, F. A., H. Browning, and U. L. Shepherd. 1998. The influence of climate change on the body mass of woodrats *Neotoma* in an arid region of New Mexico. *Ecography* 21: 140–148.

Smith, R. C., D. G. Ainley, K. Baker, E. Domack, S. D. Emslie, W. R. Fraser, J. Kennet, A. Leventer, E. Mosley-Thompson, S. E. Stammerjohn, and M. Vernet. 1999. Marine ecosystem sensitivity to historical climate change in the Antarctic Peninsula. *BioScience* 49: 393–404.

Smith, R. I. L. 1994. Vascular plants as bioindicators of regional warming in Antarctica. *Oecologia* 99: 322–328.

Smith, T. M., H. H. Shugart, G. B. Bonan, and J. B. Smith. 1992. Modeling the potential response of vegetation to global climate change. Pages 93–116 in F. I. Woodward, ed., *Advances in Ecological Research: The Ecological Consequences of Global Climate Change.* New York: Academic Press.

Sparks, T. H. 1999. Phenology and the changing pattern of bird migration in Britain. *International Journal of Biometeorology* 42: 134–138.

Sparks, T. H., and P. D. Carey. 1995. The responses of species to climate over two centuries: An analysis of the Marsham phenological record. *Journal of Ecology* 83: 321–329.

Stewart, M. M. 1995. Climate drives population fluctuations in rainforest frogs. *Journal of Herpetology* 29: 437–446.

Stouffer, R. J., S. Manabe, and K. Bryan. 1989. Interhemispheric asymmetry in climate response to a gradual increase of atmospheric CO_2. *Nature* 342: 660–662.

Thomas, C. D., and J. J. Lennon. 1999. Birds extend their ranges northwards. *Nature* 399: 213.

Thompson, D. B. A., P. S. Thompson, and D. Nethersole-Thompson. 1986. Timing of breeding and breeding performance in a population of greenshank. *Journal of Animal Ecology* 55: 181–199.

Titus, J. G., and V. Narayanan. 1995. *The Probability of Sea-Level Rise: Climatic Change.* Washington, D.C.: U.S. Environmental Protection Agency, Office of Policy, Planning and Evaluation, Climate Change Division, Adaptation Branch.

Torn, M. S., and S. J. Fried. 1992. Predicting the impacts of global warming on wildland fire. *Climate Change* 21: 257–274.

Trenberth, K. E. 1993. Northern Hemisphere climate change: Physical processes and observed changes. Pages 35–59 in H. A. Mooney, E. R.

Fuentes, and B. I. Kronberg, eds., *Earth System Responses to Global Change*. New York: Academic Press.

Vegetation/Ecosystem Modeling and Analysis Project. 1995. Vegetation/ Ecosystem Modeling and Analysis Project (VEMAP): Comparing bio- geography and biogeochemistry models in a continental-scale study of terrestrial ecosystem responses to climate change and CO_2 doubling. *Global Biogeochemical Cycles* 9: 407–437.

Visser, M. E., A. J. Vannoordwijk, J. M. Tinbergen, and C. M. Lessells. 1998. Warmer springs lead to mistimed reproduction in great tits (*Parus major*). *Proceedings of the Royal Society of London* B 265: 1867–1870.

Vitousek, P. M. 1993. Global dynamics and ecosystem processes: Scaling up or scaling down? Pages 169–177 in J. R. Ehleringer and C. B. Field, eds., *Scaling Physiological Processes: Leaf to Globe*. New York: Academic Press.

Walkovszky, A. 1998. Changes in the phenology of the locust tree (*Robinia pseudoacacia*) in Hungary. *International Journal of Biometeorology* 41: 155–160.

Washington, W. M., and G. H. Meehl. 1989. Climate sensitivity due to increased CO_2: Experiments with a coupled atmosphere and ocean gen- eral circulation model. *Climate Dynamics* 4: 1–38.

Washington, W. M., and C. L. Parkinson. 1986. *An Introduction to Three- dimensional Climate Modeling*. New York: Oxford University Press.

Wigley, T. M. L. 1985. Impact of extreme events. *Nature* 316: 106–107.

Wigley, T. M. L., and S. C. B. Raper. 2001. Interpretations for global-mean warming. *Science* 293: 451–454.

Wigley, T. M. L., and D. S. Schimel, eds. 2000. *The Carbon Cycle*. Cam- bridge: Cambridge University Press.

Wilson, E. O. 1992. *The Diversity of Life*. New York: W. W. Norton.

Winkel, W., and H. Hudde. 1996. Long-term changes of breeding para- meters of nuthatches *Sitta europaea* in two study areas of northern Ger- many. *Journal Fuer Ornithologie* 137: 193–202.

———. 1997. Long-term trends in reproductive traits of tits (*Parus major, P. caeruleus*) and pied flycatchers (*Ficedula hypoleuca*). *Journal of Avian Biology* 28: 187–190.

Wright, H. E., J. E. Kutzbach, T. Webb III, W. E Ruddiman, F. A. Street- Perrott, and P. J. Bartlein, eds. 1993. *Global Climates Since the Last Glacial Maximum*. Minneapolis: University of Minnesota Press.

Ye, D. 1989. Sensitivity of climate model to hydrology. Pages 101–108 in A. Berger, R. E. Dickinson, and J. W. Kidson, eds., *Understanding Climate Change*. Washington, D.C.: American Geophysical Union.

Zhou, X., R. Harrington, I. P. Woiwod, J. N. Perry, J. S. Bale, and S. J. Clark. 1995. Effects of temperature on aphid phenology. *Global Change Biology* 1: 303–313.

CHAPTER I

Climate Change and Its Effect on Species Range Boundaries: A Case Study of the Sachem Skipper Butterfly, *Atalopedes campestris*

LISA CROZIER

Atalopedes campestris is a butterfly on the move. While the temperature has been warming in the Pacific Northwest, the typically southern *A. campestris* has advanced its northern range edge ~700 kilometers in 35 years (Dornfeld 1980, Hinchliff 1994, 1996; pers. obs.). In the warmest decade on record, the 1990s (IPCC 2001), the butterflies established their first known populations in Washington State. In the hottest year of the century, 1998 (NCDC 1999a), *A. campestris* dispersed another 125 kilometers into new territory (pers. obs.). Is the warm weather responsible for the range expansion? If so, *A. campestris* may provide a good example of the biological consequences of global warming. Many insects, including pests and disease vectors, are prime candidates to expand their ranges in response to warming, given their sensitivity to temperature, rapid generation times, and high mobility (Walker 1991, Lawton 1995, Sutherst et al. 1995). But constraints on distribution are neither simple, nor fully understood. This species presents an unusual opportunity to study the process of range shift during climate change.

Many recent studies have documented changes in animal distribution or phenology associated with climate change (Grabherr et al. 1994, Barry et al. 1995, Parmesan 1996, Crick et al. 1997, Holbrook et al. 1997, Mantua et al. 1997, Myneni et al. 1997, McCleery and Perrins 1998, Alward et al. 1999, Brown et al. 1999, Parmesan et al. 1999, Post and Stenseth 1999, Sagarin et al. 1999, Thomas and Lennon 1999). However, a gap exists between observations of an association between distributions and various environmental factors, and a mechanistic understanding of what physiological, ecological, and evolutionary processes maintain range edges, especially in animals (Bartholomew 1958, Hoffmann and Parsons 1991). While correlation studies have great value in suggesting multiple hypotheses, experimental tests are necessary to pin down actual range-limiting factors. Understanding the true underlying mechanisms is crucial for predicting the biological outcomes of a changing climate. Unfortunately, in very few cases have hypothesized range-limiting mechanisms actually been tested.

This chapter presents my investigation into what constrains the distribution of the native, northwardly moving *Atalopedes campestris*. The range expansion occurred during relevant changes in both habitat and temperature, which led to multiple hypotheses regarding the causal mechanism of the range shift. My approach involves three steps: (1) Develop hypotheses of range-limiting factors based on correlation with the range change and the biology of the species. (2) Identify the physiological constraints that could constitute a mechanism by which climate could directly limit the range of *Atalopedes campestris*. (3) Test hypotheses experimentally with field transplantation to determine whether the proposed mechanism is acting on the current range boundary. This type of case study into the actual mechanisms of range limits in a geographically dynamic species should provide a model from which to make testable predictions about the biological consequences of climate change for a variety of species.

Analyzing Range Shifts:
From Global Change to Local Experiments

Anticipating specific biogeographical consequences of climate change requires understanding how species borders depend on climatic conditions. Our current understanding of this relationship is limited by the "scale gap" (Root and Schneider 1995). An average increase in global mean temperature is a landscape-scale prediction based on climate models using 500 × 500 km grids (Schneider 1993). These climate model predictions apply to "the big picture" of a species' range: large-scale patterns of abundance and distribution. Analyses of distributions at this scale show that combinations of temperature and precipitation criteria are frequently good predictors of species' ranges (e.g., Pigott 1975, Hengeveld 1985, Caughley et al. 1987, Root 1988, Cammell and Knight 1992). This association suggests that a change in environmental conditions may lead to a corresponding change in species' ranges.

Unfortunately, large-scale correspondence cannot differentiate essential and coincidental environmental associations. There are a large number of potential environmental variables that may yield spurious associations with species ranges. Many environmental factors are also correlated with each other in today's climate, obfuscating their relative importance for organisms (Dennis 1993). However, paleontological evidence shows that temperature, precipitation, and seasonality all vary independently during climate change (Brubaker 1988, Overpeck et al. 1992). Climates have existed in the past that do not occur today; thus it is likely that future regimes will differ from today's. If "no-analog" conditions result, present ranges do not necessarily predict a species' future distribution. Furthermore, species' response times to climate change can vary by orders of magnitude, leading to strong transient effects of rapid climate change (Davis 1976, Huntley 1991, Lawton 1995, Huntley et al. 1997). Response times are difficult to predict from static distributions due to the important but stochastic role of long-distance dispersal (Pitelka and Plant Migration Workshop 1997, Clark et al. 1998). Therefore, current large-scale associations alone are insufficient for predicting biogeographical consequences of climate change.

Ecological experiments are designed to identify causal mechanisms underlying broader patterns. For logistical reasons most ecological experiments occur on a relatively small scale, studying a single population or locality, usually for a short period of time (Kareiva and Anderson 1988). Unfortunately, experimental results cannot necessarily be extrapolated to larger scales (Levin 1992, Root and Schneider 1995). There are three potential problems with scaling up. First, larger-scale constraints may not be apparent at a smaller spatial or temporal scale or level of organization. For example, a short-term experiment would be unlikely to detect a range limit set by occasional extreme events, or subtle differences in extinction probability. Second, constraints appearing at a small scale may not be very important at a larger scale. We know, for example, that the key factor limiting population growth in one part of a species' range may differ from that in another location (e.g., Pollard 1979, Shaw 1981, Dempster 1983, Thrush et al. 2000). Although there is increasing interest in this problem (e.g., Connolly and Roughgarden 1999, Thrush et al. 2000), it is not yet clear how to predict when certain classes of factors will dominate either within a species or across species and thus how small-scale processes may generate large-scale patterns.

Third, all levels of organization from individual, to population, to species, to multiple species—and a corresponding range of spatial and temporal scales—interact to shape a species' distribution. For example, physiological limits to tolerance of environmental stress certainly exist and correlate broadly with geographic distribution. However, it is not necessarily clear whether this relationship demonstrates a fundamental physiological constraint in the species, or is contingent on population dynamics that inhibit local adaptation. Genetic variation in stress tolerance exists within natural populations (Hoffmann and Parsons 1991), suggesting that species could adapt to different climatic regimes. Intraspecific comparisons of populations from climatically different regions (e.g., Lamb 1977) complemented by many artificial selection experiments demonstrate that physiological tolerance can evolve rapidly (e.g., White et al. 1970, Tucic 1979, Huey and Kingsolver 1989). However, a comparison of the environmental correlates of the ranges of closely related species suggests that these niches tend to be conserved in evolutionary time (Peterson et al. 1999).

We currently know relatively little about range edge populations and how they may differ from central populations (Hoffmann and Parsons 1991, Hoffmann and Blows 1993). Little is known, for example, about the extent to which marginal populations are locally adapted, whether they are more likely to go extinct than equal-sized central populations, or what percentage of edge populations are demographic sinks, dependent on migration from central populations for persistence. There may be trade-offs between stress tolerance and fitness (Mongold et al. 1996, Bradshaw et al. 1998) such that migration from central-range populations opposes local adaptation (Stearns and Sage 1980, Kirkpatrick and Barton 1997). Thus population-level and species-level processes could interact with a physiological process to define a range limit.

One solution to the scaling problem is to alternate between scales interactively (Root 1991, Root and Schneider 1995). For example, an initial comparative analysis may reveal a pattern consistent with a particular hypothesized mechanistic relationship between factors. This hypothesis should then be explicitly tested at a local level to explore its mechanistic basis (e.g., Gilbert 1980, Muth 1980). Once the mechanism is understood, additional experiments in different localities or with different species can be targeted to more efficiently detect that mechanism at a larger scale and to predict under what circumstances it should be most important (e.g., Root 1991).

The approach I present in this chapter fits into this paradigm at the mechanistic level. After presenting some background information on what we already know about the relationship between butterflies and climate, I will focus on the mechanism controlling the leading edge of an advancing range in the sachem skipper. Understanding the direct impact of an environmental gradient in a natural population along the range edge is a first step toward integrating the individual- and population-level dynamics. Studying a population that may already be responding to rapid environmental change will further provide insight into the transient effects of rapid climate change, and hence the population- and species-level phenomenon of range change. Further experiments involving other locations and species will add to the results of this study and build a better bridge across the scale gap.

Butterfly Ranges and Climate

Research at many scales clearly shows that climate is important for butterfly abundance and distribution. The basic association between climate and insect distribution and abundance has been studied for at least 70 years (e.g., Uvarov 1931, Andrewartha and Birch 1954, Birch 1957, Dennis 1993). The challenge now is to focus on experiments that will integrate our mechanistic understanding of individual and population processes with distribution limits and change.

Large-scale patterns in butterfly distribution are generally consistent with the hypothesis that environmental conditions may constrain the northern boundaries of many species' ranges, although alternative hypotheses have been proposed (see discussion in Dennis 1993). As with many taxa, butterfly species richness characteristically declines along latitudinal and elevational gradients (Scriber 1973, Gieger 1987, Dennis et al. 1991, Sanchez-Rodriguez 1995). Dennis (1993) conducted a multifactorial analysis of species richness within Great Britain based on a systematic 10-km grid database on butterfly abundance (from the Butterfly Monitoring Program, Pollard 1977). Dennis (1993) recorded that butterfly species richness correlates with July temperature (correlation coefficient, $r = 0.85$), but also with January temperature ($r = 0.59$), the number of frost-free days ($r = 0.58$), and precipitation ($r = -0.49$). This analysis demonstrates a strong correspondence with environmental factors, particularly summer temperature.

If most species' ranges are constrained by temperature, then during a century of warming the prevailing direction of range shifts should be toward higher latitude and elevation. The 0.7°C warming trend in Europe since the 1890s should favor butterfly range expansion (Dennis 1993), based on physiological and ecological principles of butterfly biology. Parmesan et al. (1999) analyzed butterfly range shifts throughout Europe, and found that out of 35 species with data on their whole range, 63% have shifted northward, whereas only 3% have shifted southward. However, this analysis excluded species for which nonclimatic factors seemed to affect the range. In a separate analysis of the ranges of all British butterflies at a finer resolution, Pollard and Eversham (1995) discovered that many more species have contracted their ranges overall than have shifted. Of 59 resident British butterflies, 30 have experienced major range contractions and only 3 have expanded overall. Pollard and Eversham (1995) and Warren (1992) attribute this discrepancy to widespread habitat loss and degradation in Britain. It is not clear to what extent habitat degradation is sufficiently widespread to affect these species' entire ranges.

Importantly, all expanding British species but one (*Ladoga camilla*) are common and widespread within their ranges (Pollard and Eversham 1995), thus there may be a connection between local abundance and distribution change. Pollard and Eversham (1995) define as common and widespread those species found at > 80% of survey sites within their range. Fifty-five percent of all common species have expanded their range over the past three decades, while only 3% (a single species) of all localized species have expanded their range in Britain. Autecological studies have revealed a large number of associations between a species' population abundance over time in a given location, and climatic variation (Pollard 1979, Ehrlich 1980, Pollard and Lakhani 1985, Pollard 1988, Pollard 1991, Pollard and Yates 1993, Thomas et al. 1998). Multiple regression analyses show a significant correlation between population numbers and summer temperature for 6 out of 11 species that have expanded range since the 1940s (Pollard and Eversham 1995). Thus common butterfly species seem to track climatic fluctuations in both population size and range size better than other butterflies.

One possible mechanism of range shift resulting from climate change is that populations in an increasingly less favorable location will be more likely to go extinct and less likely to colonize new sites than populations in an increasingly favorable part of the range. This will lead to a shift in abundance, and over time, may lead to a range shift. Parmesan (1996) documented evidence of this type of shift in a North American species, Edith's checkerspot (*Euphydryas editha*). Population surveys since the 1930s documented the locations of *E. editha* populations from Canada to Mexico. Parmesan resurveyed populations, recording the number of populations that no longer survived. Significantly more population extinctions had occurred in the southern part of the range and at lower elevations than in the north or at higher elevations, resulting in a 92 km northward and 105 m upward shift in mean population location. This magnitude and direction of shift closely matched mean annual temperature isotherm shift (105 km northward, 105 m upward).

While these results support the hypothesis that climate trends influence the abundance and distribution of many species, there are very few cases where the mechanistic nature of the relationship between climate and distribution change is known. The white admiral butterfly, *Ladoga camilla*, is one of the best examples in which a detailed historical record is complemented by study of an edge population to attribute a causal mechanism to the association between distribution and climate in Britain. Using key factor analysis, Pollard (1979) determined that the limiting factor for a *Ladoga* population near the range edge is late larval and pupal development time in June. A cool June yields slow development, prolonged vulnerability to predation, and lower adult population size. Furthermore, a cool July reduces the available flight time for oviposition. The net result is that in cool years mortality is higher and fecundity is lower than in warm years. Pollard compared population sizes and appearance at new sites with May, June, and July temperature since 1900. He found a strong correlation between anomalously warm years and an increase in both new site records and population abundance in existing sites. It is not currently known whether this is a common mechanism linking distributions to climate.

Experimental Techniques

The crucial component that is missing in most analyses of animals' distributions is experimental validation that the relevant conditions outside the current range actually reduce performance or survival sufficiently to explain the range edge. Although this type of information may be unattainable for many species, the necessary manipulation is possible for many invertebrates and plants. Experimental introductions or transplants can provide direct evidence for the mechanism of range-limiting factors (Hoffmann and Blows 1994). Evidence for a significant difference between performance inside and outside a range under particular conditions is a much stronger basis for inference than correlation alone.

Transplant techniques have not often been used for the express purpose of ascertaining range-limiting factors in animals, and the results must be interpreted with caution due to the stochastic nature of interannual variability. The technique is difficult because it requires a great deal of natural history knowledge for site selection and minimal handling effects. Nonetheless, in some cases, range limiting factors have been clarified using this technique. Holdren and Ehrlich (1981) introduced individuals from a Wyoming butterfly population (*Euphydryas gillettii*) to Colorado, and the new population was viable for several generations. They conclude that this species is not present in Colorado because the Wyoming basin constitutes a dispersal barrier. Botanical examples are more plentiful. Davison (1977) found that the growth, development, and fecundity of *Hordeum murinum*, a ruderal plant in Britain, are directly inhibited by the adversity of climate outside its current range. Pigott (1975) found that the competitive advantage of two species of *Sedum* is reversed with altitude, leading to competitive exclusion of the respective species as an indirect effect of climate. The link between climate and plant distributions is better understood as a result of these and many complementary experiments (Woodward 1988, Chapin et al. 1992, Woodward 1993, Harte et al. 1995).

In some cases, researchers detect no declines in individual performance and move up to a higher level of organization to explain range limits. In a set of experiments by Prince and Carter (Carter and Prince 1985, Prince and Carter 1985, Prince et al. 1985), individual performance of the ruderal plant *Lactuca serriola* was not significantly reduced outside the range. The authors argue (Carter and Prince 1988) that in this species the balance between population extinction and recolonization is more sensitive to climate than individual performance; a slight deterioration of climatic conditions shifts the balance below a threshold such that population extinctions outnumber colonizations. Interestingly, the *E. editha* evidence (Parmesan 1996) discussed above is consistent with this hypothesized process. It is not clear whether this distinction between individual and population level sensitivity might affect the rate at which a species responds to climate change.

The logic underlying transplant experimentation is that performance should be reduced outside the range because the impact of the range-limiting factor should increase along some gradient across the range boundary. We can extend that logic further from the edge of the range to its center. We might predict that some range-limiting factors would affect the species across a relatively large area within their range, gradually decreasing in intensity from the range edge. Other range-limiting factors may only be important if they cross some threshold, causing a step in the response gradient at the range edge. Caughley et al. (1988) proposed that patterns of change in population growth rate, density, and individual body condition along a cross-section of the range should reflect the underlying mechanism of the range-limiting factor. This is a very interesting extension of the motivation behind transplant studies that may be applicable to a large number of species and could theoretically provide much needed quantification of the real impact of range-limiting factors on natural populations.

How do we move from global change to local experiments to the future? The correlation between climate and butterfly abundance and distribution suggests that we may be able to understand one potential response to climate change by studying range-limiting factors in this taxon. A mechanistic understanding of range limits is

important for predictive power in the context of rapid environmental change. I use experimental manipulations to probe the mechanisms underlying the association between climate and range limits in a case study of the skipper butterfly, *Atalopedes campestris*. One of the largest effects of rising concentrations of greenhouse gases is an increase in minimum temperature (IPCC 2001). *Atalopedes campestris* seems to be reacting specifically to a change in this climatic variable. Thus an understanding of the dynamics at the current range boundary gives us an edge in predicting the transient responses to climate change.

Atalopedes campestris: A Case Study

Atalopedes campestris, the sachem skipper, is a common, generalist butterfly in the family Hesperiidae (subfamily Hesperiinae: grass-feeding skippers) (Scott 1986). It is the only member of its genus that extends beyond the neotropics. *A. campestris* ranges from south of the equator into the United States (Burns 1989). In the central and eastern United States it frequently disperses northward in mid to late summer, but it dies out over winter and contracts to a permanent range in the southern states each year (Shapiro 1966, 1974, Scott 1986, Burns 1989, Opler and Bartlett Wright 1999). In the western United States, which may constitute a separate subspecies (Mattoon, pers. comm. 1997), *A. campestris* disperses to higher elevations in California over the summer (A. Shapiro, pers. comm., 1997), but is not found far northward beyond its overwintering range (Crozier 2001). In this chapter I define the range as the overwintering distribution, which is determined by the location of the spring flight period in April or May. In the Pacific Northwest *A. campestris* is typically rare in spring, but over three progressively larger and longer flight periods (corresponding to generations) from May through October it can achieve prodigious numbers (Crozier 2001). The final flight period extends from August to October. Eggs laid in September hatch and grow through a variable number of instars before winter (pers. obs., and J. Emmel, pers. comm. 1997).

A. campestris caterpillars eat grasses in the family Poaceae, including typical weed and lawn grasses. Larvae sew grass blades together with silk to make a protective tent at the base of the stem or undersides of leaves. *A. campestris* does not burrow underground, even in winter (Crozier, pers. obs.). Caterpillars pupate on top of the dirt, matting together bits of organic matter to blend in with the thatch. Adults nectar on many common flowers, favoring thistles (*Cirsium* spp.), clover (*Trifolium* spp.), and alfalfa (*Medicago sativa*) in Washington. *A. campestris* is a relatively large and aggressive skipper with territorial males that occupy both nectar and oviposition sites (pers. obs.). There is no indication that it is excluded from any habitat due to interspecific competition.

Range Expansion

Atalopedes campestris has a dynamic biogeographic history (Fig. 1.1). Its historical northwestern range is documented with collection

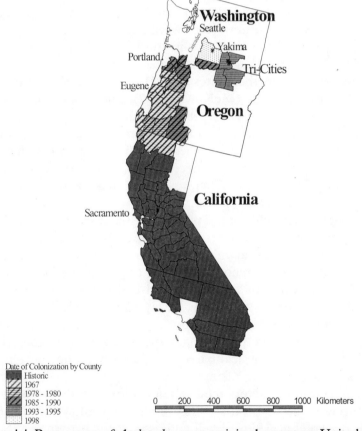

Figure 1.1. Range map of *Atalopedes campestris* in the western United States. The stages of range expansion are depicted by shading. The range map is modified (shading and dates added) from Opler (1995) by incorporating collection records from J. Hinchliff, J. Emmel, A. Shapiro, J. Pelham, and L. Crozier.

records since the mid-1800s, with which I have re-created the history of the western part of the range. In the first half of this century, the northern range edge lay in the Sacramento Valley region in California. Oregonian lepidopterists documented the invasion of Oregon in the 1960s (Dornfeld 1980). *A. campestris* first established at the base of the Willamette Valley (Eugene and Corvallis, Oregon) in 1967; it appeared in Salem, Oregon, in 1978, and finally Portland, Oregon, in 1986 (J. Hinchliff, collection records). Instead of continuing directly northward west of the Cascade Mountain Range, *A. campestris* moved eastward 335 km along both sides of the Columbia River (via White Salmon, 1990) to the Tri-Cities,[1] Washington (first record of a population was in 1993). Only two areas in Washington maintained populations at the beginning of my research in 1996: Vancouver in southwestern Washington contiguous with Portland, Oregon, and the Tri-Cities in southeastern Washington (pers. obs., Hinchliff 1996). Vagrant individuals appeared for the first time in many other far northern sites in 1991, including Wisconsin, Minnesota, North Dakota (McKown 1992), and Manitoba, Canada (Taylor 1993), but failed to establish populations. Evidence of long-distance travel is particularly important in the context of rapid climate change, because the paleoecological record shows that long jumps to new colonies may be necessary to explain rapid migrations that occurred in the past (Shapiro 1993, Pitelka and Plant Migration Workshop 1997, Clark et al. 1998).

In September 1998, at the peak of the flight season, I conducted a survey for *Atalopedes campestris* along the 125 km route northwest from the Tri-Cities to Yakima, Washington. Establishment in Yakima would mark the next major advance in the range expansion. Starting from a known population in the Tri-Cities, I stopped every 8 to 16 kilometers along Interstate Highway 82 to search for *Atalopedes*. I found substantial numbers of both males and females dispersing along the entire route, producing many first records of this species in Yakima County. Physical barriers apparently do not prevent this species from colonizing Yakima.

Range-limiting factors tend to fall within several categories (for recent reviews see Gaston 1990, Hoffmann and Blows 1994, Brown 1995, Brown et al. 1996). I will address specific hypotheses in each of the three major classes: geographical barriers to dispersal, biotic interactions, and abiotic factors.

Dispersal Limitations

Atalopedes campestris has demonstrated a capacity for both long-distance dispersal and dispersal by diffusion beyond its current range, as already described. Thus physical barriers are unlikely to have played a dominant role in the recent history of spread up the West Coast. The Rocky Mountains seem to be a barrier between the western and midwestern subspecies. Evidence of continuing spread suggests that the current range may not be in equilibrium with its environment. Many native species may be in disequilibrium if there is a lag time before changes in habitat or temperature cause a range to shift. Whether *Atalopedes* survives in this new location depends on the suitability of the habitat and the climate beyond its current range. If this lag time is operating, then fitness outside the range should not be significantly lower than fitness inside the range, which I test experimentally.

Biotic Limitations

Atalopedes' range expansion has occurred over a time period in which human land use has intensified and human population density has increased. *Atalopedes* thrives in abandoned lots, lawns, and gardens at densities that are probably higher than they would be in natural prairie habitat. Urbanization is not a sufficient explanation for the range expansion by itself, however. The expansion did not follow the main pattern of urbanization, which would have proceeded along the corridor between Portland and Seattle. Furthermore, the human presence in western Oregon has been substantial since the turn of the century and is not unique to the last 30 years. Nonetheless, understanding the role of land use changes is a very important consideration in predicting the consequences of climate change over the next century because human population growth and development may impact many terrestrial species, some negatively and some positively.

A change in habitat quality and quantity could affect a species' range by a number of possible mechanisms. To the extent that habitat has improved, higher average population density in what was originally marginal habitat would reduce the likelihood of extinction from stochastic population fluctuations. Enhanced food resources could raise individual survivorship and fecundity. Irrigation greatly increases reliability and density of grasses (larval food source), and it reduces the risk of desiccation stress. Agriculture and grazing may have improved adult food resources by facilitating the spread of native and exotic nectar sources, such as alfalfa and the late-blooming and drought-tolerant yellow star thistle (*Centaurea solstitialis*). In addition to enhancing summer survivorship, augmenting food resources may allow greater energetic reserves necessary to survive stress through long winters. Other anthropogenic impacts have degraded habitat, however. For example, pesticides, the spread of invasive cheat grass, changes in natural riparian processes (including the extirpation of beaver and the damming of the Columbia), and even radiation from the Hanford Nuclear Reservation may have impacted many species negatively, including *Atalopedes.*

At the population level, higher population sizes could expand the range by either a classic metapopulation model (Hanski 1996) or a source-sink dynamic (Pulliam 1996). Increased connectedness and decreased distance between secondary habitat patches, such as along roads, facilitates greater dispersal and colonization rates of empty patches. Since metapopulation persistence depends on the ratio between colonization and extinction rates, increasing colonization rates will enhance persistence even if there is no change in extinction rates. In a source-sink dynamic, populations in good habitat patches support populations in poor habitat via immigration. Larger source populations could support more marginal sink populations (i.e., populations that would not be sustained without replenishment by immigrants). In other words, new sink populations could appear if previously isolated populations become source populations. Enhanced population replacement and turnover, therefore, can cause the range to expand even if the conditions at the edge of the range are very poor.

Any of these factors could cause the edge of the range to shift northward either until it reaches the end of suitable habitat or until another limiting factor overwhelms the advantages of improved habitat. The key question is whether habitat is less suitable outside the current range.

I explored the suitability of habitat beyond the range by testing specifically for differences in larval growth or survivorship that could be attributed to host-plant quality or predation. Lawns are maintained in all cities of Washington State and British Columbia,

Canada, but neither the grass species planted nor the management of turf is geographically uniform. Over the course of the invasion of Oregon and Washington, the dominant grass in *Atalopedes'* habitat switched from primarily Bermuda grass (*Cynodon dactylon*) and other southern grasses, to perennial rye (*Lolium perenne*) in Oregon, and Kentucky bluegrass (*Poa pratensis*) in eastern Washington. In addition to changes in the dominant species, the relative proportion of species varies. Other potentially important factors include the use of pesticide directly on lawns or accidentally from nearby agricultural land. The butterfly might not be expanding further because some aspect of hostplant or chemical composition beyond the range is unfavorable. There may also be important differences in natural predators or parasitoids that inhibit *Atalopedes* outside the current range. If any of these possibilities are important, then transplanted caterpillars outside the current range should demonstrate significantly lower survivorship during the crucial developmental stages in summertime than control transplants within the range.

Transplant Methodology

My experiments focus on the section of *Atalopedes campestris*'s range that lies west of the Rocky Mountains. I focus on four localities, consisting of two pairs of sites at the margin (controls) and beyond the current overwintering range (as of 1998). The sites outside the range were chosen because the habitat is similar to that of marginal populations, and they are geographically the next logical step for the range expansion. The pairs differ qualitatively, representing a temperate, maritime pair west of the Cascade Range (Portland and Seattle), and a pair with a continental climate east of the Cascade Range (Tri-Cities and Yakima) (Fig. 1.1).

There are no obvious barriers to dispersal to either of the extralimital sites. Portland and the Tri-Cities currently support the northernmost populations of *Atalopedes campestris*. Both sites are on the Columbia River but are separated by the Cascade Mountain Range. Portland and Seattle are linked by the I-5 interstate with numerous towns, clearings, and native prairie habitats along the way. Dispersal to Yakima from the Tri-Cities could follow either riparian habitat along the Yakima River or rangeland up the Yakima Valley.

All areas are sufficiently urbanized to provide comparable larval and adult habitat, and are colonized by related and ecologically similar grass skippers (e.g., *Ochlodes sylvanoides, Polites sabuleti,* and *Hes-*

peria juba). Transplant sites were chosen to be as typical of the general locality as possible. For example, sites in the Tri-Cities and Yakima were irrigated during the growing season, which is necessary to maintain turf in these cities. Irrigation is much less common in Portland and Seattle, however, where grasses tend to go dormant during the summer. Transplant sites in Portland and Seattle did not receive irrigation over summer.

Portland has a diffuse population along the current range boundary and has a maritime climate comparable to Seattle's but warmer in the summertime. Portland is 280 km south of Seattle and is, on average, 1.9°C warmer in the summer but the same as Seattle ± 0.2°C over winter (WRCC 1999a; 1960–90 normals). Both western Cascade sites receive substantially more rain than the sites east of the Cascades, particularly during the La Niña year of 1998–99 (WRCC 1999a). November 1998 was the wettest on record in Seattle with 290 mm of rain (double the 30-year average of 145 mm).

The Tri-Cities currently support the northernmost overwintering population of *Atalopedes campestris,* with much more extreme winter lows and summer highs than Portland or Seattle. Yakima represents seemingly suitable habitat that is currently outside sachem range. Yakima is about 215 m higher in elevation than the Tri-Cities, and has a correspondingly lower temperature (by 2.5 to 3°C) throughout the year (WRCC 1999a).

I studied the behavior of these caterpillars within the range prior to transplant experimentation. In all transplant experiments, caterpillars were constrained within a given area with effective barriers, and plots were destructively sampled to prevent release into a novel location. Surrounding areas were also searched to ensure no caterpillars had left the plots.

To determine whether summer growth and survivorship were lower outside the range, I conducted two transplant experiments in the summer of 1998. Adult females collected in the Tri-Cities produced larvae in the lab, which I transplanted back out to the field in the designated developmental stage. I covered half of the plots with mosquito-netting to exclude predators. I censused plots by excavating the top layer of dirt, all grass, thatch, and caterpillars within. Plots were spread out across each city to span the range of local variation in physical exposure and host plant. On 7 August I transplanted a total of 1600 first instar larvae to the Tri-Cities, Yakima,

Portland, and Seattle (four plots covered and four uncovered per city); I evaluated development and survivorship on 13 July and 3 September, respectively. I analyzed arcsine-square root transformed survivorship in an Analysis of Variance (ANOVA) as a function of city and predator exclusion status.

Although survivorship was higher in covered plots, there were no significant differences between cities either in overall survivorship, or in the effect of the cages ($p = 0.05$). In the second experiment, Seattle had the lowest survivorship of the four cities (10% versus 25–34%), suggesting that some aspect of summer growth may be less favorable in Seattle than at other sites. Cooler summer temperatures or poorer food quality may be limiting west of the Cascades. However, Yakima had the highest survivorship overall (34 percent). I found similar results after a two-and-a-half-week experiment, in which I transplanted 200 second instar and 160 fourth instar caterpillars to the Tri-Cities and Yakima areas. Survivorship was higher in Yakima than the Tri-Cities for both developmental stages. Other factors may affect colonization success, such as adult fecundity. Experiments comparing oviposition across the range edge revealed no significant differences (Crozier 2001). These results contradict expectations if habitat or summer thermal conditions in Yakima are unsuitable.

Abiotic Limitations

Climatic conditions constitute the third class of range-limiting factors I will consider in this chapter. Favorable climatic periods mark each step in *Atalopedes* advance. Consideration of physiological constraints and current conditions at the range edge suggests a possible mechanism by which rising minimum temperature may affect this species. I explore physiological constraints with laboratory experiments on cold tolerance and then test their relevance to the range edge with field transplant experiments and environmental chamber simulations.

Correlation

Since the 1950s, *Atalopedes campestris* has expanded its range from northern California into the Willamette Valley in western Oregon and up the Columbia River into southeastern Washington (Fig. 1.1). This advance entails a total latitudinal shift of 5.75°N. The annual mean temperatures of all three states, California, Oregon, and Washington, have warmed from 1 to 2°C this century (Karl et al. 1995). The hottest summer in the Willamette Valley in the last 100 years occurred in 1967, the year *Atalopedes* colonized Oregon— maximum monthly temperature[2] at the colonized sites was 4.4°C above average (NCDC 1999b). The butterflies spread further in the Willamette only during a warm and dry phase of the Pacific Decadal Oscillation (1976–95; OCS 1999), which has a strong impact on northwest Oregon (Mantua et al. 1997, Taylor 1999). The weather in southeastern Washington has been much more favorable for *Atalopedes* since its establishment than it was prior, given the lack of extremely cold winters in the Tri-Cities since 1979 (Fig. 1.2). The summer of 1998 stands out as exceedingly warm in the Tri-Cities (Tmax 3.3°C above normal), coincident with *Atalopedes'* dispersal to Yakima.

Thus each leap in the expansion into Oregon and Washington occurred during a particularly warm period. However, initial expansion is not the most important phenomenon for this butterfly, as evidenced by its repeated pattern of seasonal expansion upward in elevation in California, which is not maintained over winter (A. Shapiro, unpublished data from 1972 to present), and unsuccessful colonizations of the northern midwestern United States and Canada. What stands out in the Pacific Northwest is the sustained presence over winter of populations well north of *Atalopedes'* historical range. I propose that rising winter minimum temperature may be crucial for the persistence of the northernmost population.

While mean temperature has risen globally about 0.5°C over the

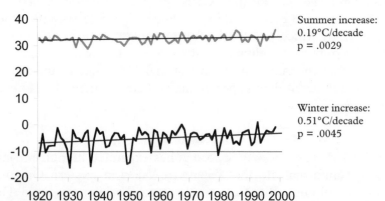

Figure 1.2. Average minimum and maximum monthly temperature in Kennewick, Washington, from 1920 to 1999 (NCDC 1999).

last century, minimum temperatures have changed at twice the rate of maximum temperature (Karl et al. 1993, Kukla et al. 1994, Karl et al. 1995, Easterling et al. 1997). Minimum temperature has also risen more at higher latitudes than at lower latitudes (CSIRO 2000) where it may have a disproportionate effect on range boundaries (MacArthur 1972). The natural consequences of a change specifically in minimum temperature are beginning to be recognized (Epstein et al. 1998, Alward et al. 1999, Brown et al. 1999), and will be important in anticipating rapid responses to global warming. A high sensitivity to the climatic parameter that has changed the most and is stereotypical of the phenomenon of a "greenhouse effect" could trigger very efficient responses to modern climate change.

I have compared the average minimum temperature (Tmin) in the coldest month each winter (December, January, or February), with the average maximum temperature (Tmax) in the warmest month (July or August) in Tri-Cities since 1920 (Fig. 1.2). Linear regressions of temperature against time are highly significant for both Tmax and Tmin (p-Tmax = .0029; p-Tmin = 0.0045), and show a strong trend toward warming. Maximum temperature has increased 0.19°C/decade. Minimum temperature has risen much more dramatically, at an average rate of 0.514°C/decade.

In summary, the overall pattern of *Atalopedes'* range expansion suggests that warm weather is important for colonization of new territories. A study of the climatic trends locally at the edge of the range reveals a long-term warming trend, particularly in minimum temperature. Evidence from failed colonizations elsewhere suggests that winter may be a critical factor for persistent populations.

Mechanism

Rising minimum temperature is the climatic factor likely to be particularly important for those species poorly adapted to cold. Most insects exposed regularly to harsh winters tolerate these conditions by entering hibernal diapause and cold hardening, or are restricted to insulated microhabitats (Uvarov 1931, Danks 1978, Bale 1987, Leather et al. 1993). Diapause is characterized by an extraordinarily low metabolic rate, which is associated with enhanced stress resistance (Hoffmann and Parsons 1991) and successful long-term dormancy (Tauber et al. 1986). Cold-adapted species are either freeze tolerant or they undergo extensive chemical changes to depress the temperature at which they freeze (i.e., the supercooling point) (Salt and James 1947, Lee and Denlinger 1991, Storey and Storey 1996).

Despite these adaptations, several boreal species examined at the edge of their range still run the risk of population extinction over winter, such as the praying mantid, *Mantis religiosa* (Salt and James 1947) and the corn earworm, *Heliothis zea* (Danks 1978).

The Tri-Cities is both cold enough to kill most species that are not cold-adapted (air temperature commonly dips below –15°C) and dry (annual precipitation of only 196 mm), offering no thermal insulation from snow (WRCC 1999a). The sachem skipper may be unique among resident butterflies in Washington State in lacking the capacity for true hibernal diapause (Scott 1979). If *Atalopedes* is typical of most non–cold adapted insects and is not freeze tolerant, has a relatively high supercooling point, and is not in well-protected microhabitats over winter, it may be close to its lower lethal temperature at the edge of its range. Given this situation, minimum temperature, specifically, is likely to directly impact persistence at the northern range limit of this species.

Experiments

I have conducted a number of experiments at different temporal scales to explore the cold tolerance of *Atalopedes campestris*. Minimum temperature tolerance experiments document the extreme temperatures below which this species cannot survive. Simulated-winter experiments characterize the relative cost of a typical day in winter within and beyond the range. Finally, field transplant experiments document mortality that actually occurs during winter along the range edge and beyond the range over the entire winter (see Table 1.1).

Table 1.1. Outline of Experiments Showing Their Relative Duration.

Experiment	Duration
MINIMUM TEMPERATURE TOLERANCE	
Supercooling Point	6 hours
Critical Minimum Temperature	2 days
SIMULATED WATER	
January Simulation	5–7 weeks
Whole Winter	6 months
FIELD TRANSPLANTS	
Half Winter	3 months
Whole Winter	6 months

Minimum Temperature Tolerance

I reared caterpillars in conditions they would likely experience if born in the final flight period (a cool 12°C night/20°C day, and a short photoperiod of 10 hrs), then conducted multiple assays of cold tolerance. *Atalopedes campestris* has a supercooling point typical of species that lack cryoprotectants or are tested during nondiapausing stages (–6.64°C, cooling rate: 6°C/hr). If they are wet, they freeze at a higher temperature (–4.42°C), as expected from the ice-nucleating property of water. Therefore, they are particularly vulnerable to rapid freezes after rain, even if the temperature is not extremely cold.

This species does not diapause. I have documented this by leaving caterpillars outside all fall and testing their ability to resume development in December and January. The caterpillars all began eating immediately upon warm-up, and many molted within 24 hours. Nor does *Atalopedes* have a developmental stage that is more cold tolerant (there is no significant difference in the supercooling points of different instars). Freezing proved fatal in all cases, even if freezing occurred above the average supercooling point. These traits are unusual for most boreal species, but typical of many southern insects. They demonstrate the lack of any special cold-adaptation in this species.

CRITICAL MINIMUM TEMPERATURE
The typical techniques for comparative cold tolerance described above are unlike exposure in the field, particularly in the rate of cooling. I therefore explored cold tolerance with a slower cooling rate (3°C/hr) and typical nighttime exposure (12 hrs) to a set minimum temperature. I assessed survival 24 hours after return to room temperature. The results show (Crozier 2001) that mortality of *Atalopedes campestris* dramatically increases from –5°C (50% mortality) to –10°C (100% mortality). This range is consistent with the supercooling point results, reflecting the increasing probability of freezing at temperatures below –5°C.

SIMULATED WINTER
To study the consequences of repeated exposure to cool temperature above the supercooling point, I studied the effects of a 6-week winter in environmental chambers. The simulated winters included

3 weeks of exposure to typical January daily temperatures within and beyond the range. Treatments explored the effects of daily minima ranging from –4 to +4°C (Crozier 2001). Survivorship was significantly lower in the –4°C treatment than any of the warmer treatments.

Taken together, these laboratory results predict that microhabitat with a long-term average minimum temperature below –4°C, or a short-term extreme minimum of –10°C is beyond *Atalopedes'* physiological tolerance and thus should define a range-limiting factor. This predicted threshold is remarkably close to the observed January minimum temperature isotherm that corresponds with the actual range edge, –4°C (Crozier 2001).

FIELD TRANSPLANTS

I used the field transplant methodology described above to explore winter mortality in the field from October to March 1997–98 and 1998–99 within and beyond the range. The summer experiment demonstrated high survival using this technique, indicating that the transplant process itself is not too stressful and the habitat is suitable. Within the range, plots were placed proximate to where I had observed wild females laying eggs late in the season. I transplanted second- to fourth-instar caterpillars, which are in the predominant overwintering stage. I conducted this experiment twice; the more modest version of 400 caterpillars/city in three cities (Corvallis, Tri-Cities, and Yakima) occurred in the El Niño year 1997–98. I greatly expanded the experiment in 1998–99 (a La Niña year) to 1400 caterpillars/city, and tested four cities (Portland, Seattle, Tri-Cities, and Yakima).

No caterpillars survived through March the first year I tried this experiment, apparently because my sample size was too small. In the second year, I assessed midwinter survivorship in January. Survivorship was significantly higher inside the range (Portland and Tri-Cities) than outside the range (Seattle and Yakima), although the absolute number of survivors was very low (Crozier 2001). The extremely high mortality observed even within the range (95%), when compared with summer transplant survival, does suggest that winter is enormously hazardous, and that range edge populations could go extinct in some years.

Natural oviposition might yield higher survivorship than transplantation for a variety of reasons. The subset of the wild population that survives the winter could somehow behave qualitatively differently than the transplanted caterpillars. More likely, local topographical variation may create sheltered sites with higher survivorship than was observed in my experiments. Nonetheless, since many eggs are laid in the "cold" sites, average mortality must be very high. Furthermore, the experimental mortality rate was consistent with changes in adult abundance from fall to spring, as assessed by population surveys (Crozier 2001).

A natural experiment tested the conclusion that winter temperature is the predominant range-limiting factor for *Atalopedes campestris*. The skippers colonized Yakima in the summer of 1998 (pers. obs.). The probability of success of the colonization should depend on the severity of the winter. If overwinter temperature limits the current range, then during a cold winter most of the colonists would probably die. This occurred over the 1998–99 winter. An unusually warm winter, however, should facilitate a successful colonization, as occurred in their second attempt in 1999–2000. When supplemented with further experiments to identify the key aspect of winter that is most important, a mechanistic explanation for how climate change can directly affect this species is possible.

Microclimatic Conditions

The microhabitat of most animals buffers them somewhat from extreme fluctuations in temperature. Standard meteorological techniques recorded at long-term weather monitoring stations typically do not measure conditions in the habitat where most animals live. It is these standardized measurements, however, that are most often used to assess the importance of climate for range limits. Comparisons between microclimatic conditions and readily available weather station databases are essential to causally link a correlation between a range limit and a climatic variable, and the local conditions that determine the mechanism of range limitation. I studied microclimatic conditions within my experimental field plots to answer three questions:

1. Do conditions at the ground level differ systematically across the range boundary, as standardized meteorological records predict?
2. Does *A. campestris* actually approach its critical thermal minimum within its microhabitat at the edge or beyond its range?
3. What aspects of cold exposure may be important other than the absolute minimum temperature?

To determine how microhabitat conditions relate to recorded air temperature, I compared my field measurements with National Climatic Data Center records from established meteorological stations in each city over the winter of 1997–98 (NCDC 1999b). I measured the temperature in my plots at five different sites in 1997–98 (one in Corvallis, two in the Tri-Cities, and two in Yakima), and at ≥ 3 sites per city from 1998 to 2001. The Stowaway XTI dataloggers record temperature every 30 minutes throughout the winter. Temperature probes were inserted into the thatch layer, where caterpillars are typically found.

CITY COMPARISONS
The temperature in the thatch differs systematically between cities. Average monthly temperature in the plots was significantly colder (2.8°C) in Yakima than in the Tri-Cities, which was in turn 3.6°C colder than Corvallis. The differences between the Tri-Cities and Yakima were significantly greater than differences within a city for both average and minimum temperature (Crozier 2001). Furthermore, mean daily NCDC temperature was strongly correlated with mean daily plot temperature (correlation coefficient = 0.8) suggesting that extrapolation from standard historical temperature records is justified. Thus the expansion of the range from Oregon to the Tri-Cities to Yakima does represent a substantial change in their thermal environment, and this microenvironment has almost definitely experienced directional change over time.

CRITICAL MINIMUM TEMPERATURE
Over the 1997–98 winter, absolute minimum temperature was not extremely cold (–3.9°C) in any of these sites relative to either caterpillar supercooling point or their critical minimum temperature. The minimum temperature the following winter (1998–99), however, was very cold despite being a warmer winter on average (–11.8°C in Yakima). Plot temperatures outside the Tri-Cities surpassed the critical minimum temperature measured in the laboratory. Based on this result, I would predict that the colonization of Yakima in the fall of 1998 will not be successful. And, in fact, no *Atalopedes* could be found through weekly population surveys in the spring of 1999 (Crozier 2001). The Tri-Cities was consistently at least 2°C warmer (average difference: 2.8°C), which would cause

significantly higher survivorship near *Atalopedes* critical minimum temperature. Thus a proportion of the potential overwintering sites immediately beyond the range do, in some years, approach the critical minimum temperature measured in the laboratory.

OTHER CHARACTERISTICS OF WINTER
Laboratory experiments showed that additional mortality results from prolonged exposure to cool temperatures (above caterpillar supercooling point). Thus the duration of exposure to temperatures below 0°C may be important for population persistence. Yakima plot temperature was below 0°C in January of 1998 for 26 days. This may be a sufficiently long period of time to cause high mortality even if caterpillars do not actually freeze. The Tri-Cities plots, however, were only below 0°C for 8 to 10 days. This type of difference could be important in differentiating tolerable and intolerable conditions.

Other Considerations

One question that is still unresolved is to what extent *Atalopedes'* persistence in a marginal location over winter hinges on how favorable conditions were for population growth the previous summer. If overwinter survivorship is a function of fall population size in addition to winter temperature, winter and summer temperature may have an interactive effect on the range boundary. I am pursuing this possibility in my continued work on the population dynamics at the range-edge.

It is natural to ask whether the marginal populations are actually sink populations, which are maintained by migration from further south. This possibility cannot be ruled out, but I have found no evidence to support it and believe it is unlikely. In 1998 the Tri-Cities population appeared earlier in spring than the nearest neighboring population in Portland (Crozier, unpub. data). Although this may not always be true, it does indicate that at least in some years the Tri-Cities population is self-sustaining.

Additional experiments verifying that experimental mortality rates accurately reflect the true cost of winter for the natural population, and clarifying how marginal populations are maintained in the face of such high mortality will provide much needed additional insight into the mechanics of an actual range edge. But the fact remains that winter temperature is the best explanation at this point for the current range boundary in eastern Washington.

Conclusion

During the past 100 years, many regions of the United States have warmed 1 to 3°C (Karl et al. 1995). We now have the opportunity to explore how the range boundaries of species that have moved over this interval may depend on temperature. The leading edge of a species' distribution during range change is an excellent place to test our understanding of the relationship between animals and their environment. Directly studying range changes in progress will yield additional information on transient effects of rapid environmental change, and how anthropogenic factors such as habitat domestication and fragmentation affect this process.

Employing a hierarchical approach, from correlation, to development of mechanistic hypotheses, to experimentation in both natural and controlled environments, I have begun to probe the relationship between climate change and species range boundaries. This research explores experimentally a set of factors that are commonly correlated with species range edges. This approach has proven successful in eliminating several hypotheses of range-limiting factors. I have shown that *Atalopedes campestris* can disperse, survive, and complete development efficiently beyond the current range edge. Although not all of the potentially relevant factors can be tested experimentally, habitat outside the range in all probability is suitable for population establishment.

Winter temperature, however, stresses the physiological limits of this species at and beyond the range edge. Like many southern insects, this species does not diapause, is not freeze tolerant, and has a relatively high supercooling point. Laboratory-based studies of cold tolerance characterize the absolute thermal constraints that could define a limit to the northern range edge. *Atalopedes campestris* suffers from extremely high mortality after either a short exposure to −10°C or repeated daily exposure to temperatures below −4°C. This predicted threshold matches the −4°C isotherm of January minimum temperature that correlates well with the actual range edge. A study of field microclimatic conditions over winter shows that these limits are surpassed near the range edge, explaining the exceptionally high mortality observed in field transplant experiments as well as natural populations over winter.

My research suggests that the recent warming trend in the Pacific Northwest is likely to have directly affected the range boundary of this species, converting previously uninhabitable territories into habitable ones. Average January minimum temperature in the Tri-Cities has risen above the critical –4°C isotherm only in the late twentieth century, coinciding with colonization by *Atalopedes campestris*. Interestingly, each step in the range expansion of this species occurred during an unusually warm summer. The factors that facilitate dispersal may be different from those that facilitate persistence, and may need to act in concert for the most rapid response to climate change. Despite dispersal to Yakima over the summer of 1998, the colonists did not appear to survive the relatively cool winter of 1998–99. Establishment did occur during the summer of 1999, which was followed by an especially warm winter (Crozier 2001). Population growth has continued until the present (2001).Whether the key factor is winter alone, or winter and summer temperature in concert, *Atalopedes campestris* appears to have the capacity to react directly and rapidly to global warming.

These butterflies benefit from human-associated habitat and are ecological generalists. These traits allow them to expand without habitat limits until they reach climatic constraints. Although generalists may have the greatest capacity to express this response to warming, a great many other species are reacting to environmental change in more subtle ways (see other chapters in this volume). Human commensals show us what endangered species might have done if habitat were not so degraded, or populations so depleted. They provide visible proof that climate change can cause large-scale impacts on wildlife. May we heed the warning that the living Earth is a sensitive system, and recognize the risks of destroying that balance.

Notes

1. The Tri-Cities refers to Richland, Kennewick, and Pasco, which are separated by the Columbia, Yakima, and Snake Rivers. *Atalopedes campestris* is present in all three cities, which form a single population.
2. Maximum monthly temperature (Tmax) is the average of all daily

maximum temperatures in a single month. Similarly, Tmin represents the average daily minimum temperature for a particular month. For simplicity, I use this parameter for all historical comparisons.

Literature Cited

Alward, R. D., J. K. Detling, and D. G. Milchunas. 1999. Grassland vegetation changes and nocturnal global warming. *Science* 283: 229–231.

Andrewartha, H. G., and L. C. Birch. 1954. *The Distribution and Abundance of Animals.* Chicago: University of Chicago Press.

Bale, J. S. 1987. Insect cold-hardiness: Freezing and supercooling—an eco-physiological perspective. *Journal of Insect Physiology* 33: 899–908.

Barry, J. P., C. H. Baxter, R. D. Sagarin, and S. E. Gilman. 1995. Climate-related, long-term faunal changes in a California rocky intertidal community. *Science* 267: 672–675.

Bartholomew, G. A. 1958. The role of physiology in the distribution of terrestrial vertebrates. Pages 81–95 in C. L. Hubbs, ed., *Zoogeography*. American Association for the Advancement of Science Publication, no. 51. AAAS: Washington, D.C.

Birch, L. C. 1957. The role of weather in determining the distribution and abundance of animals. Pages 203–218 in *Population Studies: Animal Ecology and Demography*. Cold Spring Harbor Symposia on Quantitative Biology, 22. The Biological Laboratory: Cold Spring Harbor, L.I. New York.

Bradshaw, W. E., P. A. Armbruster, and C. M. Holzapfel. 1998. Fitness consequences of hibernal diapause in the pitcher-plant mosquito, *Wyeomyia smithii. Ecology* 79(4): 1458–1462.

Brown, J. H. 1995. *Macroecology.* Chicago: University of Chicago Press.

Brown, J. H., G. C. Stevens, and D. M. Kaufman. 1996. The geographic range: Size, shape, boundaries, and internal structure. *Annual Review of Ecology and Systematics* 27: 597–623.

Brown, J. L., S. H. Li, and N. Bhagabati. 1999. Long-term trend toward earlier breeding in an American bird: A response to global warming? *Proceedings of the National Academy of Science* 96: 5565–5569.

Brubaker, L. B. 1988. Vegetation history and anticipating future vegetation change. Pages 41–60 in J. K. Agee and D. R. Johnson, eds., *Ecosystem Management for Parks and Wilderness.* Seattle: University of Washington Press.

Burns, J. M. 1989. Phylogeny and zoogeography of the bigger and better genus *Atalopedes* (Hesperiidae). *Journal of the Lepidopterists' Society* 43: 11–32.

Cammell, M. E., and J. D. Knight. 1992. Effects of climatic change on the population dynamics of crop pests. *Advances in Ecological Research* 22: 117–162.

Carter, R. N., and S. D. Prince. 1985. The geographical distribution of prickly lettuce (*Lactuca serriola*) I. A general survey of its habitats and performance in Britain. *Journal of Ecology* 73: 27–38.

Carter, R. N., and S. D. Prince. 1988. Distribution limits from a demographic viewpoint. Pages 165–184 in A. J. Davy, M. J. Hutchings, and A. R. Watkinson, eds., *Plant Population Ecology*. London: Blackwell Scientific.

Caughley, B., J. Short, G. C. Grigg, and H. Nix. 1987. Kangaroos and climate: An analysis of distribution. *Journal of Animal Ecology* 56: 751–761.

Caughley, G., D. Grice, R. Barker, and B. Brown. 1988. The edge of the range. *Journal of Animal Ecology* 57: 771–785.

Chapin, F. S., J. F. Reynolds, R. L. Jefferies, G. R. Shaver, J. Svoboda, and E. W. Chu, eds. 1992. *Arctic Ecosystems in a Changing Climate*. New York: Academic Press.

Clark, J. S., C. Fastie, G. Hurtt, S. T. Jackson, C. Johnson, G. A. King, M. Lewis, J. Lynch, S. Pacala, C. Prentice, E. W. Schupp, T. Webb, and P. Wyckoff. 1998. Reid's paradox of rapid plant migration. *Bioscience* 48: 13–24.

Connolly, S. R., and J. Roughgarden. 1999. Theory of marine communities: Competition, predation, and recruitment-dependent interaction strength. *Ecological Monographs* 69: 277–296.

Crozier, E. G. 2001. Climate Change and Species' Range Boundaries: A Case Study at the Northern Range Limits of *Atalopedes campestris* (Lepidoptera: Hesperiidae), the Sachem Skipper. Ph.D. thesis. Seattle: University of Washington.

Crick, H. Q. P., C. Dudley, D. E. Glue, and D. L. Thomson. 1997. UK birds are laying eggs earlier. *Nature* 388: 526.

CSIRO, (Australia) 2000. Temperature changes since the nineteenth century. Available online at http://www.dar.csiro.au/publications/Holper_1999a.htm, Commonwealth Scientific and Industrial Research Organization (Australia).

Danks, H. V. 1978. Modes of seasonal adaptation in the insects. I. Winter survival. *Canadian Entomologist* 110: 1167–1205.

Davis, M. B. 1976. Pleistocene biogeography of temperate deciduous forests. Pages 13–26 in R. C. West and W. G. Haag, eds., *Ecology of the Pleistocene: A Symposium*. Geoscience and Man 13. Baton Rouge: School of Geoscience, Louisiana State University.

Davison, A. W. 1977. The ecology of *Hordeum murinum* L. III. Some effects of adverse climate. *Journal of Ecology* 65: 523–530.

Dempster, J. P. 1983. The natural control of populations of butterflies and moths. *Biological Reviews of the Cambridge Philosophical Society* 58: 461–481.

Dennis, R. L. H. 1993. *Butterflies and Climate Change*. Manchester, N.Y.: Manchester University Press.

Dennis, R. L. H., W. R. Williams, and T. G. Shreeve. 1991. A multivariate approach to the determination of faunal structures among European butterfly species (Lepidoptera: Rhopalocera). *Journal of the Linnean Society. Zoology* 101: 1–49.

Dornfeld, E. J. 1980. *Butterflies of Oregon.* Forest Grove, Ore.: Timber Press.

Easterling, D. R., B. Horton, P. D. Jones, T. C. Peterson, T. R. Karl, D. E. Parker, M. J. Salinger, V. Razuvayev, N. Plummer, P. Jamason, and C. K. Folland. 1997. Maximum and minimum temperature trends for the globe. *Science* 277: 364–367.

Ehrlich, P. R. 1980. Extinction, reduction, stability and increase: The responses of checkerspot butterfly (*Euphydryas*) populations to the California drought. *Oecologia* 46: 101–105.

Epstein, P. R., H. F. Diaz, S. Elias, G. Grabherr, N. E. Graham, W. J. M. Martens, E. Mosley-Thompson, and J. Susskind. 1998. Biological and physical signs of climate change: Focus on mosquito-born diseases. *Bulletin of the American Meteorological Society* 79: 409–417.

Gaston, K. J. 1990. Patterns in the geographical ranges of species. *Biological Reviews of the Cambridge Philosophical Society* 65: 105–129.

Gieger, W. 1987. *Les papillons de jour et leurs biotopes.* Bale: Ligue Suisse pour la Protection de la Nature.

Gilbert, N. 1980. Comparative Dynamics of a Single-Host Aphid. I. The Evidence. *Journal of Animal Ecology* 49: 351–369.

Grabherr, G., M. Gottfried, and H. Pauli. 1994. Climate effects on mountain plants. *Nature* 369: 448.

Hanski, I. 1996. Metapopulation ecology. Pages 13–43 in O. E. J. Rhodes, R. K. Chesser, and M. H. Smith, eds., *Population Dynamics in Ecological Space and Time.* Chicago: University of Chicago Press.

Harte, J., M. S. Torn, F. Chang, B. Feifarek, A. P. Kinzig, R. Shaw, and K. Shen. 1995. Global warming and soil microclimate: Results from a meadow-warming experiment. *Ecological Applications* 5: 132–150.

Hengeveld, R. 1985. Dynamics of Dutch beetle species during the twentieth century (Coleoptera: Carabidae). *Journal of Biogeography* 12: 389–411.

Hinchliff, J. 1994. *An Atlas of Oregon Butterflies.* Corvallis, Ore.: The Evergreen Aurelians.

———. 1996. *An Atlas of Washington Butterflies.* Corvallis, Ore.: The Evergreen Aurelians.

Hoffmann, A. A., and M. W. Blows. 1993. Evolutionary genetics and climate change: Will animals adapt to global warming? Pages 165–178 in P. Kareiva, J. Kingsolver, and R. Huey, eds., *Biotic Interactions and Global Change.* Sunderland, Mass.: Sinauer.

———. 1994. Species borders: Ecological and evolutionary perspectives. *Trends in Ecology and Evolution* 9: 223–227.

Hoffmann, A. A., and P. A. Parsons. 1991. *Evolutionary Genetics and Environmental Stress.* Oxford: Oxford University Press.

Holbrook, S. J., R. J. Schmitt, and J. S. J. Stephen. 1997. Changes in an assemblage of temperate reef fishes associated with a climate shift. *Ecological Applications* 7: 1299–1310.

Holdren, C. E., and P. R. Ehrlich. 1981. Long range dispersal in checkerspot butterflies: transplant experiments with *Euphydryas gillettii. Oecologia* 50: 125–129.

Huey, R. B., and J. G. Kingsolver. 1989. Evolution of thermal sensitivity of ectotherm performance. *Trends in Ecology and Evolution* 4: 131–135.

Huntley, B. 1991. How plants respond to climate change: Migration rates, individualism and the consequences for plant communities. *Annals of Botany* 67: 15–22.

Huntley, B., W. Cramer, A. V. Morgan, H. C. Prentice, and J. R. M. Allen, eds. 1997. *Past and Future Rapid Environmental Changes: The Spatial and Evolutionary Responses of Terrestrial Biota.* New York: Springer-Verlag.

Intergovernmental Panel on Climate Change (IPCC). 2001. *Climate Change 2001: The Scientific Basis,* eds., J. T. Houghton, Y. Ding, D. J. Griggs, M. Noguer, P. J. van der Linden, and D. Xiaosu. Contribution of working group I to the third assessment report of the IPCC. Cambridge: Cambridge University Press.

Kareiva, P. M., and M. Anderson. 1988. Spatial aspects of species interactions: The wedding of models and experiments. Pages 35–50 in A. Hastings, ed., *Community Ecology. Proceedings of a Workshop Held in Davis, California, April 1986.* New York: Springer-Verlag.

Karl, R. R., R. W. Knight, D. R. Easterling, and R. G. Quayle. 1995. Indices of climate change for the United States. *Bulletin of the American Meteorological Society* 77: 279–292.

Karl, T. R., P. D. Jones, R. W. Knight, G. Kukla, M. Plummer, V. Razuvayev, K. P. Gallo, J. Lindseay, R. J. Charlson, and T. C. Peterson. 1993. Asymmetric trends of daily maximum and minimum temperature. *Bulletin of the American Meteorological Society* 74: 1007–1023.

———. 1995. Asymmetric trends of daily maximum and minimum temperature. Pages 80–96 in N. R. C. Climate Research Committee. *Natural Climate Variability on Decade-to-Century Time Scales.* Washington, D.C.: National Academy Press.

Kirkpatrick, M., and N. H. Barton. 1997. Evolution of a species' range. *American Naturalist* 150: 1–23.

Kukla, G., T. Karl, and M. Riches, eds. 1994. Asymmetric Change of Daily Temperature Range: Proceedings of the International MINIMAX WORKSHOP, College Park, Md., 27 to 30 September 1993. Washington D.C.: U.S. Department of Energy.

Lamb, H. H. 1977. *Climate Present, Past and Future.* Vol. 2: Climatic History and the Future. London: Methuen.

Lawton, J. H. 1995. The response of insects to environmental change. Pages 3–26 in R. Harrington and N. E. Stork, eds., *Insects in a Changing Environment.* New York: Academic Press, Harcourt Brace & Company.

Leather, S. R., K. F. A. Walters, and J. S. Bate. 1993. *The Ecology of Insect Overwintering.* Cambridge: Cambridge University Press.

Lee, R. E., and D. L. Denlinger, eds. 1991. *Insects at Low Temperatures.* New York: Chapman & Hall.

Levin, S. 1992. The problem of pattern and scale in ecology. *Ecology* 73: 1943–1967.

MacArthur, R. H. 1972. *Geographical Ecology: Patterns in the Distribution of Species.* New York: Harper & Row.

Mantua, N. J., S. R. Hare, Y. Zhang, J. M. Wallace, and R. C. Francis. 1997. A Pacific interdecadal climate oscillation with impacts on salmon production. *Bulletin of the American Meteorological Society* 78(6): 1069–1079.

McCleery, R. H., and C. M. Perrins. 1998. . . . .temperature and egg-laying trends. *Nature* 391: 30–31.

McKown, S. E. 1992. 1991 season summary. *News of the Lepidopterists' Society,* No. 2(Mar/Apr).

Mongold, J. A., A. F. Bennet, and R. E. Lenski. 1996. Evolutionary adaptation to temperature. IV: Adaptation of *Escherichia coli* at a niche boundary. *Evolution* 50(1): 35–43.

Muth, A. 1980. Physiological ecology of desert iguana (*Dipsosaurus Dorsalis*) eggs: temperature and water relations. *Ecology* 61: 1335–1343.

Myneni, R. B., C. D. Keeling, C. J. Tucker, G. Asrar, and R. R. Nemani. 1997. Increased plant growth in the northern high latitudes from 1981 to 1991. *Nature* 386: 698–702.

NCDC. 1999a. Climate of 1998 Annual Review. Available online at http://www.ncdc.noaa.gov, National Oceanic and Atmospheric Administration, National Climatic Data Center.

———. 1999b. Global Historical Climatological Network Data. Available online at http://www.ncdc.noaa.gov, National Oceanic and Atmospheric Administration, National Climatic Data Center.

OCS, 1999. Portland Water Year Precipitation, 1920–2020. Available online at www.ocs.orst.edu/reports/Pdxprcp. Oregon Climate Service.

Opler, P., and A. B. Bartlett Wright. 1999. *A Field Guide to Western Butterflies.* Boston, New York: Houghton Mifflin Company.

Opler, P. A., H. Pavulaan, and R. Stanford. 1995. *Butterflies of the United States.* Jamestown, N.D.: Northern Prairie Wildlife Research Center Home Page, available online at http://www.npsc.nbs.gov/resource/distr/lepid/bflyusa.html. Version 16Jul97.

Overpeck, J. T., R. S. Webb, and T. Webb III. 1992. Mapping eastern North American vegetation change of the past 18ka: No-analogs and the future. *Geology* 20: 1071–1074.

Parmesan, C. 1996. Climate and species' range. *Nature* 382: 765–766.

Parmesan, C., N. Ryrholm, C. Stefanescu, J. K. Hill, C. D. Thomas, H. Descimon, B. Huntley, L. Kaila, J. Kullberg, T. Tammaru, W. J. Tennent, J. A. Thomas, and M. Warren. 1999. Poleward shifts in geographical ranges of butterfly species associated with regional warming. *Nature* 399: 579–583.

Peterson, A. T., J. Soberon, and V. Sanchez-Cordero. 1999. Conservatism of ecological niches in evolutionary time. *Science* 285: 1265–1267.

Pigott, C. D. 1975. Experimental studies on the influence of climate on the geographical distribution of plants. *Weather* 30: 82–90.

Pitelka, L. F., and Plant Migration Workshop. 1997. Plant migration and climate change. *American Scientist* 85: 464–473.

Pollard, E. 1977. A method for assessing changes in the abundance of butterflies. *Biological Conservation* 12: 115–134.

———. 1979. Population ecology and change in range of the white admiral butterfly *Ladoga camilla* L. in England. *Ecological Entomology* 4: 61–74.

———. 1988. Temperature, rainfall, and butterfly numbers. *Journal of Applied Ecology* 25: 819–828.

———. 1991. Changes in the flight period of the hedge brown butterfly *Pyronia tithonus* during range expansion. *Journal of Animal Ecology* 60: 737–748.

Pollard, E., and B. C. Eversham. 1995. Butterfly monitoring 2—interpreting the changes. Pages 23–36 in A. S. Pullin, ed., *Ecology and Conservation of Butterflies*. London: Chapman & Hall.

Pollard, E., and K. H. Lakhani. 1985. Butterfly Monitoring Scheme: Effects of Weather on Abundance. Institute of Terrestrial Ecology, Annual Report 1984. N. E. R. Council. Huntington, England: Institute of Terrestrial ecology.

Pollard, E., and T. Yates. 1993. *Monitoring Butterflies for Ecology and Conservation*. London: Chapman and Hall.

Post, E., and N. C. Stenseth. 1999. Climatic variability, plant phenology, and northern ungulates. *Ecology* 80: 1322–1339.

Prince, S. D., and R. N. Carter. 1985. The geographical distribution of prickly lettuce (*Lactuca serriola*). III. Its performance in transplant sites beyond its distribution limit in Britain. *Journal of Ecology* 73: 49–64.

Prince, S. D., R. N. Carter, and K. J. Dancy. 1985. The geographical distribution of prickly lettuce (*Lactuca serriola*). II. Characteristics of populations near its distribution limit in Britain. *Journal of Ecology* 73: 39–48.

Pulliam, H. R. 1996. Sources and sinks: Empirical evidence and population consequences. Pages 45–70 in O. E. J. Rhodes, R. K. Chesser and M. H. Smith, eds., *Population Dynamics in Ecological Space and Time*. Chicago: University of Chicago Press.

Root, T. 1988. Environmental factors associated with avian distributional boundaries. *Journal of Biogeography* 15: 489–505.

———. 1991. Positive correlation between range size and body size: A possible mechanism. *Proceedings of the XXth International Ornithological Congress* 2: 817–825.

Root, T. L., and S. H. Schneider. 1995. Ecology and climate: Research strategies and implications. *Science* 269: 334–341.

Sagarin, R. D., J. P. Barry, S. E. Gilman, and C. H. Baxter. 1999. Climate-related change in an intertidal community over short and long time scales. *Ecological Monographs* 69: 465–490.

Salt, R. W., and H. G. James. 1947. Low temperature as a factor in the mortality of eggs of *Mantis religiosa* L. *Canadian Entomologist* 79: 32–36.

Sanchez-Rodriguez, J. F. and A. Baz. 1995. The effects of elevation on the butterfly communities of a mediterranean mountain, Sierra de Javalambre, central Spain. *Journal of the Lepidopterists' Society* 49: 192–207.

Schneider, S. H. 1993. Scenarios of global warming. Pages 9–23 in P. Kareiva, J. Kingsolver, and R. Huey, eds., *Biotic Interactions and Global Change*. Sunderland, Mass.: Sinauer.

Scott, J. A. 1979. Hibernal diapause of North American Papilionoidea and Hesperioidea. *Journal of Research on the Lepidoptera* 18(9): 171–200.

———. 1986. *The Butterflies of North American: A Natural History and Field Guide*. Stanford, Calif.: Stanford University Press.

Scriber, J. M. 1973. Latitudinal gradients in larval feeding specialization of the world Papilionidae (Lepidoptera). *Psyche* 80: 355–373.

Shapiro, A. M. 1966. *Butterflies of the Delaware Valley*. Special Publications of the American Entomological Society. Philadelphia: The American Entomological Society.

———. 1974. Butterflies and skippers of New York State. *Search* 4(3): 1–60.

———. 1993. Long-range dispersal and faunal responsiveness to climatic change: A note on the importance of extralimital records. *Journal of the Lepidopterists' Society* 47(3): 242–244.

Shaw, M. R. 1981. Parasitism by Hymenoptera of larvae of the white admiral butterfly, *Ladoga camilla* (L.) in England. *Ecological Entomology* 6: 333–335.

Stearns, S. C., and R. D. Sage. 1980. Maladaptation in a marginal population of the mosquito fish, *Gabusia affinis*. *Evolution* 34: 65–75.

Storey, K. B., and J. M. Storey. 1996. Natural freezing survival in animals. *Annual Review of Ecology and Systematics* 27: 365–386.

Sutherst, R. W., G. F. Maywald, and D. B. Skarratt. 1995. Predicting insect distributions in a changing climate. Pages 60–91 in R. Harrington and N. E. Stork, eds., *Insects in a Changing Environment*. New York: Academic Press, Harcourt Brace & Company.

Tauber, M. J., C. A. Tauber, and S. Masaki. 1986. *Seasonal Adaptations of Insects.* New York; Oxford: Oxford University Press.

Taylor, G. 1999. Long-Term Wet-Dry Cycles in Oregon. Available online at http://www.ocs.orst.edu/reports/wet-dry.html.

Taylor, P. 1993. The Sachem: A new skipper for Manitoba and the prairie provinces. *Blue Jay* 51: 193–195.

Thomas, C. D., and J. J. Lennon. 1999. Birds extend their ranges northwards. *Nature* 399: 213.

Thomas, J. A., D. J. Simcox, J. C. Wardlaw, G. W. Elmes, M. E. Hochberg, and R. T. Clarke. 1998. Effects of latitude, altitude and climate on the habitat and conservation of the endangered butterfly *Maculinea arion* and its *Myrmica* ant hosts. *Journal of Insect Conservation* 2: 39–46.

Thrush, S. F., J. E. Hewitt, V. J. Cummings, M. O. Green, G. A. Funnell, and M. R. Wilkinson. 2000. The generality of field experiments: Interactions between local and broad-scale processes. *Ecology* 81: 399–415.

Tucic, N. 1979. Genetic capacity for adaptation to cold resistance at different developmental stages of *Drosophila melanogaster. Evolution* 33: 350–358.

Uvarov, B. P. 1931. Insects and climate. *Transactions of the Entomological Society of London* 79: 1–255.

Walker, B. H. 1991. Ecological consequences of atmospheric and climate change. *Climatic Change* 18: 301–316.

Warren, M. S. 1992. The conservation of British butterflies. Pages 246–274 in R. L. H. Dennis, ed., *The Ecology of Butterflies in Britain.* Oxford: Oxford University Press.

White, E. B., P. DeBach, and M. J. Garber. 1970. Artificial selection for genetic adaptation to temperature extremes in *Aphytic lingnanensis* Compere (Hymenoptera: Aphelinida). *Hilgardia* 40(6): 161–190.

Woodward, F. I. 1988. Temperature and the distribution of plant species. Pages 59–75 in S. P. Long and F. I. Woodward, eds., *Plants and Temperature,* vol. 42. Symposia of the Society for Experimental Biology. Cambridge: Company of Biologists, Ltd., Department of Zoology, University of Cambridge.

———. 1993. A review of the effects of climate on vegetation: Ranges, competition, and composition. Pages 105–123 in R. L. Peters and T. E. Lovejoy, eds., *Global Warming and Biological Diversity.* New Haven: Yale University Press.

WRCC. 1999a. Western U.S. Climate Historical Summaries. Western Regional Climate Center, available online at http:\\www.wrcc.sage.dri.edu.

———. 1999b. El Niño, La Niña and the Western U.S., Alaska and Hawaii. Western Regional Climate Center, available online at http://www.wrcc.dri.edu/enso/enso.html.

CHAPTER 2

Butterflies as Model Systems for Understanding and Predicting Climate Change

JESSICA J. HELLMANN

The earth's atmosphere regulates the climate and weather and therefore determines the conditions of life for many organisms (IPCC 1996a, b). If human-driven changes in atmospheric composition alter weather and climate processes, the planet's biota will be affected. Climate studies strongly suggest that atmospheric alteration is already occurring and that the global mean temperature is rising (IPCC 1996a, Schneider 1990). Hence, global, regional, and local studies of biotic responses to climate change are necessary to understand the consequences of this change for biodiversity and, ultimately, humanity.

Research on the biological effects of climate change is needed for two fundamental reasons. First, we rely on the living resources of our planet for many essential goods and services (Daily 1997). To ensure the preservation of these ecosystem services, it is important to understand how climate change will alter the distribution, function, and viability of the biotic systems that provide for humans. In some cases, we may need to take steps to buffer service-delivering species and populations against change (Hughes et al. 1997). Second, climate change will place additional stresses on already-endangered and threatened species. Sensitive species found only in reserves or small areas of relict habitat are especially likely to suffer negative effects of global warming because they cannot escape climate impacts by dispersal (Peters and Darling 1985). To prevent endangered population and species losses to global warming, we may need to improve management of threatened species and their habitat to decrease the new extinction risks brought by climate change.

To take action in the face of climate change we need knowledge of the mechanisms that drive biotic responses to climatic shifts. We also need to compare the relative magnitude of climatic impacts in a variety of systems to set conservation priorities. Yet relatively little is known about the diversity of ways that climate change can act in communities or the degree to which individual species will be affected by the changes, and there is little time to address these deficiencies. We are already in the midst of a nearly unprecedented extinction event, and species may disappear before we are able to study them (May et al. 1995). Rapid climate change is beginning to influence abiotic conditions around the globe, and many species are already responding to this alteration (IPCC 1996a, Parmesan et al. 1999, Thomas and Lennon 1999). Hence, immediate conservation action may be required for some species.

To understand the potential for climate change to affect biodiversity, we need to learn quickly and efficiently how climate change affects species and communities. This demand for rapid but mechanistic assessment argues for detailed research in a few select taxa. Information from these model systems can then be used to generate hypotheses of impacts in other groups.

Butterflies and butterfly communities are good candidates for such models (Dennis 1993). We already have extensive knowledge of butterfly ecology, providing a unique opportunity to understand the implications of climate change and predict the magnitude of climate change effects. In addition, we know that individual butterflies and butterfly populations are strongly influenced by the abiotic environment (e.g., Pollard 1988, Roy and Sparks 2000), suggesting that if climate change can alter the dynamics of species, impacts should appear in this taxon. A thorough exploration in butterflies provides a starting point for research on the impacts of climate change for a wide variety of herbivorous[1] species.

To illustrate the potential for butterflies as climate change models, I briefly outline the qualitative responses we expect populations to express under climate change and discuss why it is important to understand impacts on insect populations. I review how climate can alter the population dynamics of butterflies via direct influences of abiotic change and via indirect influences caused by interactions with other climate-sensitive species. And I discuss factors that could influence the ecological responses of butterfly populations to climate change (i.e., evolution and other, nonclimate, environmental stresses). In many cases, I use my own study organism, *Euphydryas editha bayensis*, or the Bay checkerspot butterfly, as an extended example.

Qualitative Responses to Climate Change

In the absence of evolutionary change, populations have four possible responses to climate change: expand, decline, move, or go extinct (see Cohn 1989, Peters and Lovejoy 1992, Gates 1993, Kareiva et al. 1993). If climate change improves the quality or total area of suitable habitat (e.g., removing abiotic or biotic barriers to occupancy), populations can expand or increase in density and new populations can be established. Alternatively, if climate change impacts a species negatively, populations can decline or individuals can disperse from affected areas to non- or less-affected habitat (including unoccupied sites or areas that were uninhabitable prior to climate change). Declines can also lead to extinction, especially in those species with limited dispersal capabilities or strict habitat requirements.

We already have observed evidence of species range shifts due to climate change (MacDonald and Brown 1992, Root and Schneider 1993, Pollard and Eversham 1995, Parmesan 1996, Hill et al. 1999, Parmesan et al. 1999; in mountain regions see IPCC 1996b, Murphy and Weiss 1992, Fleishman et al. 1998), and theory suggests that climate change could also drive range contractions (Lawton 1995). Central populations in the core of a species' range tend to have the highest abundance, and population sizes at the edges of a range are often smaller and more variable than those at the core. This phenomenon may place edge populations at greater risk of extinction than core populations (Watt et al. 1979, Brown 1984). Local extinction and range contraction can lead to population isolation. Genetic isolation, divergence, or inbreeding depression can also result from limited dispersal between isolated populations (Hewitt 1996, Saccheri et al. 1998).

One critical fact we have learned from basic ecology, paleoecological records, and new evidence of shifting populations is that species often respond to climate change independently of other community members (Graham and Grimm 1990, Overpeck et al. 1992, Harrison 1993, Coope 1995). Therefore, even if every species were able to shift its range in response to global warming, climate change would impact the structure and composition of species assemblages. In addition, chains of climatic effects can cascade through a community, affecting a wide range of taxa (Pimm 1980, Coley 1998). For this reason, it is important to understand not only the direct impacts of climate change on a species of interest, but also how a species may be indirectly affected via impacts on other community members.

Insects and Climate Change

Insects compose a majority of all described biodiversity and are members of many important service-providing communities. A conservative estimate of the total number of insects inhabiting the planet is 8 million, or 60% of the estimated total number of species (Hammond 1995)! Insects consume vegetation, pollinate many species of flowering plants, and strongly alter the biotic and abiotic environment (Wilson 1987).

Entomologists have long studied the influence of weather and climate on an important subgroup of the insect fauna—agricultural pests (Uvarov 1931, Landsberg and Smith 1992, Huffaker and Gutierrez 1999). For example, empirical work has documented the impacts of precipitation (Wellington et al. 1999), temperature (Logan et al. 1976), and weather extremes (Gutierrez et al. 1977) on pest population size. Other studies report the importance of insect parasites and predators in regulating pest population dynamics, and temperature or weather can drive predatory interactions. For example, ladybird beetle predation on pea aphids in British Columbia is known to increase under elevated temperature because of increased beetle activity (Frazer and Gilbert 1976, Kingsolver 1989).

A large literature on chemical ecology and plant–insect interactions also shows how climate can affect insects via their food resources. Insect populations respond to changes in host plant quality, timing, and chemical defenses, and these plant properties change in response to weather (see Mattson and Haack 1987). For example, larvae of many herbivorous insect species feed on plants with short phenological windows for larval development (e.g., Rockwood 1974). If the temporal window of plant suitability shifts out of phase with larval development, larvae starve.

Given the role insects play in ecosystems, their economic importance as pollinators, pests, and biotic forms of pest control, and the potential risks posed to them by climate change, we may need to increase activities to manage them (Ayres and Lombardero 2000). To the degree that butterfly ecology and physiology are similar to other herbivorous species, butterflies can be used as models to explore climate sensitivity in insects. Our knowledge of how, and how much, butterfly populations can be affected by climate shifts may inform management for a large number of insect taxa.

Butterflies and Climate Change

Butterflies already have been extremely useful test systems in population biology (Vane-Wright and Ackery 1984, Boggs et al. in press). For example, butterfly studies have played a key role in the development of metapopulation theory (Ehrlich and Murphy 1981, Harrison et al. 1988, Thomas et al. 1996, Hanski 1994, 1999) and prescriptions for reserve design (Ehrlich and Murphy 1987, Kremen 1994, Schultz 1998). Butterflies also have taught us about molecular mechanisms of evolution (Watt et al. 1996), speciation (Sperling 1994, McMillan et al. 1997), hybridization (Turner 1971, Hagen 1991), and fitness differences in the wild (Watt 1992). Butterflies have been model organisms in the study of development, plasticity, norms of reaction (Brakefield et al. 1996, Nylin et al. 1996, Schlichting and Pigliucci 1998, Gotthard et al. 1999), population differentiation (McKechnie 1975, Bossart and Scriber 1995), coevolution (Ehrlich and Raven 1965, Gilbert 1971), and foraging behavior (Boggs 1987, Hill and Pierce 1989, Singer 1994) among others.

Based largely upon this extensive literature, several examples of how butterfly populations may be altered by climate change are given here (see also Crozier, this volume). It is not my intent to cover all of the facets of climate and thus all ways that climate change could affect butterflies (for detailed coverage of this sort see Dennis 1993). Instead, I focus on generalizable impacts and eco-logical factors that commonly control population abundance to illustrate the utility of butterflies as model systems (see also Dennis 1992, Warren 1992). Specifically, I discuss the potential impacts of changes in temperature, the amount of sunshine, and the frequency of climatic extremes. (The effects of rainfall are discussed only in the context of plant growth and senescence.) Examples focus on temporal impacts of climate change and the effects on larvae and adults. Although not treated here, climate change also could alter the environmental cues butterflies use for timing their life cycle (e.g., induction of diapause or pupation) (Hunter and McNeil 1997) or affect other population properties such as voltinism and sex ratio (Dennis 1993).

Direct Effects of Climate Change in Butterflies

The impacts of climate change on butterflies can be divided into direct and indirect effects. Direct effects are those impacts of abiotic change that alter the physiology, performance, or mobility of indi-vidual butterflies, in turn impacting population-level processes. Indirect impacts, treated later here, affect individual butterflies and butterfly population dynamics via interactions with other species, where either the biotic association or the interacting species is mod-ified by climate or weather.

Degree Days

The rate of butterfly larval growth and level of larval activity are determined by body temperature (Taylor 1981, Ratte 1984). A widely used model of growth in insects is a linear summation model of larval development expressed in terms of "degree days." The degree day concept assumes that development rate is proportional to temperature (above some minimum temperature). To a first-order approximation, this model is fairly realistic. Moderate increases in temperature increase the pace at which larvae grow (Casey 1976), potentially decreasing the amount of time larvae are exposed to sources of mortality (e.g., predation) and, if rates of mortality do not decline in other life stages, possibly leading to pop-ulation expansion (Pollard 1979). The primary mechanism of accelerated growth is increased foraging rates under elevated temperature (see Mathavan and Pandian 1975, Scriber and Lederhouse 1983, and references therein). In contrast, moderate temperature declines slow larval development, increasing the time of exposure to predators, and possibly leading to increased mortality rates and population depression.

A linear model of larval development, however, does not hold at temperature extremes. At very high temperatures, efficiency and activity gains are lost, rate of larval development declines, and mortality greatly increases (Sherman and Watt 1973, Sharpe and DeMichele 1977, Taylor 1981). At very low temperatures, development is not possible or is severely slowed such that development cannot be completed before the onset of the cold (or dry) season (Pullin and Bale 1989). Extremely cold temperatures can kill larvae.

Climate Surprises

If global climate change causes local warming, some butterfly larvae may be able to develop and survive at higher latitudes or elevations (Brakefield 1987). Climate change could allow northerly expansion into regions where prechange temperatures were too low for larval development. However, climatologists are predicting that climate change will alter the frequency of extreme events or climate "surprises" (e.g., harsh freezes or drought) in addition to changing mean conditions (e.g., average temperature or precipitation) (Overpeck 1996, Schneider and Root 1996). The occurrence of surprises may be more important in determining range limits than mean climate because surprises can kill individuals and increase the probability of population extinction (see also Kingsolver and Watt 1983). Therefore, surprises, rather than site suitability determined by degree days, may control the potential occupancy of habitat on the edges of a butterfly species' range.

For example, extreme cool weather in Michigan led to population declines of northerly expanding species of *Papilio* in 1992 (Scriber and Gage 1995). Several southern *Papilio* species had colonized Michigan in a warming period extending from 1986 to 1991, but extreme cold weather and a late frost in 1992 led to large larval mortality and population die-back. The effects of this extreme event overrode the impacts of the warming trend with respect to *Papilio* persistence in the region.

Topography can play an important role in mediating these regional climate surprises (Wellington et al. 1999). The ends of topographic gradients can experience qualitatively different weather, and some habitats within a region can be protected from climatic extremes. At the ends of a gradient or in protected habitat pockets, individuals may be buffered from the mortality effects of extreme temperatures or catastrophic weather events. Protected populations could then recolonize the entire landscape when favorable conditions return. On a regional or landscape scale, we might observe exaggerated or more frequent population dieback and expansion under climate change, and complex topography may play an important role in protecting species from extinction under altered weather regimes.

For example, the Bay checkerspot butterfly, *Euphydryas editha bayensis*, inhabits topographically diverse areas of native grassland in California. Regular shifts in butterfly density on various slope types were observed in the mid-1980s; the occurrence of individuals on warmer slopes increased when the total population size grew, and the population retreated to cool slopes when total abundance declined (Murphy and Weiss 1988). Larval survivorship differences by slope type drive this spatial expansion and contraction, suggesting that cool slopes offer an important refuge in years of extreme heat or drought. If the frequency of these extremes increases, the availability of cool slopes will be critical to the persistence of *E. e. bayensis*.

Adult Flight

Temperature and weather also affect the physiology and activity of adult butterflies. Mating, feeding, and oviposition require flight, and flight is strongly affected by ambient temperature and solar radiation. Butterflies thermoregulate to achieve body temperatures within a small thermal window for flight (Clench 1966, Watt 1968, Heinrich 1974, 1986, Pivnick and McNeil 1987). Individuals orient themselves toward the sun for maximal absorption of solar heat when their body temperatures are below the thermal window and, either by basking or by using the reflectivity of their wings, elevate their body temperatures (Douwes 1976, Kingsolver 1985). (Several butterfly species also generate heat by shivering; e.g., Kammer 1970.) Above the thermal window, butterflies orient away from the sun to minimize additional thermal load. Thermal stress occurs when ambient temperatures are above suitable body temperatures, and body temperatures too low for flight can occur during periods of cloudiness or cooling by convection (i.e., wind) (Digby 1955). The number of days or hours available for flight during a flight season is inversely related to the number of extreme temperature days, the number of cloudy days, and the frequency and duration of weather systems.

For those species that spend an approximately constant portion of their life laying eggs, total available flight time directly equates to total reproductive output. Females of many species spend a majority of their flight time searching for suitable oviposition sites (Wiklund and Persson 1983). For example, studies on *Colias* populations showed that weather can constrain fecundity to only 20 to 50% of potential fecundity, suggesting that poor weather leads to a depressed population size in the following generation (Kingsolver 1983). Failure to lay a full complement of eggs as a function of limited sunny days also drives population fluctuations in the British butterfly, *Anthocharis cardamines* (Courtney and Duggan 1983). In *A. cardamines*, the number of eggs deposited per season is directly proportional to the number of sunshine hours. Therefore, climate change could lead to either population increases or declines depending on whether climate change limits or expands flight time (i.e., searching and egg-laying time) and, therefore, reproductive output.

Climate change also could impact butterfly dispersal and migration. First, by changing the total time available for adult flight, changes in weather could depress the frequency with which individuals colonize unoccupied patches and disperse to other established populations (i.e., frequency of gene flow). Alternatively, increased flight time due to decreased frequency of cold, very hot, or cloudy days could lead to more frequent dispersal events. Second, climate and weather changes could impact seasonal migratory species (Johnson 1969, Baker 1978). Research on migration in other insects suggests that migratory species can be blown off course by extreme weather events (Southwood 1981, Johnson 1995, Dingle 1996). Butterflies, however, may avoid flight during strong winds and heavy weather or fly near the ground to maintain control of their flight path (Williams 1930, Taylor 1974, Gilbert and Singer 1975). To the extent that the former is true, increased frequency of extreme events under climate change could lower the number of individuals that successfully arrive at the migration destination. In cases where the latter pertains, more numerous extreme events could slow or delay migration time. Migratory species are also susceptible to climate changes in multiple habitats along their flight path (Malcolm 1987). The conservation of migratory species in the face of climate change will require cooperative efforts in several locations.

Indirect Effects of Climate Change in Butterflies

The indirect effects of climate change on butterflies arise from interactions with other species. Butterflies use other species for food, are involved in mutualistic associations (e.g., lycaenids and ants), and are prey for other taxa. Interactions can have a strong influence on the range of butterfly responses to climate change. In some cases, the indirect consequences of climate change may exceed direct effects in magnitude and importance. Here, I focus on how the relationship between butterflies, their plant resources, and their predators may be altered by changes in atmospheric composition and climatic variables.

Associations with Plants

Associations between Lepidoptera and their larval host plants are remarkably conserved. Related butterfly species are almost always found on related host plants, suggesting that colonization of new hosts is an evolutionary hurdle (Ehrlich and Raven 1965). The number of plant species that a butterfly can consume is thought to be limited by the physiological machinery needed to detoxify plant chemicals (Berenbaum 1995; see Bernays and Graham 1988). Specialized associations with host species create an opportunity for climate change to impact larval feeding (Dennis and Shreeve 1991). Weather may change the butterfly–host plant interaction by altering the temporal overlap between larval development and plant availability (Harrington et al. 1999). To illustrate this phenomenon, an extended example is drawn from my own research on *E. e. bayensis.*

E. e. bayensis has a sensitive temporal interaction with its host plants that is controlled by weather and climate. This butterfly is univoltine (one generation per year) and inhabits patches of annual grassland on serpentine soils in the region of the San Francisco Bay.

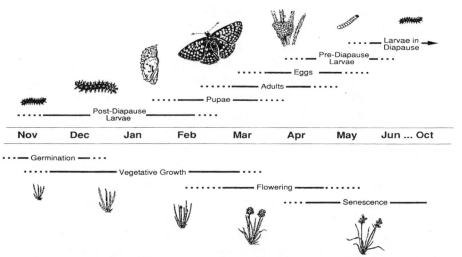

Figure 2.1. Life cycle diagram of *Euphydryas editha bayensis* and one of its two larval host plants, *Plantago erecta*. Dotted lines refer to the among-year variance in the timing of each organism. The location and length of the bars is determined by climatic conditions in a single year. In years of drought or deluge, larval development is delayed and can push beyond the date of senescence for *P. erecta*. A second host plant, *Castilleja* spp., matures up to 2 weeks later than *P. erecta*, and can serve as an important resource to extend the period of larval development. (Drawing by Lawrence Lavendel.)

Obligatory hibernation, or diapause, of larvae is required during the grassland's dry season (Fig. 2.1).

An important driver of total population size in *E. e. bayensis* is annual variation in larval mortality. Larvae experience food shortages and die in large numbers when host plants senesce before larvae are sufficiently developed to diapause (Ehrlich 1965, Singer 1972, Ehrlich et al. 1980, Dobkin et al. 1987, Cushman et al. 1994). Food shortages occur when plant growth and death are accelerated by high temperature or drought or when larval hatching and growth are delayed by cloudy days or low temperatures. In years with moderate temperatures and precipitation, in contrast, host plants can sustain sufficient numbers of larvae to diapause, and populations remain stable or expand.

The oligophalous nature of *E. e. bayensis* and the properties of the butterfly's host species complicate the larval race to diapause in extreme years. Pre-diapause larvae feed on two types of host plants: *Plantago erecta* and species in the genus *Castilleja* (Singer 1971, Hickman 1993), and the timing of senescence in these two hosts is distinct. In particular, *Castilleja* develops and dies approximately 1 to 2 weeks later than *P. erecta*, and larvae that forage in *Castilleja* patches have been noted to have a higher probability of survival than larvae outside these patches in extreme years (Singer 1971, 1972).

To investigate the relationship between larvae and their hosts, larvae were given the option to feed on both host plants over the length of larval development in field trials, and their diet choice was tracked through time (Hellmann 2000). In separate enclosures, larvae hatched either from eggs on *P. erecta* or from eggs on *Castilleja*. Approximately half of the larvae from *P. erecta* eggs moved to *Castilleja* shortly after hatching. Similarly, approximately half of all larvae that hatched on *Castilleja* moved to *P. erecta* early in the study period. However, the fraction of larvae feeding on *P. erecta* quickly decreased as time passed while the corresponding fraction of larvae foraging on *Castilleja* increased. As *P. erecta* quality declined, proportional use of *Castilleja* grew. This suggests that *Castilleja* sustains feeding longer than *P. erecta*, a quality that enables it to partially buffer larvae against climatic extremes. If these extremes occur with greater frequency under climate change, *Castilleja* could play an increasingly important role in larval diet.

Further evidence of the importance of *Castilleja* under extreme conditions can be seen in experiments of plant and larval performance under elevated temperature (Hellmann 2000). Persistently elevated temperature accelerates both the senescence of hosts and the growth of larvae. However, larval survivorship under these conditions is greater when *Castilleja* is present than when it is absent. If regional climate brings temperatures that are consistently higher than prechange conditions, this result suggests that the availability of a long-lasting host (i.e., *Castilleja*) is key to maintaining butterfly populations (Hellmann 2000).

Exactly how much *Castilleja* can help sustain checkerspot populations under change is limited by the plant's total abundance. Unfortunately, *Castilleja* is patchily distributed and highly variable among years. At low abundance (with small patches and large distances between patches), the fraction of larvae that are able to feed on this species is relatively small. *P. erecta*, in contrast, is widely dispersed throughout the grassland, and its abundance fluctuates little annually.

Further, it is not yet known how climate change may influence the distribution and abundance of *Castilleja*, but several scenarios are possible (Hellmann 2000). If *Castilleja* abundance is high under a future climate, the likelihood of butterfly population persistence under warming also may be high because many larvae could feed on this key resource. Conversely, if the changing climate significantly decreases *Castilleja* abundance, few larvae will be able to exploit its buffering potential, and butterfly populations could decline. To account for this latter possibility, management activities that promote *Castilleja* would be helpful in preserving Bay checkerspot populations.

This type of temporal interaction between *E. e. bayensis* larvae and their host plants is not unique, and many host-restricted species may be prone to climate impacts by changes in plant dynamics. For example, large larval mortality has also been observed when hosts and larvae fall out of phase in *Pieris virginiensis*. Cappuccino and Kareiva (1985) investigated interannual differences in the availability of hosts for *Pieris virginiensis* and found that in some years, host plants senesced before larvae had completed foraging. During years with early plant death, many larvae were left stranded with no alternative hosts. Sensitivity to weather-driven phenology also extends to butterfly species whose larvae feed on woody species because larval foraging is closely linked to leaf flush, leaf age, and the timing of leaf chemical defense (Feeny 1970, Coley 1980, Holliday 1977, Watt and McFarlane 1991, Hunter and Elkington 2000). Whether temporal shifts in woody species will lead to increases or decreases in lepidopteran numbers is dependent on the direction of the shift in tree phenology versus the timing of larval development (Buse and Good 1996, Visser and Holleman 2000, Buse et al. 1998, Dewar and Watt 1992).

Timing is not the only factor of host accessibility and quality. Changes in soil moisture, soil temperature, cloud cover, precipitation, and air temperature can each affect the nutrient content or palatability of host plants. A change in food quality can be harmful to herbivores if it forces them to increase foraging time, thereby exposing them to predators or increasing the potential for food limitation (Ayres 1993). Extreme weather events also can drive changes

in host plant suitability and availability and can even lead to severe plant death and die-back (Inouye 2000). An example of how the impact of weather on plants can translate to high butterfly mortality was observed in Colorado in 1969 (Ehrlich et al. 1972). An unseasonably warm spring was followed by a brief, late, and severe snowstorm that killed or degraded a large majority of lupine inflorescences, the oviposition and larval foraging site of the butterfly *Glaucopsyche lygdamus*. A local population extinction of this butterfly species was observed at elevations where lupine was severely damaged and the eggs they housed were killed. In a second example, the host plants of a single subpopulation in a source-sink metapopulation of *E. editha* were killed in a severe summer frost (Thomas et al. 1996). As a result, larvae starved, and the butterfly subpopulation went extinct. A nearby subpopulation also declined from the loss of immigrants.

Increases in the atmospheric concentration of CO_2 also can affect the relationship between butterflies and their larval host plants. Under elevated CO_2, the nutrient concentrations of plants change, increasing the ratio of carbon to other essential nutrients such as nitrogen and water. The general pattern emerging from studies on the impacts of elevated CO_2 on arthropod herbivores is that individuals grow more slowly, eat more plant biomass, and have an extended development time on plants grown under elevated CO_2 (Ayres 1993, Watt et al. 1995, Coley 1998). Lengthened development time can result in greater exposure to predators and parasites, and decreased food quality increases the potential for food to be limiting, thereby decreasing total population sizes. Studies of the impacts of elevated CO_2 on plant community composition also indicate that plant species respond differently to increased atmospheric carbon, resulting in possible changes in plant dominance (see Bazzaz 1990, Vitousek 1994). If vegetation composition changes, altering the availability of host resources, butterfly populations would be affected in turn.

In a study with the butterfly *Junonia coenia*, researchers found decreased larval growth and survivorship on plants grown under elevated CO_2 (Fajer et al. 1989). Larval growth was slowed from decreased water and nutrient concentrations in the host plant; the CO_2 treatment, however, did not affect the concentration of defensive compounds. Studies of moths also have shown increased feeding under elevated CO_2 (Lincoln et al. 1986, 1993), and one study has seen a change in the digestibility of leaf material due to increased leaf toughness and a change in plants' ability to defend themselves (Dury et al. 1998).

In the *E. e. bayensis* system, larvae may need to feed longer to meet nutritional demands under elevated CO_2, and these foraging needs could delay development to diapause beyond the date of plant senescence. This effect could increase larval mortality at the onset of the grassland dry season and exacerbate phenological shifts driven by climate alone (as already discussed). Elevated CO_2 also might alter competitive interactions among host plants on the serpentine, either positively or negatively impacting the availability of larval food resources. Experiments on community composition in the grassland indicate that CO_2 can affect the density of native forbs. The primary influence, however, is on late-season annuals; CO_2 impacts on early-season forbs (including larval host plants) are restricted to soil types where native forbs co-occur with invasive grasses and are less abundant (Chiariello and Field 1996). The direct impacts of CO_2 on plant quality and senescence date of larval plants are not known.

Impacts on plants can indirectly affect adults as well as larvae. Butterflies feed as adults, and most use some form of plant material for essential sugars and nutrients (e.g., nectar, pollen, sap, and rotting fruit; Boggs 1987). Just as weather and CO_2 concentration can affect the quality, availability, and timing of larval host plants, they also can change the quality or availability of adult food resources. For those species that allocate a large fraction of adult resources to reproduction and have little larval stores for egg production, the effect of climate change on adult food could substantially impact total reproductive output (Ehrlich and Gilbert 1973, Murphy et al. 1983, Boggs 1986, 1997, Boggs and Ross 1993).

Predation and Parasitism

Evidence suggests that in some systems, predation and parasitism are important drivers of population dynamics (Dempster 1983, Hassell 1985). (Competition among insect herbivores, in contrast, is thought to be infrequent [Strong et al. 1984, but see Denno et al. 1995, Ohgushi 1997].) Direct effects of climate change on the development rates and activity of insect predators can affect butterfly predation (Porter 1983, Courtney 1986). For example, under local cooling, the food requirements of endothermic predators increase, leading to the consumption of more prey (Gilbert and Singer 1975). Weather and climate also can alter the exposure of butterflies to predators by influencing butterfly activity or total time of susceptibility (Lederhouse 1983, Shreeve 1986). For example, low temperatures make both larvae and adults relatively immobile, but some predators (e.g., birds and some small parasitoids) can successfully attack at low air temperatures (see Dennis 1993 and references therein). Similarly, at high temperatures, adult butterflies may be vulnerable to predation while standing in heat avoidance posture (Kingsolver and Watt 1983). Weather-driven declines in habitat quality also can increase the visibility of larvae as they search for food (Sato 1979, Dennis 1993).

Generating Prediction in Butterfly Systems

While a great deal of uncertainty remains in the field of butterfly ecology, the knowledge we do have can serve as an important starting point for exploring the causes and magnitudes of climate impacts in insects and other climate-sensitive species. We can use many of the direct and indirect impacts of climate change in butterflies listed here to build predictions of how much butterfly populations might be altered by human influences on the climate (Roy et al. 2001). Specifically, we can translate our knowledge of potential climate change mechanisms in butterflies into mathematical models and build quantitative predictions with relatively minor, but focused, additional study (Hellmann 2000). Such an approach would be useful, but has been explored very little.

Several key goals could be addressed with a mathematical modeling framework built from mechanisms in sample butterfly systems. First, quantitative models can be used to explore the sensitivity of butterfly populations to a variety of climate scenarios. Second, mathematical models of several butterfly or other insect systems could be compared to identify common factors that are important determinants of population sensitivity. Third, models could be used to indicate which management prescriptions might be effective at reducing the impacts of climate change. Fourth, models may allow us to piece together climate impacts with other pressures on biotic diversity, such as pollution and anthropogenic habitat destruction. Finally, mathematical descriptions may help us understand how evolutionary processes may modify ecological predictions.

Butterfly Evolution and Climate Change

The possibility of evolutionary responses to climate change renders the preceding discussion of ecological climate consequences incomplete. Many assume that the pace of climate change will be faster than potential rates of evolutionary adaptation for many organisms, causing species intolerant of change to decline or go extinct (Peters and Darling 1985, Hoffmann and Blows 1993). If adaptation to rapid climate change is possible, however, we might expect to find it in short-lived species such as insects (Smith et al. 1987). Adaptation is likely to spread most quickly in species with short, overlapping generations and large population sizes. All else being equal, adaptation would be slower in insects that have long generation times or are rare.

I briefly discuss two examples of how climate shifts may induce population genetic changes in butterflies. (For a thorough discussion of climate change and adaptation see Hoffmann and Parsons [1997] or Travis and Futuyma [1993] and the references therein.) It remains to be determined, however, how phenomena such as changes in gene frequency will affect a species' ecology—its population dynamics, risk of extinction, or geographic distribution (e.g., Huey and Kingsolver 1993). Further research is needed to explore the potential for evolution to prevent declines in population size or distribution brought on by climate change (Rodriguez-Trelles et al. 1998, Armbruster et al. 1999).

Adaptation to climate change requires genetic variation at loci controlling climate-influenced characters. Flight is one critical element of butterfly performance that is strongly impacted by climate (Hill et al. 1999). As an ectothermic animal, a butterfly flies only when its body temperature lies within a narrow temperature range, and body temperature is a function of environmental temperature and thermoregulatory mechanisms (Clench 1966, Watt 1968, Heinrich 1974, 1986, Pivnick and McNeil 1987). Because flight translates directly to mating, feeding, and egg laying, flight is essential for reproduction. We would expect genetic differences among individuals for flight capacity to play an important role in determining species survivorship under climate change because of the relationship between flight, temperature, and fitness. Genetic variation in flight capacity could facilitate evolution under climate change because natural selection could select among existing genotypes.

In species of the butterfly genus *Colias,* there are genetic differences among individuals with respect to flight capacity (Watt et al. 1983). These differences are the result of a polymorphism at the gene that codes for phosphoglucose isomerase (PGI) synthesis. PGI controls an important component of the metabolic, or energy processing, pathway, and is therefore strongly related to flight. The mechanism behind flight performance differences among PGI genotypes involves a biochemical trade-off between enzymatic thermal stability and energy processing throughput that is particularly pronounced in homozygotes. To the extent that this trade-off holds, an individual's genotype determines whether it has a narrow window of flight but high energy throughput or a broad window of flight but lower kinetic power. In ordinarily variable temperature ranges typical of pre–climate change conditions, individuals of the kinetically favored (i.e., high throughput) enzyme genotypes are more likely to fly, sustain flight, and thus have greater reproductive output than heat resistant, but kinetically disfavored, individuals. When exposed to extreme heat stress, however, the flight capacity (and hence fecundity) of most kinetically favored genotypes is damaged, whereas heat resistant genotypes retain more of their original flight capability.

While it seems likely that individuals with the thermally stable PGI genotypes will be selected under a warmer climate, the existence of heat resistant genotypes may not protect *Colias* populations from declines in abundance under climate change (Watt 1992). Because resistant individuals are less fecund than other PGI genotypes (Hayes 1985), total population reproduction could significantly decline under conditions of warming if the frequency of heat resistant, but kinetically inferior and less fecund, genotypes increases. With an understanding of these population genetic effects in *Colias,* one could build predictions of population size changes under numerous scenarios of climate warming. Population genetic modeling in this system could be used to further our understanding of how genetic variation in combination with biological trade-offs may or may not enable species to escape the risks of extinction brought on by climate change.

Changes in PGI gene frequency in *Colias* would be a direct response to climate change. Other types of evolution might occur as a result of indirect effects. For example, if climate change altered host plant quality, palatability, or availability and led to increased mortality or decreased survivorship on historical hosts, new diets might be strongly selected. Species may colonize new hosts under such selection if genetic variation for diet breadth exists in a population or if fortuitous mutations occur (Singer et al. 1994).

Evidence from *E. e. monoensis* suggests that the evolution of new host use is possible (Singer et al. 1993). A population of this subspecies recently colonized a widespread invasive species (*Plantago lancelolata*) and preference for its native host (*Collinsia parviflora*) has declined. Such shifts may enable species to colonize new habitats, possibly expanding their population distribution and size. If climate change drives genetic-based changes in host preference, it could trigger population expansion rather than decline.

It is unclear how common host shifts under climate change could be (Singer et al. 1994). In the case of *E. e. bayensis*, the selective pressure for colonization of new hosts may already be strong. Again, *E. e. bayensis* inhabits small pockets of native grassland widely surrounded by invasive grasses. If an *E. e. bayensis* individual were able to colonize an invasive host, even in the absence of climate change, it should have high fitness because its offspring could use resources far more abundant and widespread than the native hosts. It seems unlikely that changes in host plant suitability driven by climate change would increase the likelihood of host shifts in this subspecies because the butterfly has not responded to an already-strong pressure over hundreds of generations to shift to invasive plants.

Additional Impacts on Butterfly Systems

Many human stresses on the environment are superimposed on climate change, and these factors complicate our ability to predict the impacts of the latter. Conversely, any prediction that considers only climate change is narrow—organisms will be impacted by many human influences at once. The way that climate and other impacts interact with and feed back on each other can be pieced together only with an understanding of the relative effects of each. By understanding how climate and other factors alter the ecology of organisms, we can join factors together, building more realistic predictions and better informing management in the face of multidimensional environmental change. To illustrate the potential of nonclimate stresses to influence the biota, I briefly discuss examples of how habitat loss and nitrogen deposition might interact with some of the preceding impacts of climate change in butterflies.

Habitat Loss

Habitat alteration limits the potential for species to migrate in response to climate change. Many species that once existed in continuous habitat are now found in small habitat patches, separated by degraded habitat, urbanization, and agriculture. Post-Pleistocene patterns of movement suggest that range shifts of hundreds of kilometers may be necessary to keep pace with modern climate change and avoid extinction (Quinn and Karr 1993). Because of habitat fragmentation, however, this scale of movement may not be possible for many species (Hill et al. 1999).

E. e. bayensis is an example of a habitat-restricted species that may not be able to shift its range. Because habitat patches are separated by a human-dominated matrix of invasive plants and urbanization, dispersal events between habitat sites within the species' region of occupancy are infrequent. For example, in one year of intensive mark-release-recapture study, researchers found only eight transfers between sites separated by less than 300 meters (Ehrlich 1961). Dispersal over longer distances, therefore, is even less frequent. If E. e. bayensis host plants were widespread in California, as they perhaps once were (Murphy and Ehrlich 1989), range shifts in response to climate alteration might have been feasible. Human land use changes and the introduction of exotics, however, have probably eliminated that possibility.

Nutrient Deposition

Regional pollution impacts ecosystem nutrient cycling and therefore affects herbivores. Nitrogen, for example, is a limiting nutrient in many ecosystems, and competition among plant species for nitrogen determines the structure of many communities (Vitousek et al. 1997). Nitrogen deposition (e.g., deposition from urban pollution or fertilizer application) can change plant quality and the dominance hierarchy of vegetation, potentially impacting the suitability of a habitat for an herbivorous species.

In the *E. e. bayensis* system, nitrogen addition promotes the growth of invasive grasses, and these grasses exclude larval host plants (Hobbs et al. 1988, Chiarello and Field 1996, Weiss 1999). Local butterfly extinction or population decline from food scarcity can result from the failure to manage for grass reduction. This phenomenon of native grassland degradation in the face of nitrogen deposition is common to other grassland communities (e.g., Bobbink 1991) and may have implications for many open-habitat butterflies (Thomas et al. 1986).

Conclusion

Several studies have established that climate change is affecting biotic systems (Barry et al. 1995, Parmesan 1996, Myneni et al. 1997, Parmesan et al. 1999, Thomas and Lennon 1999, Roy and Sparks 2000). These studies have laid the groundwork for an emerging set of climate change research challenges. We now need to uncover the diversity of ways in which local and regional climatic effects are altering population dynamics. If we learn the mechanisms of biotic responses to change, we can anticipate possible negative impacts on endangered species and in communities of value to humans. Detailed climate study in every ecological system, however, is a daunting and unrealistic task. Instead, we can begin to explore the possible range of biotic outcomes to climate change in model communities and taxa. If models are chosen carefully and systematically, we may be able to relate climate change sensitivity to particular ecological characteristics in a quantitative way.

I have outlined several mechanisms of direct and indirect climatic effects in butterflies with the goal of providing a base of information from which we can begin to build predictions of population and species response to climate change in this group. Butterflies might serve as effective models for climate effects among insects, and an existing literature on butterfly ecology provides an opportunity to translate known mechanisms into predictions.

Ecologists interested in global change and conservation ultimately wish to predict which species risk extinction from the barrage of human-caused stresses on the environment. We also would like to understand what steps can be taken to mitigate any negative effects these global impacts might bring. Mechanistic climate change studies, among other approaches, are needed to achieve these prediction and management goals. Mechanistic studies also aid efforts to integrate evolutionary mechanisms into ecological impacts and may enable us to piece together the many interactive and multidimensional influences humans have on the biosphere.

Acknowledgments

I am indebted to the scientists and field workers who have developed the classic E. *editha* research system, particularly the efforts of P. Ehrlich and M. Singer. I also thank my own field assistants, especially S. Fallon and W. Knight. Thanks to C. Boggs, G. Daily, M. C. Devine, P. Ehrlich, E. Fleishman, J. Hughes, J. Kingsolver, T. Ricketts, B. Saenz, W. Watt, C. Wheat, and one anonymous reviewer for helpful discussion and manuscript comments. Lastly, thank you to T. Root and S. Schneider for the inspiration to work on issues of climate change. During the composition of this review, my doctoral work was supported by a National Science Foundation graduate fellowship.

Notes

1. A great majority of Lepidoptera larvae feed on plants; a small minority are mycophagous, predatory, or parasitic (see Scoble 1992, Pierce 1995).

Literature Cited

Armbruster, P., W. E. Bradshaw, A. L. Steiner, C. M. Holzapfel. 1999. Evolutionary responses to environmental stress by the pitcher-plant mosquito, *Wyeomyia smithii*. *Heredity* 83: 509–519.

Ayres, M. P. 1993. Plant defense, herbivory, and climate change. Pages 75–94 in P. M. Kareiva, J. G. Kingsolver, and R. B. Huey, eds., *Biotic interactions and global change*. Sunderland, Mass.: Sinauer.

Ayres, M. P., and M. J. Lombardero. 2000. Assessing the consequences of global change for forest disturbance from herbivores and pathogens. *Science of the Total Environment* 262: 263–286.

Baker, R. R. 1978. *The Evolutionary Ecology of Animal Migration*. London: Hodder and Stoughton.

Barry, J. P., C. H. Baxter, R. D. Sagarin, and S. E. Gilman. 1995. Climate-related long-term faunal changes in a California rocky intertidal community. *Science* 267: 672–678.

Bazzaz, F. A. 1990. The response of natural ecosystems to the rising global CO_2 levels. *Annual Review of Ecology and Systematics* 21: 167–196.

Berenbaum, M. R. 1995. Chemistry and oligophagy in the Papilionidae. Pages 27–38 in J. M. Scriber, Y. Tsubaki, and R. C. Lederhouse, eds., *Swallowtail Butterflies: Their Ecology and Evolutionary Biology*. Gainesville, Fla.: Scientific.

Bernays, E., and M. Graham. 1988. On the evolution of host specificity in phytophagous arthropods. *Ecology* 69: 886–892.

Bobbink, R. 1991. Effects of nutrient enrichment in Dutch chalk grassland. *Journal of Applied Ecology* 28: 28–41.

Boggs, C. L. 1986. Reproductive strategies of female butterflies: Variation in and constraints on fecundity. *Ecological Entomology* 11: 7–15.

———. 1987. Ecology of nectar and pollen feeding in Lepidoptera. Pages 369–391 in F. Slansky and J. G. Rodriguez, eds., *Nutritional Ecology of Insects, Mites, Spiders, and Related Invertebrates*. New York: John Wiley and Sons.

———. 1997. Dynamics of reproductive allocation from juvenile and adult feeding: Radiotracer studies. *Ecology* 78: 192–202.

Boggs, C. L., and C. L. Ross. 1993. The effect of adult food limitation on life history traits in *Speyeria mormonia* (Lepidoptera: Nymphalidae). *Ecology* 74: 433–441.

Boggs, C. L., W. B. Watt, and P. R. Ehrlich. In press. *Ecology and Evolution Taking Flight: Butterflies as Model Systems*. Chicago: University of Chicago Press.

Bossart, J. L., and J. M. Scriber. 1995. Maintenance of ecologically significant genetic variation in the tiger swallowtail butterfly through differential selection and gene flow. *Evolution* 49: 1163–1171.

Brakefield, P. M. 1987. Geographical variability in, and temperature effects on, the phenology of *Maiola jurtina* and *Pyronia tithonus* (Lepidoptera, Satyrinae) in England and Wales. *Ecological Entomology* 12: 139–148.

Brakefield, P. M., J. Gates, D. Keys, F. Kesbeke, P. J. Wijngaarden, A. Monteiro, V. French, and S. B. Carroll. 1996. Development, plasticity and evolution of butterfly eyespot patterns. *Nature* 384: 236–242.

Brown, J. H. 1984. On the relationship between abundance and distribution. *American Naturalist* 124: 255–279.

Buse, A., and J. E. G. Good. 1996. Synchronization of larval emergence in winter moth (*Operophtera brumata* L.) and budburst in pedunculate oak (*Quercus robur* L.) under simulated climate change. *Ecological Entomology* 21: 335–343.

Buse, A., E. G. Good, S. Dury, and C. M. Perrins. 1998. Effects of elevated temperature and carbon dioxide on the nutritional quality of leaves of oak (*Quercus robur* L.) as food for the winter moth (*Operophtera brumata* L.). *Functional Ecology* 12: 742–749.

Cappuccino, N., and P. Kareiva. 1985. Coping with a capricious environment: A population study of a rare Pierid butterfly. *Ecology* 66: 152–161.

Casey, T. M. 1976. Activity patterns, body temperature and thermal ecology in two desert caterpillars (Lepidoptera: Sphingidae). *Ecology* 57: 485–497.

Chiariello, N. R., and C. B. Field. 1996. Annual grassland responses to elevated CO_2 in multiyear community mesocosms. Pages 139–157 in C. Körner and F. Bazzaz, eds., *Carbon Dioxide, Populations, and Communities*. San Diego: Academic.

Clench, H. K. 1966. Behavioural thermoregulation in butterflies. *Ecology* 47: 1021–1034.

Cohn, J. P. 1989. Gauging the biological impacts of the greenhouse effect. *BioScience* 39: 142–146.

Coley, P. D. 1980. Effects of leaf age and plant life-history patterns on herbivory. *Nature* 284: 545–546.

———. 1998. Possible effects of climate change on plant/herbivore interactions in moist tropical forests. *Climatic Change* 39: 455–472.

Coope, G. R. 1995. The effects of Quaternary climatic changes on insect populations: Lessons from the past. Pages 29–48 in R. Harrington and N. E. Stork, eds., *Insects in a Changing Environment*. San Diego: Academic.

Courtney, S. P. 1986. The ecology of pierid butterflies: Dynamics and interactions. *Advances in Ecological Research* 15: 51–131.

Courtney S. P., and A. E. Duggan. 1983. The population biology of the orange tip butterfly *Anthocharis cardamines* in Britain. *Ecological Entomology* 8: 271–281.

Cushman, J. H., C. L. Boggs, S. B. Weiss, D. D. Murphy, A. W. Harvey, and P. R. Ehrlich. 1994. Estimating female reproductive success of a threatened butterfly: Influence of emergence time and host plant phenology. *Oecologia* 99: 194–200.

Daily, G. C., ed. 1997. *Nature's Services: Societal Dependence on Natural Ecosystems*. Washington, D.C.: Island Press.

Dempster, J. P. 1983. The natural control of populations of butterflies and moths. *Biological Review* 58: 461–481.

Dennis, R. L. H., ed. 1992. *The Ecology of Butterflies in Britain*. Oxford: Oxford University Press.

———. 1993. *Butterflies and Climate Change*. Manchester, England: Manchester University Press.

Dennis, R. L. H., and T. G. Shreeve. 1991. Climatic change and the British butterfly fauna: Opportunities and constraints. *Biological Conservation* 55: 1–16.

Denno, R. F., M. S. McClure, and J. R. Ott. 1995. Interspecific interactions in phytophagous insects: Competition reexamined and resurrected. *Annual Review of Entomology* 40: 297–331.

Dewar, R. C., and A. D. Watt. 1992. Predicted changes in the synchrony of larval emergence and budburst under climatic warming. *Oecologia* 89: 557–559.

Digby, P. S. B. 1955. Factors affecting the temperature excess of insects in sunshine. *Journal of Experimental Biology* 32: 279–298.

Dingle, H. 1996. *Migration: The Biology of Life on the Move.* Oxford: Oxford University Press.

Dobkin, D. S., I. Olivieri, and P. R. Ehrlich. 1987. Rainfall and the interaction of microclimate with larval resources in the population dynamics of checkerspot butterflies (*Euphydryas editha*) inhabiting serpentine grassland. *Oecologia* 71: 161–166.

Douwes, P. 1976. Activity in *Heodes virgaureae* (Lep: Lyncaenidae) in relation to air temperature, solar radiation, and time of day. *Oecologia* 22: 287–298.

Dury, S. J., J. E. G. Good, C. M. Perrins, A. Buse, and T. Kaye. 1998. The effects of increased CO_2 and temperature on oak leaf palatability and the implications for herbivorous insects. *Global Change Biology* 4: 55–61.

Ehrlich, P. R. 1961. Intrinsic barriers to dispersal in checkerspot butterfly. *Science* 134: 108–109.

———. 1965. The population biology of the butterfly, *Euphydryas editha.* II. The structure of the Japer Ridge colony. *Evolution* 19: 327–336.

Ehrlich, P. R., and L. E. Gilbert. 1973. Population structure and dynamics of the tropical butterfly *Heliconius ethilla*. *Biotropica* 5: 69–82.

Ehrlich, P. R., and D. D. Murphy. 1981. The population biology of checkerspot butterflies (*Euphydryas*). *Biologisches Zentralblatt* 100: 613–629.

———. 1987. Conservation lessons from long-term studies of checkerspot butterflies. *Conservation Biology* 1: 122–131.

Ehrlich, P. R., and P. H. Raven. 1965. Butterflies and plants: A study in coevolution. *Evolution* 18: 586–608.

Ehrlich, P. R., D. E. Breedlove, P. F. Brussard, and M. A. Sharp. 1972. Weather and the "regulation" of subalpine populations. *Ecology* 53: 243–247.

Ehrlich, P. R., D. D. Murphy, M. C. Singer, C. B. Sherwood, R. R. White, and I. L. Brown. 1980. Extinction, reduction, stability and increase: The responses of checkerspot butterfly (*Euphydrays*) populations to the California drought. *Oecologia* 46: 101–105.

Fajer, E. D, M. D. Bowers, and F. A. Bazzazz. 1989. The effects of enriched carbon dioxide atmospheres on plant–insect herbivore interactions. *Science* 243: 1198–1200.

Feeny, P. 1970. Seasonal changes in oak leaf tannins and nutrients as a cause of spring feeding by winter moth caterpillars. *Ecology* 51: 565–581.

Fleishman, E., G. T. Austin, and A. D. Weiss. 1998. An empirical test of Rapoport's rule: Elevational gradients in montane butterfly communities. *Ecology* 79: 2482–2493.

Frazer, B. D., and N. Gilbert. 1976. Coccinellids and aphids: A quantitative study of the impact of adult ladybirds preying on field populations of pea aphids. *Journal of the Entomological Society of British Columbia* 73: 33–56.

Gates, D. M. 1993. *Climate Change and its Biological Consequences.* Sunderland, Mass.: Sinauer.

Gilbert, L. E. 1971. Butterfly–plant coevolution: Has *Passiflora adenopoda* won the selectional race with Heliconiine butterflies? *Science* 172: 585–586.

Gilbert, L. E., and M. C. Singer. 1975. Butterfly ecology. *Annual Review of Ecology and Systematics* 6: 365–397.

Gotthard, K., S. Nylin, and C. Wiklund. 1999. Seasonal plasticity in two satyrine butterflies: State-dependent decision-making in relation to day length. *Oikos* 84: 453–462.

Graham, R. W., and E. C. Grimm. 1990. Effects of global climate change on the patterns of terrestrial biological communities. *Trends in Ecology and Evolution* 5: 289–292.

Gutierrez, A. P., G. D. Butler, Y. Wang, and D. Westphal. 1977. The interaction of pink bollworm (Lepidoptera: Gelichiidae), cotton, and weather: A detailed model. *Canadian Entomologist* 109: 1457–1468.

Hagen, R. H. 1991. Population structure and host use in hybridizing subspecies of *Papilio glaucus* (Lepidoptera: Papilionidae). *Evolution* 44: 1914–1930.

Hammond, D. L. 1995. The current magnitude of biodiversity. Pages 113–138 in V. H. Heywood, ed., *Global Biodiversity Assessment.* Cambridge: Cambridge University Press.

Hanski, I. 1994. A practical model of metapopulation dynamics. *Journal of Animal Ecology* 63: 151–162.

———. 1999. *Metapopulation Ecology.* New York: Oxford University Press.

Harrington, R., I. Woiwod, and T. Sparks. 1999. Climate change and trophic interactions. *Trends in Ecology and Evolution* 14: 146–150.

Harrison, S. 1993. Species diversity, spatial scale, and global change. Pages 388–401 in P. M. Kareiva, J. G. Kingsolver, and R. B. Huey, eds., *Biotic Interactions and Global Change.* Sunderland, Mass.: Sinauer.

Harrison, S., D. D. Murphy, P. R. Ehrlich. 1988. Distribution of the bay checkerspot butterfly, *Euphydryas editha bayensis:* Evidence for a metapopulation model. *American Naturalist* 132: 360–382.

Hassell, M. P. 1985. Insect natural enemies as regulating factors. *Journal of Animal Ecology* 54: 323–334.

Hayes, J. L. 1985. Egg distribution and survivorship in the pierid butterfly, *Colias alexandra. Oecologia* 66: 495–498.

Heinrich, B. 1974. Thermoregulation in endothermic insects: Body temperature is closely attuned to activity and energy supplies. *Science* 185: 747–756.

———. 1986. Thermoregulation and flight activity of a satyrine, *Coenonympha inornata* (Lep: Satyridae). *Ecology* 67: 593–597.

Hellmann, J. J. 2000. The Role of Environmental Variation in the Dynamics of an Insect–Plant Interaction. Ph.D. dissertation, Stanford University.

Hewitt, G. M. 1996. Some genetic consequences of ice ages, and their role in divergence and speciation. *Biological Journal of the Linnean Society* 58: 247–276.

Hickman, J. C., ed. 1993. *The Jepson Manual of Higher Plants of California.* Berkeley: University of California Press.

Hill, C. J., and N. E. Pierce. 1989. The effects of adult diet on the biology of butterflies. 1. The common imperial blue, *Jalmenus evagoras. Oecologia* 81: 249–257.

Hill, J. K., C. D. Thomas, D. S. Blakeley. 1999. Evolution of flight morphology in a butterfly that has recently expanded its geographic range. *Oecologia* 121: 165–170.

Hill, J. K., C. D. Thomas, B. Huntley. 1999. Climate and habitat availability determine twentieth century changes in a butterfly's range margin. *Proceedings of the Royal Society of London* B 266: 1197–1206.

Hobbs, R. J., S. L. Gulmon, V. J. Hobbs, and H. A. Mooney. 1988. Effects of fertilizer addition and subsequent gopher disturbance on a serpentine annual grassland community. *Oecologia* 75: 291–295.

Hoffmann, A. A., and M. W. Blows. 1993. Evolutionary genetics and climate change: Will animals adapt to global warming? Pages 165–178 in P. M. J. Kareiva, G. Kingsolver, and R. B. Huey, eds., *Biotic Interactions and Global Change.* Sunderland, Mass.: Sinauer.

Hoffmann, A. A., and P. A. Parsons. 1997. *Extreme Evolutionary Change and Evolution.* Cambridge: Cambridge University Press.

Holliday, N. J. 1977. Population ecology of winter moth (*Operophtera brumata*) on apple in relation to larval dispersal and time of bud burst. *Journal of Applied Ecology* 14: 803–813.

Huey, R. B., and J. G. Kingsolver. 1993. Evolution of resistance to high temperature in ectotherms. *American Naturalist* 142: S21–S46.

Huffaker, C. B., and A. P. Gutierrez, eds., 1999. *Ecological Entomology.* New York: John Wiley and Sons.

Hughes, J. B., G. C. Daily, and P. R. Ehrlich. 1997. Population diversity: Its extent and extinction. *Science* 278: 689–692.

Hunter, A. F., and J. S. Elkinton. 2000. Effects of synchrony with host plant on populations of a spring-feeding Lepidopteran. *Ecology* 81: 1248–1261.

Hunter, M. D., and J. N. McNeil. 1997. Host-plant quality influences diapause and voltinism in a polyphagous insect herbivore. *Ecology* 78: 977–986.

Inouye, D.W. 2000. The ecological and evolutionary significance of frost in the context of climate change. *Ecology Letters* 3: 457–463.

Intergovernmental Panel on Climate Change (IPCC). 1996a. *Climate Change 1995: The Science of Climate Change*, eds., J.T. Houghton, L. G. Meira Filho, B. A. Callandar, N. Harris, A. Kattenberg, and K. Maskell. Contribution of working group I to the second assessment report of the IPCC. Cambridge: Cambridge University Press.

———. 1996b. *Climate Change 1995: Impacts, Adaptations, and Mitigation of Climate Change: Scientific Technical Analyses*, eds., R. T. Watson, M. C. Zinyowera, and R. H. Moss. Contribution of working group II to the second assessment report of the IPCC. Cambridge: Cambridge University Press.

Johnson, C. G. 1969. *Migration and Dispersal of Insects by Flight.* London: Methuen.

Johnson, S. J. 1995. Insect migration in North America: Synoptic-scale transport in a highly seasonal environment. Pages 31–66 in V. A. Drake and A. G. Gatehouse, eds., *Insect Migration: Tracking Resources through Space and Time.* Cambridge: Cambridge University Press.

Kammer, A. E. 1970. Thoracic temperature, shivering and flight in the monarch butterfly, *Danaus plexippus. Zeitschrift für vergleichende Physiologie* 68: 334–344.

Kareiva, P. M., J. G. Kingsolver, and R. B. Huey, eds., 1993. *Biotic Interactions and Global Change.* Sunderland, Mass.: Sinauer.

Kingsolver, J. G. 1983. Thermoregulation and flight in *Colias* butterflies: Elevational patterns and mechanistic limitations. *Ecology* 64: 534–545.

———. 1985. Butterfly thermoregulation: Organismic mechanisms and population consequences. *Journal of Research on the Lepidoptera* 24: 1–20.

———. 1989. Weather and the population dynamics of insects: Integrating physiological and population ecology. *Physiological Zoology* 62: 314–334.

Kingsolver, J. G., and W. B. Watt. 1983. Thermoregulatory strategies in *Colias* butterflies: Thermal stress and the limits to adaptation in temporally varying environments. *American Naturalist* 121: 32–55.

Kremen, C. 1994. Biological inventory using target taxa: A case study of the butterflies of Madagascar. *Ecological Applications* 4: 407–422.

Landsberg, J., and M. S. Smith. 1992. A functional scheme for predicting the outbreak potential of herbivorous insects under global atmospheric change. *Australian Journal of Botany* 40: 565–577.

Lawton, J. H. 1995. The response of insects to environmental change. Pages 3–26 in R. Harrington and N. E. Stork, eds., *Insects in a Changing Environment.* San Diego: Academic Press.

Lederhouse, R. C. 1983. Population structure, residency and weather related mortality in the black swallowtail butterfly, *Papilio polyxenes. Oecologia* 59: 307–311.

Lincoln, D. E, D. Couvet, and N. Sionit. 1986. Response of an insect herbivore to host plants grown in carbon dioxide enriched atmospheres. *Oecologia* 69: 556–560.

Lincoln, D. E., E. D. Fajer, and R. H. Johnson. 1993. Plant–insect herbivore interactions in elevated CO_2 environments. *Trends in Ecology and Evolution* 8: 64–68.

Logan, J. A., D. J. Wollkind, S. C. Hoyt, and L. K. Tanigoshi. 1976. An analytic model for description of temperature dependent rate phenomena in arthropods. *Environmental Entomology* 5: 1133–1140.

MacDonald, K. A., and J. H. Brown. 1992. Using montane mammals to model extinctions due to global change. *Conservation Biology* 6: 409–415.

Malcolm, S. B. 1987. Monarch butterfly migration in North America: Controversy and conservation. *Trends in Ecology and Evolution* 2: 135–138.

Mathavan, S., and T. J. Pandian. 1975. Effect of temperature on food utilization in the monarch butterfly *Danaus chrysippus. Oikos* 26: 60–64.

Mattson, W. J., and R. A. Haack. 1987. The role of drought in outbreaks of plant-eating insects. *BioScience* 37: 110–118.

May, R. M., J. H. Lawton, and N. E. Stork. 1995. Assessing extinction rates. Pages 1–24 in J. H. Lawton and R. M. May, eds., *Extinction Rates.* Oxford: Oxford University Press.

McKechnie, S. W. 1975. Population genetics of *Euphydryas* butterflies. I. Genetic variation and the neutrality hypothesis. *Genetics* 81: 571–594.

McMillan, W. O., C. D. Jiggins, and J. Mallet. 1997. What initiates speciation in passion-vine butterflies? *Proceedings of the National Academy of Sciences* 94: 8628–8633.

Murphy, D. D., and P. R. Ehrlich. 1989. Conservation biology of California's remnant native grasslands. Pages 201–211 in L. F. Huenneke and H. Mooney, eds., *Grassland Structure and Function: California Annual Grassland.* Dordrecht: Kluwer.

Murphy, D. D., and S. B. Weiss. 1988. A long-term monitoring plan for a threatened butterfly. *Conservation Biology* 2: 367–374.

———. 1992. Effects of climate chance on biological diversity in Western North America: Species losses and mechanisms. Pages 355–368 in R. L. Peters, R. L. and T. E. Lovejoy, eds., *Global Warming and Biological Diversity.* New Haven: Yale University Press.

Murphy, D. D., A. E. Launer, and P. R. Ehrlich. 1983. The role of adult feeding in egg production and population dynamics of the checkerspot butterfly *Euphydryas editha. Oecologia* 56: 257–263.

Myneni, R. B., C. D. Keeling, C. J. Tucker, G. Asrar, and R. R. Nemani. 1997. Increased plant growth in the northern high latitudes from 1981 to 1991. *Nature* 386: 698–702.

Nylin, S., K. Gotthard, and C. Wiklund. 1996. Reaction norms for age and size at maturity in Lasiommata butterflies: Predictions and tests. *Evolution* 50: 1351–1358.

Ohgushi, T. 1997. Plant-mediated interactions between herbivorous insects. Pages 115–130 in T. Abe, S. A. Levin, and M. Higashi, eds., *Biodiversity: An Ecological Perspective*. New York: Springer-Verlag.

Overpeck, J. T. 1996. Warm climate surprises. *Science* 271: 1820–1821.

Overpeck J. T., R. S. Webb, and T. Webb III. 1992. Mapping eastern North American vegetation change of the past 18 ka: No-analogs and the future. *Geology* 20: 1071–1074.

Parmesan, C. 1996. Climate and species' range. *Nature* 382: 765–766.

Parmesan, C., N. Ryrholm, C. Stefanescu, J. K. Hill, C. D. Thomas, H. Descimon, B. Huntley, L. Kaila, J. Kullberg, T. Tammaru, J. Tennent, J. A. Thomas, and M. Warren. 1999. Poleward shifts in geographical ranges of butterfly species associated with regional warming. *Nature* 399: 579–583.

Peters, R. L., and J. D. S. Darling. 1985. The greenhouse effect and nature reserves. *BioScience* 35: 707–717.

Peters, R. L., and T. E. Lovejoy, eds., 1992. *Global Warming and Biological Diversity*. New Haven: Yale University Press.

Pierce, N. E. 1995. Predatory and parasitic Lepidoptera: Carnivores living on plants. *Journal of the Lepidopterists' Society* 49: 412–453.

Pimm, S. L. 1980. Food web design and the effect of species deletion. *Oikos* 35: 139–140.

Pivnick, K. A., and J. N. McNeil. 1987. Diel patterns of activity of *Thymelicus lineola* adults (Lep: Hisperiidae) in relation to weather. *Ecological Entomology* 12: 197–207.

Pollard, E. 1979. Population ecology and change in range of the white admiral butterfly *Ladoga camilla* L. in England. *Ecological Entomology* 4: 61–74.

Pollard, E. 1988. Temperature, rainfall and butterfly numbers. *Journal of Applied Ecology* 25: 819–828.

Pollard, E., and B. C. Eversham. 1995. Butterfly monitoring 2: Interpreting the changes. Pages 23–36 in A. S. Pullin, ed., *Ecology and Conservation of Butterflies*. New York: Chapman and Hall.

Porter, K. 1983. Multivoltinism in *Apanteles bignellii* and the influences of weather on sychronization with its host *Euphydryas aurinia*. *Entomologia experimentalis et applicata* 34: 155–162.

Pullin, A. S., and J. S. Bale. 1989. Effects of low temperature on diapausing *Aglais urticae* and *Inachis io* (Lepidoptera: Nymphalidae): Cold hardiness and overwintering survival. *Journal of Insect Physiology* 35: 277–281.

Quinn, J. F., and J. R. Karr. 1993. Habitat fragmentation and global change. Pages 451–463 in P. M. Kareiva, J. G. Kinsolver, and R. B. Huey, eds., *Biotic Interactions and Global Change.* Sunderland, Mass.: Sinauer.

Ratte, H. T. 1984. Temperature and insect development. Pages 33–66 in K. H. Hoffmann, ed., *Environmental Physiology and Biochemistry.* New York: Springer-Verlag.

Rockwood, L. L. 1974. Seasonal changes in the susceptibility of *Crescentia alata* leaves to the flea beetle, *Oedionychus* sp. *Ecology* 55: 142–148.

Rodriguez-Trelles, F., M. A. Rodriguez, S. M. Scheiner. 1998. Tracking the genetic effects of global warming: *Drosophila* and other model systems. Conservation Ecology 2: 2. Available online at http://www.consecol.org/ vol2/iss2/art2.

Root, T. L., and S. H. Schneider. 1993. Can large-scale climatic models be linked with multiscale ecological studies? *Conservation Biology* 7: 256–270.

Roy, D. B., and T. H. Sparks. 2000. Phenology of British butterflies and climate change. *Global Change Biology* 6: 407–416.

Roy, D. B., P. Rothery, D. Moss, E. Pollard, and J. A. Thomas. 2001. Butterfly numbers and weather: Predicting historical trends in abundance and the future effects of climate change. *Journal of Animal Ecology* 70: 201–217.

Saccheri, I., M. Kuussaari, M. Kankare, P. Vikman, W. Fortelius, and I. Hanski. 1998. Inbreeding and extinction in a butterfly metapopulation. *Nature* 392: 491–494.

Sato, Y. 1979. Experimental studies on parasitization by *Apanteles glomeratus* L. (Hym: Braconidae). IV. Factors leading a female to the host. *Physiological Entomology* 4: 63–70.

Schlichting, C. D., and M. Pigliucci. 1998. *Phenotypic Evolution: A Reaction Norm Perspective.* Sunderland, Mass.: Sinauer.

Schneider, S. H. 1990. The greenhouse effect: Science and policy. *Science* 10: 771–781.

Schneider, S. H., and T. L. Root. 1996. Ecological implications of climate change will include surprises. *Biodiversity and Conservation* 5: 1109–1119.

Schultz, C. B. 1998. Dispersal behavior and its implications for reserve design in a rare Oregon butterfly. *Conservation Biology* 12: 284–292.

Scoble, M. J. 1992. *The Lepidoptera: Form, Function and Diversity.* Oxford: Oxford University Press.

Scriber, J. M., and R. C. Lederhouse. 1983. Temperature as a factor in the development and feeding ecology of tiger swallowtail caterpillars, *Papilio glaucus* (Lepidoptera). *Oikos* 40: 95–102.

Scriber, J. M., and S. H. Gage. 1995. Pollution and global climate change: Plant ecotones, butterfly hybrid zones and changes in biodiversity. Pages 319–344 in J. M. Scriber, Y. Tsubaki, and R. C. Lederhouse, eds., *Swallowtail Butterflies: Their Ecology and Evolutionary Biology.* Gainesville, Fla.: Scientific.

Sharpe, P. J. H., and D. W. DeMichele. 1977. Reaction kinetics of poikilotherm development. *Journal of Theoretical Biology* 64: 649–670.

Sherman, P. W., and W. B. Watt. 1973. The thermal ecology of some *Colias* butterfly larvae. *Journal of Comparative Physiology* 83: 25–40.

Shreeve, T. G. 1986. The effect of weather on the life cycle of the speckled wood butterfly *Pararge aegeria. Ecological Entomology* 11: 325–332.

Singer, M. C. 1971. Evolution of food-plant preference in the butterfly *Euphydryas editha. Evolution* 25: 383–389.

———. 1972. Complex components of habitat suitability within a butterfly colony. *Science* 176: 75–77.

———. 1994. Behavioral constraints on the evolutionary expansion of insect diet: A case history from checkerspot butterflies. Pages 279–396 in L. A. Real, ed., *Behavioral Mechanisms in Evolutionary Ecology.* Chicago: Chicago University Press.

Singer, M. C., C. D. Thomas, and C. Parmesan. 1993. Rapid human-induced evolution of insect–host associations. *Nature* 366: 681–683.

Singer, M. C., C. D. Thomas, H. L. Billington, and C. Parmesan. 1994. Correlates of speed of evolution on host preference in a set of twelve populations of the butterfly *Euphydryas editha. Ecoscience* 1: 107–114.

Smith, R. H., R. M. Sibly, and H. Moller. 1987. Control of size and fecundity in *Pieris rapae:* Towards a theory of butterfly cycles. *Journal of Animal Ecology* 56: 341–350.

Southwood, T. R. E. 1981. Ecological aspects of insect migration. Pages 197–208 in D. J. Aidley, ed., *Animal Migration.* Cambridge: Cambridge University Press.

Sperling, F. A. H. 1994. Sex-linked genes and species differences in Lepidoptera. *Canadian Entomologist* 126: 807–818.

Strong, D. R., J. H. Lawton, and R. Southwood. 1984. *Insects on Plants: Community Patterns and Mechanisms.* Cambridge: Harvard University Press.

Taylor, F. 1981. Ecology and evolution of physiological time in insects. *American Naturalist* 117: 1–23.

Taylor, L. R. 1974. Insect migration, flight periodicity and the boundary layer. *Journal of Animal Ecology* 43: 225–238.

Thomas, C. D., and J. J. Lennon. 1999. Birds extend their ranges northwards. *Nature* 399: 213.

Thomas, C. D., M. C. Singer, D. A. Boughton. 1996. Catastrophic extinction of population sources in a butterfly metapopulation. *American Naturalist* 148: 957–975.

Thomas, J. A., C. D. Thomas, D. J. Simcox, and R. T. Clarke. 1986. Ecology and declining status of the silver-spotted skipper butterfly (*Hesperia comma*) in Britain. *Journal of Applied Ecology* 23: 365–380.

Travis, J., and D. J. Futuyma. 1993. Global change: Lessons from and for evolutionary biology. Pages 251–263 in P. M. Kareiva, J. G. Kingsolver, and R. B. Huey, eds., *Biotic Interactions and Global Change*. Sunderland, Mass.: Sinauer.

Turner, J. R. G. 1971. 2000 generations of hybridization in a Heliconius-Melpomene butterfly. *Evolution* 25: 471–482.

Uvarov, B. P. 1931. Insects and climate. *Transactions of the Entomological Society of London* 79: 1–247.

Vane-Wright, R. I., and P. R. Ackery, eds., 1984. *The Biology of Butterflies*. London: Academic Press.

Visser, M. E., and L. J. M. Holleman. 2000. Warmer springs disrupt the synchrony of oak and winter moth phenology. *Proceedings of the Royal Oak Society* B 268: 289–294.

Vitousek, P. M. 1994. Beyond global warming: Ecology and global change. *Ecology* 75: 1861–1876.

Vitousek, P. M., J. D. Aber, R. W. Howarth, G. E. Liken, P. A. Matson, D. W. Schindler, W. H. Schlesinger, and D. G. Tilman. 1997. Human alteration of the global nitrogen cycle: Source and consequences. *Ecological Applications* 7: 737–750.

Warren, M. S. 1992. Butterfly populations. Pages 73–92 in R. L. H. Dennis, ed., *The Ecology of Butterflies in Britain*. Oxford: Oxford University Press.

Watt, A. D., and A. M. McFarlane. 1991. Winter moth on Sitka spruce: Synchrony of egg hatch and budburst, and its effect on larval survival. *Ecological Entomology* 16: 387–390.

Watt, A. D., J. B. Wittaker, M. Docherty, G. Brooks, E. Lindsay, and D. T. Salt. 1995. The impact of elevated CO_2 on insect herbivores. Pages 197–217 in R. Harrington and N. E. Stork, eds., *Insects in a Changing Environment*. San Diego: Academic Press.

Watt, W. B. 1968. Adaptive significance of pigment polymorphisms in *Colias* butterflies. I. Variation of melanin pigment in relation to thermoregulation. *Evolution* 22: 437–458.

———. 1992. Eggs, enzymes, and evolution: Natural genetic variants change insect fecundity. *Proceedings of the National Academy of Science* 89: 10608–10612.

———. 1997. Accuracy, anecdotes, and artifacts in the study of insect thermal ecology. *Oikos* 80: 399–400.

Watt, W. B., D. Han, and B. E. Tabashnik. 1979. Population structure of Pierird butterflies. II. A "native" population of *Colias philodice eriphyle* in Colorado. *Oecologia* 44: 44–52.

Watt, W. B, R. C. Cassin, and M. S. Swan. 1983. Adaptation at specific loci. III. Field behavior and survivorship differences among *Colias* PGI genotypes are predictable from in vitro biochemistry. *Genetics* 103: 691–724.

Watt, W. B., K. Donohue, and P. A. Carter. 1996. Adaptation at specific loci. VI. Divergence vs. parallelism of polymorphic allozymes in molecular function and fitness-component effects among *Colias* species (Lepidoptera, Pieridae). *Molecular Biological Evolution* 13: 699–709.

Weiss, S. B. 1999. Cars, cows, and checkerspot butterflies: Nitrogen deposition and management of nutrient poor grasslands for a threatened species. *Conservation Biology* 13: 1476–1486.

Wellington, W. G., D. L. Johnson, and D. J. Lactin. 1999. Weather and insects. Pages 313–353 in C. B. Huffaker and A. P. Gutierrez, eds., *Ecological Entomology*. New York: John Wiley and Sons.

Wiklund, C., and A. Persson. 1983. Fecundity, and the relation of egg weight variation to offspring fitness in the speckled wood butterfly, *Pararge aegeria*, or why don't butterfly females lay more eggs? *Oikos* 40: 53–63.

Williams, C. B. 1930. *The Migration of Butterflies*. Edinburgh: Oliver and Boyd.

Wilson, E. O. 1987. The little things that run the world. *Conservation Biology* 1: 344–346.

CHAPTER 3

Historical Studies of Species' Responses to Climate Change: Promise and Pitfalls

RAPHAEL SAGARIN

Identifying responses of ecological systems and their biotic components to regional and global climatic changes is a critical element of climate change science. The traditional methods of experimental ecology involving manipulations of small numbers of species over small spatial and/or temporal scales may be limited in some cases when assessing the impacts of climate change on plant and animal populations and communities. Although experimental manipulations have been and will continue to be very valuable in suggesting potential targets of species, community, and even ecosystem change (Carpenter et al. 1993, Chapin et al. 1995), they are more limited in their ability to detect real changes to populations and communities due to gradual, widespread climate change. For many systems, experimental manipulations that attempt to simulate changing climatic conditions and discern species' responses will not be feasible because of the large spatial scales and ecological complexity involved. Moreover, results of small-scale manipulations may not reasonably be expected to scale up to the level of entire species ranges where many species encounter vastly different sets of climatic and biological variables. In some cases, we may be interested in the responses of rare or endangered species to climate change for which experimental manipulations are prohibited for legal and ethical reasons. The temporal scale of most experimental manipulations has traditionally been too short to assess effects of gradual climate change. Even in systems with fairly well understood dynamics, changes on a short-term scale are much different than changes in the long-term (Chapin et al. 1995), which could lead to spurious conclusions in short-term experimental studies. Finally, traditional experimental studies of ecological dynamics (e.g., species interactions, recruitment dynamics) have tended to assume a background environment that is more or less constant on a long-term (interannual or interdecadal) basis, so that results are considered representative of the system's dynamics through time (Francis and Hare 1994). This premise is clearly violated by the climate changes that are occurring and are expected to continue as greenhouse gas concentrations increase.

Experimental studies of species' responses to climate change do have an important role in studies of biotic response to climate change. Specifically, they are the only method by which mechanisms of response, be they physiological (individual organism responses), physical (due to changes in habitat, currents, stream-flow, etc.), or ecological (due to interactions between species), can be elucidated. Many studies have already used experimental methods in laboratory and field settings to illustrate how such mechanisms might affect species under climate change (see Davis et al. 1998 for a fine example; also Sanford, this volume; citations in Saavedra, this volume).

Nevertheless, for many systems we will have to rely heavily on historical observational data and attempts to correlate the physical and biotic components of these data in order to paint a broad-scale picture of species' responses to climate changes. Francis and Hare (1994) have shown, through examples of climate-related shifts in abundance of zooplankton and salmon, that historical-descriptive science is not merely a fallback position in the face of the daunting task of studying large-scale ecological processes, but is instead a sound method for identifying links between physical and biological systems.

Historical Science Defined

In Francis and Hare's (1994) definition, historical-descriptive science begins with observations of a system. At this stage, the observations may be anecdotal (e.g., fishermen noticing unusual species in their catches during warm-weather periods) or tangential to a main line of research (e.g., an investigator noticing changes in species composition over many years at a study site). Whatever the source, these observations are intriguing enough that an investigator uses them as a launching point for a more in-depth study of the system.

Simple models, or narratives, are then developed that suggest hypotheses of how the various observations fit together. For studies of how climate changes will affect biotic systems, many of these models have already been suggested in the form of predictions of population or community changes under expected warming climate. The most commonly proposed model is that species' geographic ranges will shift poleward in order to maintain the same position relative to climatic conditions. This hypothesis has been forwarded for a wide variety of biotic systems including terrestrial plants (Davis 1989, Woodward 1992), infectious disease (Shope 1991, Patz et al. 1996), invertebrates (Bhaud et al. 1995), birds (Root 1993, Taper et al. 1995), freshwater fishes (Scott and Poynter 1991, Murawski 1993), algae (Breeman 1990), marine fishes (Frank et al. 1990), and coastal marine communities (Fields et al. 1993, Lubchenco et al. 1993). Other types of narratives that have been developed for species under warming climate and that might be tested using historical studies include (but are not limited to) changes in timing of reproductive or other life-history stages (Breeman 1990, Bhaud et al. 1995, Sparks and Carey 1995), increased stratification of oceans affecting nutrient upwelling (Fields et al. 1993, Roemmich and McGowan 1995), and alterations of species interactions (Murawski 1993, Davis et al. 1998).

In the final step, historical observations from a variety of sources are examined for patterns that support or don't support the model. The physical data used typically include long-term time-series data for one or more climatic variables, such as daily temperature records or annual tallies of frost-free days. Historical data on biotic systems might include presence/absence data, population counts, or measures of community composition and diversity through time. These data may take the form of "snapshot" data taken at the same location at separate points in time, or preferably, as time series data collected for many years, such as catch records for fisheries or Christmas bird counts. The analysis of these data may be as simple as noting a gradual increase in temperature alongside gradual changes in species' abundances (Southward et al. 1995), or more rigorous attempts to correlate physical and biological variables using time-series analysis (Francis and Hare 1994, McGowan et al. 1996).

Examples of the Historical Method in Action

Recently, a handful of papers have used some form of this historical-descriptive method to demonstrate evidence of species' responses to climate changes that have occurred over the last century. Barry et al.(1995) and Sagarin et al. (1999) documented a general increase in southern invertebrate species and a decrease in northern species between two investigations of a single intertidal site conducted 60 years apart. This faunal change was concomitant with an increase in mean annual shoreline ocean temperatures. Holbrook et al. (1997) found an overall decline in abundance and shift toward dominance of southern reef fish species in the Southern California Bight following a shift to warmer ocean temperatures in the mid-1970s. California Cooperative Oceanic Fisheries Investigations (CalCOFI) data have been used to document population increases and northward range expansion of sardine populations during multidecadal ocean warming since the mid-1970s (Lluch-Belda et al. 1992). Southward (1967, 1995) found that warm-water species in the English Channel increased in abundance and cold-water species declined during periods of ocean warming (1920–1960; 1981–1995) whereas the opposite occurred during a cooling period (1960–1981). Parmesan (1996) compared her extensive surveys of the Edith's checkerspot butterfly in western North America to historical records and found that its range has contracted northward and toward higher elevation with climate warming. These studies are consistent with predictions developed in earlier papers of poleward species range shifts or range contractions from the equator, although with the exception of Parmesan (1996), they offer indirect evidence by only looking at a limited area within species' ranges.

Other evidence seems to support different models for responses of species to climate change. Roemmich and McGowan (1995) suggested that warming of the surface layer in southern California coastal waters had increased stratification and prevented upwelling of more nutrient rich water. This model was supported by long-term oceanographic records and species collections from CalCOFI which showed a 70% decline in zooplankton in southern California since 1951, concurrent with sea surface warming of up to 1.5°C. This may also help to explain a 90% decline in sooty shearwater abundance and a 40% decline in seabird abundance overall in the California current between 1987 and 1994 (Veit et al. 1996, 1997).

The science of phenology, which focuses on timing of periodic biological events such as migration, egg laying, and leafing, has reemerged as a source of models and data for historical studies of species' responses to climate change. Sparks and Carey (1995) analyzed plant phenological records collected by seven generations of an English family since 1736 and found significant correlations between temperature and dates of first leafing, flowering, and appearance. This relationship was used to predict earlier dates for the phenological variables under proposed climate warming scenarios. Similar phenological models are supported for laying dates in U.K. birds, in which 20 of 65 species studied between 1971 and 1995 showed earlier laying dates, and only 1 species showed later laying dates (Crick et al. 1997), as well as for amphibians in England which showed progressively earlier spawning dates between 1978 and 1994, a period of climatic warming (Beebee 1995).

In this chapter, I defend the use of the historical-descriptive method in climate change studies, but also discuss shortcomings of the method itself, as well as potential problems with the types of data typically encountered in historical studies. These issues are discussed within the context of my research (with collaborators James P. Barry, Sarah E. Gilman, and Charles H. Baxter) on observed long-term (multidecadal) changes to the nearshore marine communities of the Hopkins Marine Station (HMS) in Pacific Grove, California, at the southern end of Monterey Bay (36°37.3'N, 121°54.3'W) (Fig. 3.1).

Long-Term Faunal Changes at Hopkins Marine Station

Our impetus for historical studies of the intertidal zone at Hopkins Marine Station (HMS) was the observation by Charles Baxter, a Stanford University lecturer, that intertidal species composition had changed dramatically since his arrival at HMS in the late 1970s. In spring 1993, we relocated the original brass bolts that marked an intertidal transect established by W.G. Hewatt in 1931

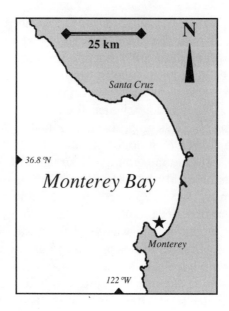

Figure 3.1. Map of Monterey Bay, California.

(Hewatt 1934, 1937) at HMS (Fig. 3.2A, B). The transect runs for
98.8 m perpendicular to shore from high to low intertidal zone.
Hewatt had recorded the abundance of invertebrates in over 100
square-yard plots along this transect. Between 1993 and 1995 we
resurveyed 57 of the square-yard (0.84 m^2) plots in the exact location of Hewatt's plots in order to compare abundances of invertebrates. We had sufficient data to quantitatively compare 62 species,
representing a range of intertidal taxa, trophic levels, and life-history strategies. Changes in these species are shown in Figure 3.3.
Almost all species changed in abundance, with 46 species showing
a significant change in abundance (paired t-test, $P < 0.05$), but
there was no overall direction of change, with 24 species increasing
and 22 decreasing.

In contrast, a very strong pattern of change appears when
species are considered by their geographic range. We assigned each
species to a geographic range category based on its range relative to
the study site in Monterey Bay as reported in the most comprehensive guides to the fauna (primarily Brusca 1980, Morris et al. 1980,

Figure 3.2. Photos of investigators on Hewatt's transect. (A) W. G. Hewatt ca. 1932. (B) Raphael Sagarin 1998.

Ricketts et al. 1985). Species with a northern limit south of Cape Mendocino, California, were considered southern species. Those with a southern limit north of Point Conception, California, were considered northern species, and those with boundaries extending

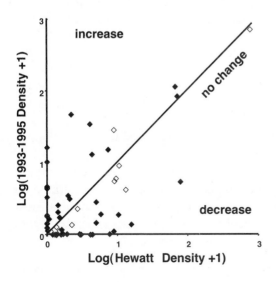

Figure 3.3. Plot of density in 1995 vs. density in Hewatt's study in 57 paired plots for 62 species. Solid symbols are species that showed a significant change in abundance between the two studies (paired t-test, $P < 0.05$).

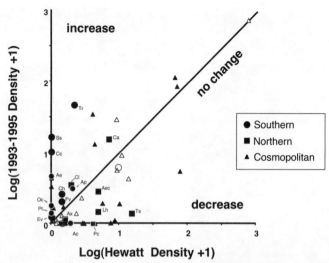

Figure 3.4. Plot of density in 1995 vs. density in Hewatt's study in 57 paired plots for 62 species. Solid symbols are species that showed a significant change in abundance between the two studies (paired *t*-test, *P* < 0.05). Open symbols are nonsignificant changes. Southern species: circles. Northern species: squares. Cosmopolitan species: triangles. Ap: *Acanthina punctulata* (snail); Ac: *Alpheus clamator* (shrimp); Ae: *Anthopleura elegantissima* (anemone); Aec: *Anthopleura elegantissima* (anemone, clonal form); Ax: *Anthopleura xanthogrammica* (anemone); Cl: *Calliostoma ligatum* (snail); Cc: *Corynactis californica* (anemone); Ca: *Crepidula adunca* (snail); Ch: *Lepidochitona* (*Cyanoplax*) *hartwegii* (chiton); Ev: *Erato vitellina* (snail); Fv: *Fissurella volcano* (limpet); Lh: *Leptasterias hexactis* (sea star); Oc: *Ocenebra circumtexta* (snail); Pc: *Petrolisthes cinctipes* (crab); Pt: *Pseudomelatoma torosa* (snail); Ss: *Serpulorbis squamigerus* (tube snail); Ts: *Tectura scutum* (limpet); Tr: *Tetraclita rubescens* (barnacle). Adapted from Sagarin et al. 1999.

beyond these points in both directions were considered cosmopolitan species.

Division of species in this manner reveals a dramatic range-related pattern of species' abundance changes (Fig. 3.4). Almost all southern species (10 of 11) increased in abundance, whereas most northern species (5 of 7) decreased. Cosmopolitan species showed no trend, with 12 increasing and 16 decreasing.

Climatic Setting

Nearshore sea temperatures have been recorded every day of the year at the HMS since 1920. These records have always been taken by hand with a standard thermometer, so the entire record is technologically consistent. Yearly mean, maximum, and minimum shoreline temperature increased significantly during the period from 1920 to 1995 (Fig. 3.5). Linear regressions for annual mean, maximum, and minimum shoreline temperature for HMS versus year indicate a significant increase of annual sea surface temperature of 0.79°C, 1.26°C, and 0.66°C, respectively, between 1931 (the time of Hewatt's study) and 1995.

In addition to the gradual long-term temperature increases, sea temperature conditions during the 13-year period preceding our study were consistently warmer than the same period preceding Hewatt's study, with a difference in mean temperature of 0.99°C, and a maximum difference in seasonal temperature of 1.94°C in late July (Fig. 3.6).

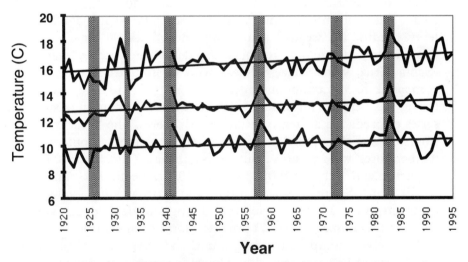

Year

Figure 3.5. Plot of annual maximum, mean, and minimum shoreline ocean temperature recorded at HMS from 1920 to 1995. Gray vertical bars indicate strong El Niño–Southern Oscillation (ENSO) years. Slopes for maximum, mean, and minimum temperature regressions are 0.0199, 0.0125, and 0.0104, respectively. Adapted from Sagarin et al. 1999.

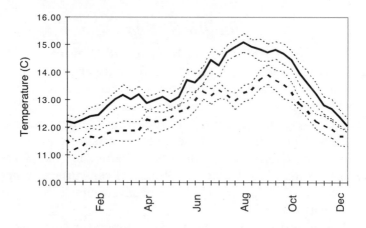

Figure 3.6. Monthly averaged shoreline ocean temperatures for HMS for the 13 years preceding this study (black line, above) and the 13 years preceding Hewatt's study (dashed line, below). Dashed lines are + or − standard error. Adapted from Sagarin et al. 1999.

Air temperature records, on the other hand, do not show a clear trend. There are no continuous air temperature records for Monterey, because the weather station was moved several times during this century. Therefore, temperature data from nearby stations (Santa Cruz, Watsonville, Salinas) with continuous records were compared to discontinuous records from the Monterey station (using least squares linear regression) to generate a continuous time series of predicted air temperature for Monterey. The linear regression slopes of the predicted minimum and maximum Monterey air temperatures were positive, but nonsignificant (0.001 and 0.006, respectively; R^2 values = 0.0363 and 0.0009, respectively) (Fig. 3.7).

El Niño–Southern Oscillation events of varying intensities occurred in the 13-year periods preceding and during each study (Fig. 3.8). Quinn et al. (1987) provide a long-term historical record with which to compare the magnitude and frequency of El Niño events. Six of 13 years preceding Hewatt's survey were classified as El Niño years including a strong two-year event (1925–1926). Five of 13 years preceding our surveys were El Niño years, including the event of 1982–83, considered to be the strongest of the century before our surveys (Hansen 1990). Both studies were conducted

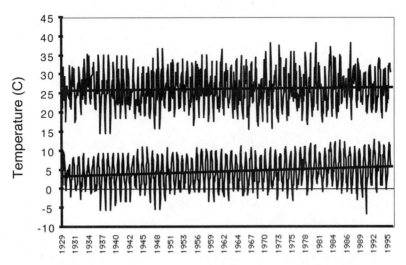

Figure 3.7. Air temperature for Monterey Bay from National Climatic Data Center data. Plot of predicted Monterey air temperature from 1929 to 1995 based on regressions from three nearby stations. Slopes of maximum and minimum temperature linear least squares regressions are 0.0017 ($R^2 = 0.0023$) and 0.0057 ($R^2 = 0.0355$), respectively. Adapted from Sagarin et al. 1999.

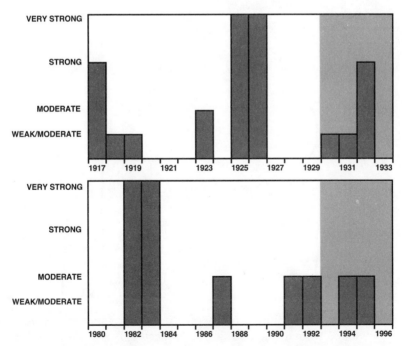

Figure 3.8. Strength of El Niño–Southern Oscillation (ENSO) events prior to Hewatt's study (top) and prior to this study (bottom), based on Quinn et al. (1987) and Vernon Kousky (pers. comm. 1998). Hatched areas represent survey periods for each study. Adapted from Sagarin et al. 1999.

during El Niño years, including weak (1931), moderate (1994, 1995 Vernon Kousky, pers. comm. 1998), and strong (1932) events.

Model: Range-Related Changes Are Climate-Related

The range-related pattern of change at this site matches expectations of animal populations shifting northward due to climatic warming. Figure 3.9 illustrates diagrammatically how northward shifts in geographic distributions of animals would result in the patterns we observed. Note both a population with a Gaussian distribution of abundance (bell-shaped curves) and a population with abrupt drops in populations at the edges of the range (steep-sided curves) can hypothetically show this pattern. This is significant because preliminary surveys (R. Sagarin, unpublished data) show

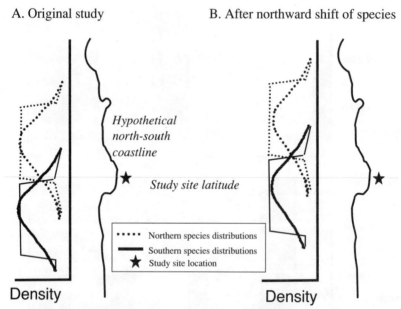

A. Original study   B. After northward shift of species

Hypothetical
north-south
coastline

★ *Study site latitude*

····· Northern species distributions
——— Southern species distributions
★ Study site location

Density   Density

Figure 3.9. Hypothetical shift in northern and southern species' distributions. As viewed from a single study site (indicated by star), the southern species abundances (solid lines) appear to increase, and the northern species abundances (dotted lines) appear to decrease as species move northward through time (e.g., from A to B).

that at least some of our species show extremely steep-sided abundance distributions.

It is difficult with only a single study site to separate actual range shifts from population changes within an existing range. Nevertheless, several of the intertidal species for which we have additional information have shown changes consistent with northern range shifts. For example, *Serpulorbis squamigerus*, a southern gastropod, which is now extremely abundant at HMS and in Monterey Bay (R. Sagarin, pers. obs.), was not recorded by Hewatt, and was recorded as rare or occurring as single, scattered individuals by investigators in 1966 and 1980 (Hadfield 1966, Morris et al. 1980). The rapid rise of *Serpulorbis* between 1980 and the first re-survey of Hewatt's transect in 1993, as well as abundance data I have collected for this species in the northern half of its range (R. Sagarin unpublished data) could indicate that *Serpulorbis* has shown a shift in range in a manner illustrated by the steep-sided distributions in Figure 3.8. *Tetraclita rubescens*, a southern barnacle that increased significantly in abundance has also shifted its range northward in recent decades (Connolly and Roughgarden 1998). Further observations of recent arrivals of southern species such as lobster (*Panulirus interruptus*), the Kellet's whelk, *Kelletia kelletia* (Herrlinger 1981, G. Villa and J. Watanabe, unpub. obs.) and the eel-grass limpet, *Tectura depicta* (Zimmerman et al. 1996) by other investigators lend additional support to the hypothesis of northward movement of animal populations. These observations, which cover either small spatial areas or limited time periods, are suggestive, but are too limited to confirm range shifts. Ideally, abundances at many sites (especially concentrated near the range edge) must be observed over several years to ensure that ranges have indeed shifted. The Partnership for Interdisciplinary Studies of Coastal Oceans (PISCO), a recent collaboration between intertidal ecologists at the Universities of California at Santa Barbara and Santa Cruz, Stanford University, and Oregon State may provide such a monitoring scheme.

An important consideration in interpreting our data from Hewatt's transect is that they are based on only two surveys separated by 60 years. They therefore can give no indication of the short-term variability in populations along the transect, suggesting the possibility that the pattern we observed was the product of a chance "snapshot" of widely fluctuating populations. To begin to address this scale of variation, we resampled 19 plots in 1996, 3 years after they were originally resampled. The comparison of these plots in 1996 versus 1993 reveals only very small changes in species' abundances (Fig. 3.10). Furthermore, no range-related pattern of change is evident.

The implication of this second, short-term study is that changes in populations are much different when viewed in short time frames (3 years) than in long time frames (60 years). Interestingly, the short-term changes, although inconsistent with the pattern of range-related changes seen over 60 years, did not erase the sharp range-related pattern seen in the long-term. This is illustrated when the 19 resampled plots are examined over the long term by comparing the 1996 survey to Hewatt's study (Fig. 3.11). Without additional intermediate data points, it is impossible to know how long the process of species composition shift takes, although the relatively small magnitude and lack of consistent direction of abun-

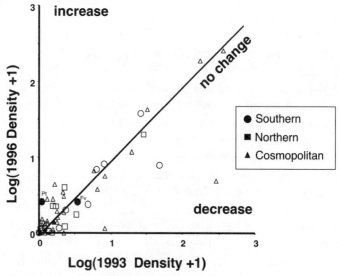

Figure 3.10. Plot of density in 19 paired plots in 1996 vs. 1993. Symbols are as in Fig. 3.4. Filled symbols represent significant changes ($P <$ 0.025). The lower alpha value represents a Dunn-Sidak correction on the paired t-test, used because the data were drawn from repeated measures of the same plot (Hewatt's study, 1993 and 1996). Adapted from Sagarin et al. 1999.

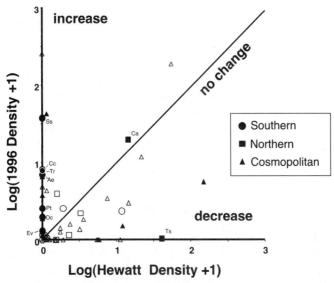

Figure 3.11. Plot of density in 19 paired plots (as in Fig. 3.9) between 1996 and Hewatt's study. Symbols are as in Fig. 3.3. Filled symbols represent significant changes ($P ** 0.025$). The lower alpha value represents a Dunn-Sidak correction on the paired t-test, used because the data were drawn from repeated measures of the same plot (Hewatt's study, 1993 and 1996). Adapted from Sagarin et al. 1999.

dance changes in the short-term suggest that the range-related pattern may take many years to appear. It is likely that species respond individually to climate change depending on thermal sensitivities, reproductive strategies, mobility, life span, and interactions with other species (Graham and Grimm 1990). The pattern we see today thus integrates changes by many species over many years. This may help to explain the failure of other investigators to find such strong range-related changes in their own reexaminations of historical data sets that date back only 20 or 30 years. For example, John Pearse and collaborators, in reexamining extensive intertidal faunal surveys originally conducted from 1971 to 1973 from Santa Cruz, California, did not find a strong range-related pattern of change overall, but did find increases in several southern species (J. Pearse, unpublished data).

Nevertheless, an intertidal community with hundreds of invertebrate and algal species is an incredibly complex system characterized by a myriad of potential direct and indirect biotic interactions that may affect populations of any given species. Over the course of 60 years, factors other than warming sea temperatures have undoubtedly changed populations of intertidal invertebrates. We examined several alternative hypotheses to explain the changes we observed (Barry et al. 1995, Sagarin et al. 1999). Our findings are summarized in Table 3.1. While any of these alternative hypotheses might explain some of the faunal changes we observed, none of them can explain the strong range-related pattern of change in its entirety.

Glynn Study and Other Algal Changes

Marine researchers have predicted that, in the absence of sea-level rise (indeed our surveys show no change in the height of intertidal benchmarks relative to sea level since 1930, see Sagarin et al. 1999), the upper limits of some intertidal species may shift toward lower tidal heights with warming climate (Lubchenco et al. 1993). Other historical studies done at HMS reveal large community-level changes that are consistent with this prediction. Beginning in spring 1994 we relocated seven small (10 cm^2) plots established by Peter Glynn (1965) within the zone characterized by high densities of the barnacle *Balanus glandula* and the turf alga *Endocladia muricata*. In surveys, we found that the vertical range of *Endocladia muricata* has shifted lower by 0.18 meters at the upper boundary and 0.40 meters at the lower boundary. These tidal height changes were reflected in widespread changes in species' abundances and species diversity at the locations of Glynn's plots. At present, we cannot identify specific mechanisms that may have caused this tidal height shift. Work by Denny and Paine (1998) showed that an 18.6-year oscillation in the moon's orbital inclination may be responsible for large changes in time of emersion in the intertidal at HMS. Although such shifts in emersion time might be expected to drive shifts in intertidal zonation (with shifts to lower tidal heights seen in times of greater emersion due to greater desiccation), analysis of Denny and Paine's data shows that both lunar inclination and emersion times were similar between Glynn's study and the more recent studies. Whether the algal height shift is related to the long-term changes in climate at HMS remains to be seen, although it is certainly consistent with predictions of species responses to climate warming.

Table 3.1. Alternative Hypotheses, as Discussed in Barry et al. (1995) and Sagarin et al. (1999)

Alternate hypotheses	Potential Influences	Supporting Evidence	Opposing Evidence	Conclusion
Anthropogenic effects	• Trampling, collecting • Pollution	• Little evidence	• Establishment of Hopkins Marine life refuge, 1931 • Localized effects of canneries	• Likely to be minor in affecting changes along the transect
Seasonal changes	• Hewatt's original sampling dates unknown, re-surveys done in spring and summer	• Shifts in sand cover may affect species' distributions (R. Paine, pers. comm. 1994)	• Both Glynn (1965) and Hewatt (1934) reported little seasonal variation in the HMS intertidal	• Seasonal changes unlikely to be responsible for large changes seen in many species
Substratum changes	• Seismic shifts affect tidal height distributions • Erosion of substratum changes primary habitat • Loss of movable boulder habitats	• Hewatt (1934) reports many small moveable boulders in transect—not apparent in recent study • Decline in abundance of 6 of 10 species which use under-rock habitat	• Surveying shows no shift in tidal height of Hewatt's transect relative to surrounding shore or other parts of the intertidal • Granitic rock at HMS resists erosion. Substratum drawn in Hewatt's maps clearly matches substratum today	• Seismic effects unlikely • Loss of boulders possibly responsible for some species declines

continues

Table 3.1. *Continued*

Alternate hypotheses	Potential Influences	Supporting Evidence	Opposing Evidence	Conclusion
Return of otters	• Predation on several intertidal invertebrate species	• Decline in 4 of 10 potential otter prey species including mussels, urchins, sea-stars, and crabs	• Otters rarely seen foraging in area of transect	• Re-surveyed transect plots do not include mussel beds where otter predation likely to be greatest • Effect not likely to be great in area of transect re-surveyed. All prey species are cosmopolitan, so pattern in northern and southern species is not affected by otters
Return of oystercatchers	• Intertidally foraging bird returned to HMS between Hewatt's study and the present	• One prey species declined in abundance	• One prey species increased in abundance, 4 species showed no change in abundance	• Effect is minimal on observed population changes
Return of harbor seals	• Seal population has increased since 1970s at HMS—may affect intertidal invertebrates through crushing, shading, and feces	• No direct evidence—studies of effects are ambiguous (Boal 1980, Horng and Hayhurst, unpublished data)	• Seals rarely haul out on rocks along the transect	• Effect is minimal on observed population changes

El Niño	• El Niño manifests in California as increased northward transport and increased sea temperatures	• Southern species (especially fish) often found north of their range during El Niño years	• Long-lived larvae (more likely to be affected by changes in transport processes) did not show greater changes (increases or decreases) than species with short or no planktonic phase • El Niño patterns similar in years preceding and during Hewatt's study as years preceding and during current study (Fig. 3.8)	• Hypothesis cannot be rejected—important, but as yet unknown feedbacks between long-term warming and El Niño events may benefit southern species • Effects of El Niños on intertidal invertebrate populations are poorly understood
Celestial mechanics	• An 18.6 y oscillation in the moon's orbital inclination changes the amount of time intertidal animals are exposed to air, on average (Denny and Paine 1998)	• Strong correlation between this oscillation and HMS water temperature suggests that changes we observed could have occurred over 18.6 y rather than 60 y	• Temperature has increased gradually throughout the 60 y period since Hewatt's study • Intertidal animals were exposed to air longer on average during Hewatt's study.	• Points out the inability of our study to establish direct mechanistic links between species' changes and sea temperature changes • Does not invalidate the finding of range-related species changes

Algal changes were also observed in our re-surveys of Hewatt's transect. Although Hewatt did not quantify algal populations directly, several references and photos in his thesis indicate that the area surrounding his transect was dominated by the rockweed *Pelvetia compressa*. Hewatt reports 100% cover of this alga in a tidal zone where we discovered it to be largely absent or rare, with maximal cover of 30% in any plot within this zone (Sagarin et al. 1999). Photographic comparison of rocks and areas of the intertidal that Hewatt photographed also reveals a dramatic decline in *Pelvetia* (Figs. 3.12 and 3.13). Reports from 1948 indicate that this decline

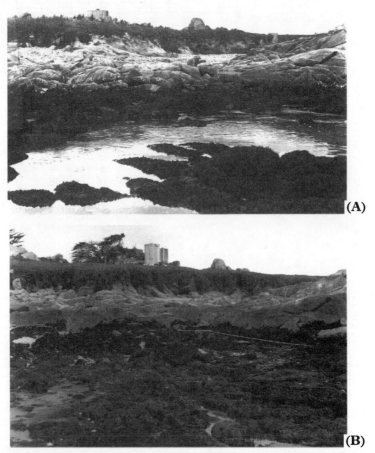

(A)

(B)

Figure 3.12. View of Hewatt's transect area. (A) ca. 1932. (B) 1998 (at lower tide than A). Dark areas in A are almost exclusively the alga *Pelvetia compressa*. Same areas in B are nearly devoid of this alga and characterized by turf algae such as *Endocladia muricata*.

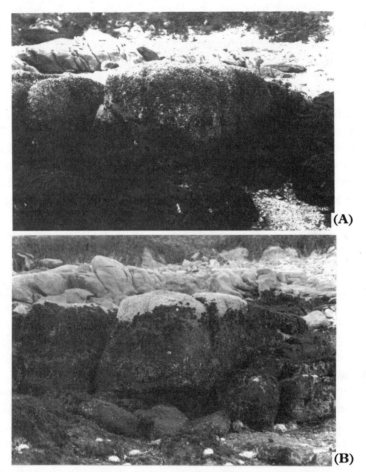

Figure 3.13. Close-up of intertidal rock near Hewatt's transect. (A) ca. 1932. (B) 1998. Note that heavy layer of *Pelvetia* in lower half of A is completely absent in B, which shows only scattered turf algae, mostly *Endocladia muricata*.

was well under way at HMS at that time (Stephenson and Stephenson 1972), although we cannot be sure what populations have done between 1948 and 1993 when our studies began. This change would not be predicted based on our division of species into northern and southern categories, as *Pelvetia* is basically southern in range, and all of our southern invertebrates increased in abundance. We currently have no strong hypothesis to explain this change. It is interesting that *Lepidochitona hartwegii*, a southern chiton that uses *Pelvetia* almost exclusively for food and shelter (Andrus and Legard 1975, DeBevoise 1975), increased in abundance during the same period that this alga declined, indicating that habitat loss did not outweigh benefits to this species, which may include the warmer sea temperatures.

Our findings suggest that changes in algal composition and zonation may be associated with dramatic changes in faunal populations. Although some of the observed algal changes match simple predictions of species' responses to climate change, others do not. In light of this, the asymmetry between the detailed faunal observations by Hewatt (1934) and the paucity of algal observations is frustrating and highlights the need in historical studies to have data on as many components of the system as possible. Even the small amount of algal information provided by Hewatt proved extremely useful.

The Good and the Bad of the Hopkins Studies

The studies at HMS illustrate both the value and the challenges of long-term, historically based research in identifying biological responses to climate change. Perhaps the most important conclusion that can be drawn from these studies is that long-term monitoring programs are essential in attempting to reject alternative hypotheses to climate change as a driver of species or community changes. The dramatic range-related shift in species' abundances we observed, as well as the vertical shift in algal zonation, would not have appeared in a short-term study. Long-term monitoring of both biotic and climatic variables is a straightforward method of tracking changes that may be too subtle to be observed in the course of a single investigator's career. The National Science Foundation's Long-Term Ecological Research (LTER) program is an excellent step in this direction. Unfortunately, the program's coverage of nearshore coastal areas is still lacking. Out of 24 LTER sites, only six are coastal, and only one includes rocky shores, which are sites of high diversity and vulnerability to climate change. Ironically, long-term monitoring programs such as CalCOFI in southern California (used, for example, by Roemmich and McGowan 1995) and several programs in the United Kingdom (used by Southward et al. 1995) have suffered drastic scale reductions and budget cutbacks even as they contributed essential data to convincing studies of biological responses to climate change. It has been estimated that 40% of the long-term marine monitoring schemes in Europe were terminated in the late 1980s (Duarte et al. 1992), precluding study of some of the warmest years on record. These and other problems faced in the collection and use of long-term monitoring data sets are discussed in detail by Gross and Pake (1995) in a report for the Ecological Society of America.

An additional point illustrated by the HMS studies is that scientific reserves can play a vital role in long-term studies and the conclusions drawn from them. The designation of the Hopkins Marine Life Refuge in 1931 not only ensured that the study sites were protected from human disturbance, it also gave us a higher degree of certainty that the biological changes we observed were not due to the direct destructive practices of humans, such as foraging, collecting, or trampling.

From a sampling standpoint, the studies benefited from being replicated in the exact location of the original studies. Spatial variation in invertebrate abundance in the Hopkins intertidal zone is extreme, so that transects through the same intertidal zone just a few meters apart can show completely different patterns (R. Sagarin, pers. obs.). Likewise, algal tidal height distributions are extremely sensitive to aspect of the rocks. By sampling *Endocladia* height on different sides of rock faces at HMS (unpublished data), we found that height of this alga is consistently lower on south-facing rock slopes than on north-facing slopes. Thus, without knowing that *Endocladia* plots were surveyed in the precise location of Glynn's original plots, it would be difficult to rule out the effects of substratum aspect in the tidal height changes we observed.

Nonetheless, resampling in the same location raises its own problems in the lack of independence between repeated-measures of the same quadrat. One solution to this problem is to adjust the alpha value of statistical comparisons (such as a paired t-test) using the Bonferoni or Dunn-Sidak corrections (Sokal and Rohlf 1995). Unfortunately, these corrections severely impair the power to detect changes. For example, resampling a plot twice requires finding significance at the 0.025 level, based on the Dunn-Sidak correction, and additional re-surveys lower the alpha value further, making this impractical for data that would be collected many times from the same location. As an alternative, multivariate methods such as the paired T^2 test (Morrison 1990) may be used, but they are computationally more complex. Other multivariate methods for dealing with repeated measures focus on response curves. For example, Scott (1993) proposed a time segment analysis for repeated measures of permanent quadrats which uses a combination of the mean of measured variable (e.g., percent cover of vegetation in a quadrat) and slope, or rate of change, between observations of the quadrat to generate curves showing the time trend in the data. Gurevitch and Chester (1986) and Potvin et al. (1990) discuss the benefits and limitations of such approaches and suggest methods for dealing with the problem of repeated measures. Lesica and Steele (1996) show through example plant populations that these methods are practical for long-term monitoring in the context of climate change studies. Thus, the benefits of permanent quadrats discussed above outweigh the statistical complexities of dealing with repeated measures data.

Our study of Hewatt's transect also illustrates the value of having quantitative data, rather than categorical data. Many ecological surveys rely on presence/absence data or categorical data (e.g., rare, common, abundant), making it difficult to determine any but the most extreme population changes. This point is illustrated dramatically by viewing our data from Hewatt's transect as if Hewatt had provided data only on the presence or absence of invertebrate species. In this analysis, the percentage of local deletions of species (found in Hewatt's study, but not in our re-surveys) is nearly identical across geographic range categories, whereas the number of local additions (found in our re-surveys, but not in Hewatt's survey) is only slightly greater for southern species (Fig. 3.14A). In other words, no range-related pattern of change could be found in this study in the absence of quantitative data. This surprising result is largely due to the effect of rare species on presence/absence changes in this comparative study. When sampling along a single transect in two separate points in time, rare species may easily be counted (or missed) by chance in one study but not the other. After removing the rare species (defined as those species with fewer than 10 individuals counted in both studies) from the presence/absence analysis, some range-related pattern of change can be observed (Fig. 3.14B), with a high percentage of southern species additions and no northern species additions. It should be stressed though, that the emergent pattern is not as clear or convincing as that shown with quantitative data (see Fig. 3.4), and furthermore, the elimination of

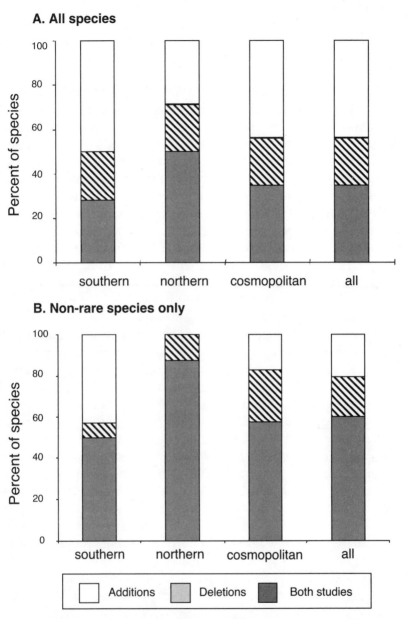

Figure 3.14. Changes in presence/absence of all species recorded by either Hewatt (1934) or Sagarin et al. (1999) in the 57 plots resurveyed between 1993 and 1995. "Additions" are species recorded by Sagarin et al., but not Hewatt. "Deletions" are species recorded by Hewatt, but not Sagarin et al. "Both studies" are species recorded by both Hewatt and Sagarin et al. (B) Changes in presence/absence of non-rare species only (defined as those with greater than 10 individuals recorded by either Hewatt or Sagarin et al.).

rare species was only possible because we had data on their abundances. Despite the fortunate contributions of well-marked study sites and quantitative data from the original authors, the Hopkins studies' data were limiting for their lack of biological information beyond simple numbers and locations of individuals. Many sessile invertebrates are difficult to count on an individual basis. Sponges and most tunicates were ignored by both us and Hewatt for this reason. Information on their percent coverage, if taken by Hewatt, would have helped us determine whether changes had taken place in these taxa. Further, some biological responses to climate change may be more subtle, and not necessarily reflected in population density changes. For example, Tegner et al. (1996) found that number of stipes per plant, rather than traditional measures of whole plant survival, was a more sensitive indicator of kelp forest response to temperature shifts in southern California.

Measures of population and community changes beyond individual counts will become necessary as model predictions of species' responses to climate change become more sophisticated. For example, there is some evidence that climate change may intensify upwelling off the California coast (Bakun 1990). Indeed, recent analysis shows that upwelling increased along the coastline surrounding Monterey Bay during the period from 1946 to 1990 (Schwing and Mendelssohn 1997). Intertidal sites adjacent to upwelling zones in South Africa and Chile have shown enhanced algal productivity and increased herbivore biomass relative to nonupwelling sites, presumably due to nutrient enrichment during some upwelling periods (Bosman et al. 1987). This suggests the hypothesis that increased upwelling should have led to an increase in biomass of intertidal herbivores. Lacking historical information on animal biomass at HMS, we had no way of testing this hypothesis.

In a wider analysis, the Hopkins studies, like most studies of biological responses to climate change to date, are correlational in nature. They have drawn their strength from how well the relationship between observed biological changes and climate changes has matched predictions of these changes. Often the predictions themselves are quite simplistic and fail to take into account species interactions, different rates of migration between species or within a species range, and other "ecological noise" that will confound interpretation of range shifts. For example, Davis et al. (1998) used microcosm experiments on insects with various preferred temperature ranges to show that species interactions during periods of simulated warming result in range shifts that are different than expected from predictions that only take into account a species' preferred temperature range. As another example, Zimmerman et al. (1996) found that a southern limpet which had recently invaded Monterey Bay was responsible for a catastrophic decline in subtidal eelgrass beds. Although the limpet's northward expansion is consistent with the simple expectation from climate warming, the decline of the eelgrass is not, as seagrasses might be expected to show enhanced photosynthetic rates and growth in an era of increased atmospheric CO_2 (Beardall et al. 1998).

Nevertheless, for the present, these correlational studies are a welcome addition to our small but growing body of data regarding biological responses to climate change. It is clear, nonetheless, that such studies would benefit from improved resolution in three important areas:

1. Temporal–Fine resolution in this area is essential to answer the question, How are both the biological and climatological signals changing through time? This calls for time series data, which are often available for climatic factors, but less common for biological variables. Notable exceptions are the work of Southward et al. (1995) and of Roemmich and McGowan (1995) (see earlier discussion). The advantages of such data are well-summarized by McGowan et al. (1996) who point out that time series data allow researchers to identify important frequencies of biological changes and to examine coherence between biological and physical events.

   The obvious weakness of our "snapshot" historical data in terms of temporal coverage was pointed out by Denny and Paine (1998) who found that the 18.6-year oscillation in the moon's orbital inclination (preceding discussion) may be correlated with nearshore sea temperatures. Because we lack regular time series data on the biological changes at HMS it is difficult to know if the species changes we observed are more strongly related to the gradual rise in near-shore sea temperatures over 60 years, the shorter-term sea temperature changes associated with the lunar oscillation, or other changes that have occurred in the last 60 years (see Table 3.1).

2. Spatial–Strong spatial resolution is important in answering the question, Are the changes occurring throughout the species' range? This is critical in separating global changes from local perturbations. Parmesan's (1996) extensive study of Edith's checkerspot butterfly population spanned the entire North American range of the species and thus was able to convincingly show that populations were more likely to have gone extinct in the southern part of the range. In our study, by contrast, it was impossible without additional sources of information to separate abundance changes of species from actual shifts in their ranges because our data were gathered at a single site.

3. Taxonomic–Strong taxonomic resolution answers the question, Are these changes happening to everything? As our study of Hewatt's transect illustrates, this is an excellent tool for eliminating alternate hypotheses. A pattern (e.g., range-related abundance changes) is more likely to be a general biological response to climatic changes if it is consistent across taxonomic groups or life-history strategies. As a counterexample, if we found that only filter feeders (e.g., barnacles, mussels, etc.) had changed in abundance, we would suspect a more specific agent of change than a general temperature increase.

It is unlikely that any one study will adequately address these three areas. Thus investigators should be aware of weaknesses in these areas as they interpret their results. More importantly, investigators from different fields and different locations should compare results frequently, amalgamating their findings in order to overcome deficiencies in the spatial, temporal, or taxonomic resolution of any one study.

Future Directions

The most important future direction these historical-descriptive studies can take (beyond the simple step of continued monitoring) is to establish mechanistic links between the observed biological and climatic changes. These mechanisms may be ecological (concerning changes in biological habitats, symbiotic relationships, or changes in species interactions), physical (concerning changes in the structure of the environment in which the species live), or physiological (concerning individuals' internal responses to changing climatic conditions).

In some cases, detailed observational studies or natural experiments may reveal likely mechanisms of climate sensitivity. For example Fraser et al. (1992) found that two closely related species of penguins showed opposite population trends since the 1970s in the Antarctic with chinstrap penguins increasing and Adelie populations declining. A closer examination found that chinstraps, in contrast to the Adelies, prefer open water habitats and thus had benefited from climate-related loss of pack ice in the Antarctic. Further, within populations of Adelies, those that lived on islands with increased snowfall and snowmelt were detrimentally affected by the increased inaccessibility of nest sites (Fraser, unpublished data). Thus habitat changes mediated by climate changes were shown to be the mechanism responsible for population changes.

Taking advantage of changing background conditions and long-term records, Hersteinsson and Macdonald (1992) found that a period of climate warming in northern latitudes since the late 1800s led to increased primary productivity and rodent abundance, facilitating a northward range shift in the predatory red fox. This range shift led in turn to a range shrinkage in the more northerly arctic fox, which had been observed to be competitively inferior to the red.

Especially when physiological mechanisms are concerned, however, direct experimental manipulation has been the primary tool for understanding of responses to climate change. Terrestrial plant ecologists, for example, have aimed experiments at identifying how individual components of climate change (such as elevated temperature, increased atmospheric carbon dioxide, changes to soil-moisture content) will affect plant growth (Bazzaz 1990, also Saavedra, this volume). In cases where these experiments have been conducted over broad spatial scales or long time scales, these effects can be linked to population and community changes (Carpenter et al. 1993, Chapin et al. 1995).

Physiological mechanisms that could be important under climate change have also been explored at the biochemical level. George Somero and collaborators have studied the ecological implications of heat-shock proteins (hsps), which are molecular chaperones that function to repair stress-damaged proteins and may be an indicator of thermal stress in organisms. Hofmann and Somero (1996) found that a northern mussel species had higher levels of hsps than its southern congener when held under identical conditions, and that the northern species induced hsps at lower temperatures when the mussels were exposed to higher temperatures, suggesting that the northern species is more sensitive to warm temperatures. Future studies of organism responses to climate change may follow these models by focusing on physiological markers as a means of identifying mechanistic causes of population changes.

Ultimately both observational and experimental studies should be linked interactively to provide the clearest picture of both patterns and processes of species' responses to climate change. Some studies have suggested such links. For example, Stillman and Somero (1996) found that a suite of physiological and morphological adaptations to temperature stress may be responsible for interspecific differences in porcelain crab distributions. They suggested that small increases in nearshore ocean temperature, similar to those documented at HMS, could lead to elevated habitat temperature which in turn could affect the latitudinal and tidal height ranges of these crabs. *Petrolisthes cinctipes*, a northern porcelain crab, declined along Hewatt's transect, although we cannot be certain if this decline was due to increased temperatures or some other factor. Other studies are clearly needed for establishing connections between small-scale experiments and large-scale observations. For example, some detailed observational studies have suggested that metabolic constraints limit the ranges of animals (see Root 1988 for birds, Welch et al. 1998 for salmon). It would clearly be worthwhile to combine these studies with well-designed experiments on metabolic responses of these species to temperature changes to confirm the observational hypotheses. Exploring these links in a way that allows continual feedback of information and hypotheses across scales of biological organization (or between experimental and observational studies) is what Root and Schneider (1995) described as the strategic cyclical scaling (SCS) paradigm. SCS approaches in the biological realm can in turn become part of an integrated assessment approach to confronting the challenges of climate change from a societal standpoint.

Conclusion

Climate change and its effects on biological systems is an extremely complex, often unpredictable process. Yet useful methods for studying the interface of climate change and biology are ironically quite simple, primarily utilizing observational data. Already the small number of historical-descriptive studies that have focused on climate change have demonstrated convincing correlations between changes to biotic systems and climate changes for a broad range of organisms including marine and terrestrial invertebrates, fish, birds, amphibians, reptiles, and plants (Hughes 2000). In most cases, these correlations match well with proposed models of species' responses to climate change. There is a clear argument to take these studies further (especially in exploring causal mechanisms behind the correlations) and to vastly expand our body of historical studies that deal with climate change.

Investigators have begun to search for existing studies that can be reexamined, as we have done with Hewatt's and Glynn's data. Often these searches can be frustrating because historical records are incomplete, poorly documented, or simply nonexistent for a particular study system. In a report on long-term data sets, Gross and Pake (1995) found that one of the most prevalent problems with these data was poorly documented "meta-data," or information about how the data were collected. In other cases, habitat loss may have effectively destroyed the historical record. Parmesan, for example, had to reject many historical records of butterfly populations in her study of extinction patterns because the habitats had been converted to other uses (Parmesan 1996). Moreover, the biological systems in question may have been so altered by human exploitation by the time records were initiated that it may have been difficult to separate natural from anthropogenic changes. Dayton et al. (1998) give excellent illustrations of this problem in the context of long-term records of southern California kelp forest communities. Pollution, severe exploitation of vertebrate and invertebrate fisheries, as well as climatic shifts and strong El Niño events have all affected this system since the start of record keeping, leaving the authors unable to define baseline "natural" conditions with which to compare current conditions. Occasionally, though, hidden historical treasures are uncovered. For instance, Terry Root discovered long-term unpublished phenological records kept by a wildlife biologist in Michigan that may show trends toward earlier migrations of North American birds (Root, unpublished data). Furthermore, the Future of Long-term Ecological Data Committee of the Ecological Society of America (Gross and Pake 1995) maintains a catalog of long-term data sets maintained by state and federal agencies, as well as individuals, that may be useful for examining climate-related trends in species' changes (http://esa.sdsc.edu/FLED/FLED.html).

Equally important to examining historical records is the commencement of new studies that will provide important historical data for the future. Ideally, these should be set up as part of a long-term monitoring program. At the very least, they should be done with sensitivity to the needs of future examiners, incorporating well-marked study sites, quantitative data, and as comprehensive coverage (spatially, temporally, and taxonomically) as possible. Additionally, as we have shown with examples here (Figs. 3.12 and 3.13), photographic and video data may be an extremely valuable source of comparative historic information.

Francis and Hare (1994) point out that historical studies assume that observations are dependent upon variable states of nature that preceded them. They contrast this to reductionist, experimental studies of ecological systems, which have typically assumed an invariant state of nature, such that interactions observed at one point in time are representative of the system through time. This distinction is especially relevant to studies of species' responses to climate change, where we know that species are interacting amidst a continually changing background environment. Moreover, when these background changes are predicted to have profound effects on species' distributions and physiologies, it becomes clear that relevant experimental or reductionist approaches must factor a changing environment into their designs and conclusions. Ultimately, there are important roles for both historical and experimental studies in our understanding of species responses to climate change. Recognizing the strengths and weaknesses of both approaches, and critically evaluating their conclusions accordingly, will be essential in painting an accurate picture of ecological changes in the modern era of climate change.

Acknowledgments

Research discussed in this chapter was supported by the National Wildlife Federation, NASA Earth Systems Science Fellowship (grant # NGT 30339), an NSF graduate student fellowship (RTG grant # BIR94-13141 and GRT grant # GER 93-54870), and grants from the Lerner-Gray Fund for Marine Research and the Myers' Oceanographic and Marine Biology Trust. I thank my coinvestigators Sarah Gilman, Jim Barry, and Chuck Baxter for allowing me to use some of our data in this analysis. Judy Thompson, Freya

Sommer, and Joe Wible of the Hopkins Marine Station provided excellent administrative and research support. This manuscript was greatly improved by thoughtful reviews from Steve Gaines, Jane Lubchenco, Terry Root, Eric Sanford, George Somero, and Erika Zavaleta. Stephen Schneider has provided excellent advice and guidance as part of the National Wildlife Federation's Climate Change Fellowship program.

Literature Cited

Andrus, J. K., and W. B. Legard. 1975. Description of the habitats of several intertidal chiton species (Mollusca: Polyplacophora) found along the Monterey Peninsula of Central California. *Veliger* 18: 3–8.

Bakun, A. 1990. Global climate change and intensification of coastal ocean upwelling. *Science* 247: 198–201.

Barry, J. P., C. H. Baxter, R. D. Sagarin, and S. E. Gilman. 1995. Climate-related, long-term faunal changes in a California rocky intertidal community. *Science* 267: 672–675.

Bazzaz, F. A. 1990. The response of natural ecosystems to the rising CO_2 levels. *Annual Reviews of Ecology and Systematics* 21: 167–196.

Beardall, J., S. Beer, and J. A. Raven. 1998. Biodiversity of marine plants in an era of climate change: Some predictions based on physiological performance. *Botanica Marina* 41: 113–123.

Beebee, T. J. C. 1995. Amphibian breeding and climate. *Nature* 374: 219–220.

Bhaud, M., J. H. Cha, J. C. Duchene, and C. Nozias. 1995. Influence of temperature on the marine fauna: What can be expected from a climatic change. *Journal of Thermal Biology* 20: 91–104.

Bosman, A. L., P. A. R. Hockey, and W. R. Siegfried. 1987. The influence of coastal upwelling on the functional structure of rocky intertidal communities. *Oecologia* 72: 226–232.

Breeman, A. 1990. Expected effects of changing seawater temperatures on the geographic distribution of seaweed species. Pages 69–76 in J. J. Beukema, W. J. Wolff, and J. J. W. M. Brouns, eds., *Expected Effects of Climatic Change on Marine Coastal Ecosystems*. Dordrecht: Kluwer.

Brusca, R. C. 1980. *Common Intertidal Invertebrates of the Gulf of California*. 2nd ed. Tucson: University of Arizona Press.

Carpenter, S. R., T. M. Frost, J. F. Kitchell, and T. K. Kratz. 1993. Species dynamics and global environmental change: A perspective from ecosystem experiments. Pages 267–279 in P. M. Kareiva, J. G. Kingsolver, and R. B. Huey, eds., *Biotic Interactions and Global Change*. Sunderland, Mass.: Sinauer.

Chapin, F. S. I., G. R. Shaver, A. E. Giblin, K. J. Nadelhoffer, and J. A. Laundre. 1995. Responses of arctic tundra to experimental and observed changes in climate. *Ecology* 76: 694–711.

Connolly, S. R., and J. Roughgarden. 1998. A range extension for the volcano barnacle, *Tetraclita rubescens*. *California Fish and Game* 84: 182–183.

Crick, H. Q. P., C. Dudley, D. E. Glue, and D. L. Thomson. 1997. UK birds are laying eggs earlier. *Nature* 388: 526.

Davis, A. J., L. S. Jenkinson, J. H. Lawton, B. Shorrocks, and S. Wood. 1998. Making mistakes when predicting shifts in species range in responses to global warming. *Nature* 391: 783–786.

Davis, M. B. 1989. Lags in vegetation response to greenhouse warming. *Climatic Change* 15: 75–82.

Dayton, P. K., M. J. Tegner, P. B. Edwards, and K. L. Riser. 1998. Sliding baselines, ghosts, and reduced expectations in kelp forest communities. *Ecological Applications* 8: 309–322.

DeBevoise, A. E. 1975. Predation on the chiton *Cyanoplax hartwegii* (Mollusca: Polyplacophora). *Veliger* 18: 47–50.

Denny, M. W., and R. T. Paine. 1998. Celestial mechanics, sea-level changes, and intertidal ecology. *Biological Bulletin* 194: 108–115.

Duarte, C. M., J. Cebrian, and N. Marba. 1992. Uncertainty of detecting sea change. *Nature* 356: 190.

Fields, P. A., J. B. Graham, R. H. Rosenblatt, and G. N. Somero. 1993. Effects of expected global climate change on marine faunas. *Trends in Ecology and Evolution* 8: 361–367.

Francis, R. C., and S. R. Hare. 1994. Decadal-scale regime shifts in the large marine ecosystems of the north-east Pacific: A case for historical science. *Fisheries Oceanography* 3: 279–291.

Frank, K. T., R. I. Perry, and K. F. Drinkwater. 1990. Predicted response of Northwest Atlantic invertebrate and fish stocks to CO_2-induced climate change. *Transactions of the American Fisheries Society* 119: 353–365.

Fraser, W. R., W. Z. Trivelpiece, D. G. Ainley, and S. G. Trivelpiece. 1992. Increases in Antarctic penguin populations: Reduced competition with whales or a loss of sea ice due to environmental warming? *Polar Biology* 11: 525–531.

Glynn, P. 1965. Community composition, structure, and interrelationships in the marine intertidal *Endocladia muricata–Balanus glandula* association in Monterey Bay, California. *Beaufortia* 12: 1–198.

Graham, R. W., and E. C. Grimm. 1990. Effects of global climate change on the patterns of terrestrial biological communities. *Trends in Ecology and Evolution* 5: 289–292.

Gross, K. L., and C. E. Pake. 1995. Final report of the Ecological Society of America Committee on the Future of Long-Term Ecological Data

(FLED). Washington, D.C.: Ecological Society of America. Available online at http://esa.sdsc.edu/FLED/FLED.html.

Gurevitch, J., and J. S. T. Chester. 1986. Analysis of repeated measures experiments. *Ecology* 67: 251–255.

Hadfield, M. G. 1966. The Reproductive Biology of the California Vermetid Gastropods *Serpulorbis squamigerus* (Carpenter, 1857) and *Petaloconchus montereyensis* (Dall, 1919). Ph.D. dissertation, Biological Sciences, Stanford University, Stanford, California.

Hansen, D. V. 1990. Physical aspects of the El Niño event of 1982–1983. Pages 1–20 in P. Glynn, ed., *Global Ecological Consequences of the 1982–1983 El Niño–Southern Oscillation*. New York: Elsevier Science.

Herrlinger, T. J. 1981. Range extension of *Kelletia kelletia*. *Veliger* 24: 78.

Hersteinsson, P., and D. W. Macdonald. 1992. Interspecific competition and the geographical distribution of red and arctic foxes *Vulpes vulpes* and *Alopex lagopus*. *Oikos* 64: 505–515.

Hewatt, W. G. 1934. Ecological Studies on Selected Marine Intertidal Communities of Monterey Bay. Ph.D. dissertation, Biological Sciences, Stanford University, Stanford, California.

———. 1937. Ecological studies on selected marine intertidal communities of Monterey Bay, California. *American Midland Naturalist* 18: 161–206.

Hofmann, G. E., and G. N. Somero. 1996. Interspecific variation in thermal denaturation of proteins in the congeneric mussels *Mytilus trossulus* and *M. galloprovincialis:* Evidence from the heat-shock response and protein ubiquitination. *Marine Biology* 126: 65–75.

Holbrook, S. J., R. J. Schmitt, and J. S. Stephens, Jr. 1997. Changes in an assemblage of temperate reef fishes associated with a climatic shift. *Ecological Applications* 7: 1299–1310.

Hughes, L. 2000. Biological consequences of global warming: is the signal already apparent? *Trends in Ecology and Evolution* 15: 56–61.

Lesica, P., and B. M. Steele. 1996. A method for monitoring long-term population trends: An example using rare arctic-alpine plants. *Ecological Applications* 6: 879–887.

Lluch-Belda, D., S. Hernandez-Vazquez, D. B. Lluch-Cota, C. A. Salinas-Zavala, and R. A. Schwartzlose. 1992. The recovery of the California sardine as related to global change. *California Cooperative Oceanic Fisheries Investigations Reports* 33: 50–59.

Lubchenco, J., S. Navarrete, B. N. Tissot, and J. C. Castilla. 1993. Possible ecological responses to global climate change: Nearshore benthic biota of northeastern Pacific coastal ecosystems. Pages 147–165 in H. A. Mooney, ed., *Earth System Responses to Global Change: Contrasts between North and South America*. San Diego: Academic Press.

McGowan, J. A., D. B. Chelton, and A. Conversi. 1996. Plankton patterns,

climate, and change in the California Current. *California Cooperative Oceanic Fisheries Investigations Reports* 37: 45–68.

Morris, R. H., D. P. Abbott, and E. C. Haderlie. 1980. *Intertidal Invertebrates of California.* Stanford: Stanford University Press.

Morrison, D. F. 1990. *Multivariate Statistical Methods.* 3rd ed. New York: McGraw-Hill.

Murawski, S. A. 1993. Climate change and marine fish distributions: Forecasting from historical analogy. *Transactions of the American Fisheries Society* 122: 647–658.

Parmesan, C. 1996. Climate and species' range. *Nature* 382: 765–766.

Patz, J. A., P. R. Epstein, T. A. Burke, and J. M. Balbus. 1996. Global climate change and emerging infectious diseases. *Journal of the American Medical Association* 275: 217–223.

Potvin, C., M. J. Lechowicz, and S. Tardif. 1990. The statistical analysis of ecophysiological response curves obtained from experiments involving repeated measures. *Ecology* 71: 1389–1400.

Quinn, W. H., V. T. Neal, and S. E. A. de Mayolo. 1987. El Niño occurrences over the past four and a half centuries. *Journal of Geophysical Research* 92: 14,449–14,461.

Ricketts, E. F., J. Calvin, and J. W. Hedgpeth. 1985. *Between Pacific Tides.* 5th ed. Stanford: Stanford University Press.

Roemmich, D., and J. McGowan. 1995. Climatic warming and the decline of zooplankton in the California current. *Science* 267: 1324–1326.

Root, T. 1988. Energy constraints on avian distributions and abundances. *Ecology* 69: 330–339.

———. 1993. Effects of climate change on North American birds and their communities. Pages 280–292 in P. M. Kareiva, J. G. Kingsolver, and R. B. Huey, eds., *Biotic Interactions and Global Change.* Sunderland, Mass: Sinauer.

Root, T. L., and S. H. Schneider. 1995. Ecology and climate: Research strategies and implications. *Science* 269: 334–341.

Sagarin, R. D., J. P. Barry, S. E. Gilman, and C. H. Baxter. 1999. Climate-related changes in an intertidal community over short- and long-time scales. *Ecological Monographs* 69: 465–490.

Schwing, F., and R. Mendelssohn. 1997. Increased coastal upwelling in the California Current System. *Journal of Geophysical Research* 102: 3421–3438.

Scott, D. 1993. Time segment analysis of permanent quadrat data: Changes in *Hieracium* cover in the Waimakariri in 35 years. *New Zealand Journal of Ecology* 17: 53–57.

Scott, D., and M. Poynter. 1991. Upper temperature limits for trout in New Zealand and climate change. *Hydrobiologia* 222: 147–151.

Shope, R. 1991. Global climate change and infectious diseases. *Environmental Health Perspectives* 96: 171–174.

Sokal, R. R., and F. J. Rohlf. 1995. *Biometry.* 3rd ed. New York: W. H. Freeman.

Southward, A. J. 1967. Recent changes in abundance of intertidal barnacles in south-west England: A possible effect of climatic deterioration. *Journal of the Marine Biological Association: U. K.* 47: 81–95.

Southward, A. J., S. J. Hawkins, and M. T. Burrows. 1995. Seventy years' observations of changes in distribution and abundance of zooplankton and intertidal organisms in the western English Channel in relation to rising sea temperature. *Journal of Thermal Biology* 20: 127–155.

Sparks, T. H., and P. D. Carey. 1995. The responses of species to climate over two centuries: An analysis of the Marsham phenological record, 1736–1947. *Journal of Ecology* 83: 321–329.

Stephenson, T. A., and A. Stephenson. 1972. *Life Between Tidemarks on Rocky Shores.* San Francisco: W. H. Freeman and Company.

Stillman, J. H., and G. N. Somero. 1996. Adaptation to temperature stress and aerial exposure in congeneric species of intertidal porcelain crabs (genus: *Petrolisthes*): Correlation of physiology, biochemistry and morphology with vertical distribution. *Journal of Experimental Biology* 199: 1845–1855.

Taper, M. L., K. Bohning-Gaese, and J. H. Brown. 1995. Individualistic responses of bird species to environmental change. *Oecologia* 101: 478–486.

Tegner, M. J., P. K. Dayton, P. B. Edwards, and K. L. Riser. 1996. Is there evidence for long-term climatic change in southern California kelp forests? *California Cooperative Oceanic Fisheries Investigations Reports* 37: 111–126.

Veit, R. R., P. Pyle, and J. A. McGowan. 1996. Ocean warming and long-term change in pelagic bird abundance within the California current system. *Marine Ecology Progress Series* 139: 11–18.

Veit, R. R., J. A. McGowan, D. G. Ainley, T. R. Wahls, and P. Pyle. 1997. Apex marine predator declines ninety percent in association with changing oceanic climate. *Global Change Biology* 3: 23–28.

Welch, D. W., Y. Ishida, and K. Nagasawa. 1998. Thermal limits and ocean migrations of sockeye salmon (*Oncorhynchus nerka*): Long-term consequences of global warming. *Canadian Journal of Fisheries and Aquatic Science* 55: 937–948.

Woodward, F. I. 1992. A review of the effects of climate on vegetation: Ranges, competition, and composition. Pages 105–123 in R. L. Peters and T. E. Lovejoy, eds., *Global Warming and Biological Diversity.* New Haven: Yale University Press.

Zimmerman, R. C., D. G. Kohrs, and R. S. Alberte. 1996. Top-down impact through a bottom-up mechanism: The effect of limpet grazing on growth, productivity and carbon allocation of *Zostera marina* L. (eelgrass). *Oecologia* 107: 560–567.

CHAPTER 4

Community Responses to Climate Change: Links between Temperature and Keystone Predation in a Rocky Intertidal System

ERIC SANFORD

A growing body of evidence suggests that human activities are dramatically increasing the concentration of greenhouse gases in the atmosphere and warming the planet (Mann et al. 1998, Pollack et al. 1998, IPCC 2001). During the twentieth century average global surface temperatures increased by about 0.6°C. Atmospheric models predict that global warming will cause average surface temperatures to rise another 1.4 to 5.8°C by the end of this century (IPCC 2001).

What effect would a change of this magnitude have on natural systems? The simplest prediction suggests that species will respond to warmer temperatures by shifting to higher latitudes or altitudes. This perspective treats species as isolated units responding directly and independently to environmental change. Although range shifts are important (e.g., Zavaleta and Royval, this volume), few studies have addressed the possibility that climate change may disrupt communities and ecosystems more immediately and dramatically by altering species interactions.

In this chapter, I review our understanding of how temperature change may impact species' ranges, species interactions, and natural communities. I suggest that experimentally testing the response of key species interactions to changes in environmental variables may be a useful way to evaluate the sensitivity of biotic communities to climate change. This approach is illustrated in rocky intertidal communities on the central Oregon coast using field and laboratory experiments. This study underscores the importance of considering the effects of climate change on species interactions and highlights the value, as well as the limitations, of using experimental approaches to predict community-level responses.

Temperature and Species' Geographic Ranges

To date, most predictions regarding the potential effects of climate change on wildlife and ecosystems have focused on shifts in species' geographic ranges. Species are expected to respond to warming temperatures by gradually shifting to higher latitudes or altitudes to remain within their thermal tolerance ranges (Breeman 1990, Davis and Zabinski 1992, Gates 1993, Lubchenco et al. 1993). Several recent studies in both marine and terrestrial systems report long-term changes in species distributions that are consistent with these predictions (Barry et al. 1995, Parmesan 1996, Pounds et al. 1999, Sagarin et al. 1999, Parmesan et al. 1999, Thomas and Lennon 1999, Reed, this volume, Sagarin, this volume).

Projected range shifts are based on the long-standing hypothesis that temperature sets the geographic range limits of most species (e.g., Orton 1920, Hutchins 1947). Although there are only a handful of studies that establish a direct physiological or experimental connection between temperature and geographic range boundaries (Sundene 1962, Nobel 1980, references in Breeman 1990), there are many strong correlations between temperature isotherms and species' range limits that suggest such links may be common (e.g., Woodward 1987, Root 1988).

These patterns imply a general model of range limits based on the relationship between temperature and physiological tolerance. An organism's performance or fitness is expected to vary as a roughly bell-shaped function of temperature (Fig. 4.1A), driven by the relationship between temperature and the efficiency of underlying physiological processes (Shelford 1913, Huey and Stevenson

1979, Cossins and Bowler 1987, Huey and Kingsolver 1989). Optimal performance occurs in the middle of this temperature range. At high and low temperatures there are critical limits, beyond which an organism is no longer able to reproduce successfully (i.e., the organism may be unable to produce offspring, or alternatively, the organism's offspring may be unable to survive; Orton 1920, Hutchins 1947, Bhaud et al. 1995). These upper and lower critical limits are thought to set the geographic boundaries of a species' range (Fig. 4.1A).

Given this model, a species' range contracts when temperature change eliminates populations living at the margin of the range. For example, a population living near its southern boundary may disappear when a small temperature increase pushes individuals beyond their upper critical limit (Fig. 4.1B). This model implies that populations living near their range limits are most vulnerable to a slight temperature change (Davis and Zabinski 1992), whereas populations living near the middle of their geographic range (i.e., close to their physiological optimum) should be relatively unaffected by a slight temperature change.

Considerable evidence suggests that physiological optima and critical limits are likely to vary among species within the same community. For example, different species of plants living within the same habitat frequently vary in their physiological responses to temperature and other environmental factors (such as light, CO_2, and soil nutrients; Pacala and Hurtt 1993). A similar pattern may apply to animals. For example, a recent study found that seven congeneric species of darkling beetles living in a short-grass prairie in Colorado showed peak activity at different ambient temperatures (Whicker and Tracy 1987). In the laboratory, beetles selected these same, species-specific temperatures from a thermal gradient. These and similar studies (e.g., Lévêque 1997, Tomanek and Somero 1999) suggest that species within a community are likely to respond differently to temperature change.

This idea is also supported by the fossil record in North America, which indicates that species' ranges shifted independently from one another in response to past changes in climate (Foster et al. 1990). For example, during the last North American glacial cycle (14,000–10,000 years ago), individual species of small mammals showed shifts in geographic range marked by different directions and rates (Graham 1992). Similarly, pollen records suggest that the

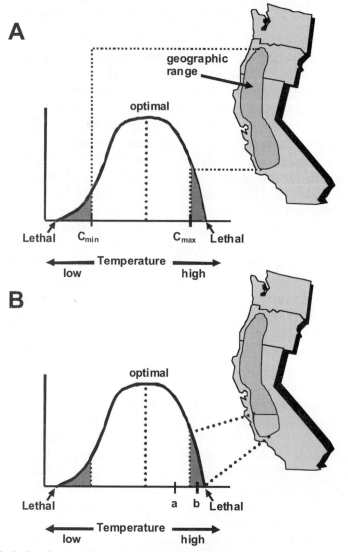

Figure 4.1. A simple model relating temperature and the geographic range of a hypothetical sedentary species in California and Oregon. (A) Individual performance varies as a species-specific function of temperature. Geographic range limits are set by a critical minimum (C_{min}) and critical maximum (C_{max}) temperature, beyond which the organism is unable to successfully reproduce (shaded regions under curve). (B) Dynamics of a hypothetical range contraction. A small temperature increase occurs throughout the geographic range. Individuals near their southern range limit formerly experienced temperature "a," but now experience a higher temperature "b" within the species' upper critical range. Individuals are unable to successfully reproduce, and southern populations ultimately disappear (cross-hatched region of geographic range). Analogous processes may expand the northern range limit, leading to a northward shift of the entire range.

distributions of tree species responded individualistically to climatic changes over the past 18,000 years (Webb 1992). To the extent that species' ranges reflect temperature tolerances, these individualistic responses suggest that communities are impermanent associations of species with different underlying physiologies.

Assuming these individualistic responses apply broadly to most species, ongoing climatic warming is likely to shift species' ranges independently of one another, and perhaps disassemble present-day communities, leading to the formation of new species associations. As a result, scientists considering climatic impacts have generally treated species as independent units, rather than considering the influence of temperature on communities and community processes. Most studies have focused on predicting range shifts of single species and have used a three-part strategy (summarized by Harte et al. 1992): (1) field-derived correlations and physiological information from laboratory studies are used to estimate a species' thermal tolerance, (2) atmospheric models are used to predict temperature changes, and (3) these data are combined to determine whether a species would need to move in order to remain in a suitable climate. This strategy has been used to project range shifts for a variety of species of marine algae, trees, insects, fish, and mammals (e.g., Breeman 1990, Scott and Poynter 1991, Davis and Zabinski 1992, Rogers and Randolph 1993, Johnston and Schmitz 1997).

This approach is conceptually simple and appealing, and predicting range shifts is a useful first step in evaluating potential impacts of climate change. However, the near exclusive focus on range shifts implies that: (1) impacts on communities can be evaluated solely by assessing the response of individual species to the direct effects of temperature change, (2) changes will be manifested mainly through lethal effects on individuals, and (3) populations near geographic range limits will be most vulnerable to temperature change, whereas those near the central portion of a range will be relatively unaffected.

Temperature and Species Interactions

The potential effects of climate change on natural communities may be underestimated by a focus on the direct, lethal effects of temperature. Many studies have concluded that, at least in the short-term, anticipated warming will have little impact on the ability of a species to survive within its current range (e.g., Johnston and Schmitz 1997, Sætersdal and Birks 1997). This conclusion is based on the likelihood that available behavioral and physiological plasticity will minimize the risk of extinction to most species (Dawson 1992). Although range shifts may have important long-term effects, changes in species distributions (in and of themselves) may not significantly disrupt communities during this century (Paine 1993, but see Zavaleta and Royval, this volume).

However, an emphasis on range shifts neglects other important, and potentially more immediate, effects. In particular, few studies have addressed the possibility that climatic changes may impact local populations more rapidly and dramatically by altering the strength of interactions between a species and its competitors, mutualists, predators, prey, or pathogens (Lubchenco et al. 1991). This hypothesis follows logically from the idea that performance varies as a species-specific function of temperature (Fig. 4.1A). For example, following a slight temperature increase, an individual that was previously living near its physiological optimum may be forced to spend more time operating at a reduced level of performance. Such sublethal effects may change the way that one species interacts with another, for example by changing competitive ability, rates of consumption, or susceptibility to predators or pathogens. Moreover, these effects may occur throughout a species' range, not just in populations living near range boundaries.

The recognition that physical conditions may mediate the strength or outcome of species interactions is not a new idea. This concept was the focus of several foundational papers in modern experimental ecology. For example, Tansley (1917) determined that the outcome between two competing plant species was dependent on soil type. Similarly, Park's (1954) classic flour beetle studies demonstrated that competition between different species of *Tribolium* was mediated by temperature and humidity. Although these and other early experiments clearly suggested that biotic processes could be modified by abiotic factors, community ecology has passed through phases emphasizing the primary importance of either abiotic (e.g., Andrewartha and Birch 1954) or biotic factors (e.g., Paine 1966) in determining patterns of abundance and distribution.

In recent years, however, there has been a strong movement toward integrating the roles of biotic and abiotic factors into an understanding of community dynamics (Menge and Sutherland 1976, 1987, Tilman 1980, Kingsolver 1989, Dunson and Travis 1991, Hunter and Price 1992, Bertness and Callaway 1994, Bruno and Bertness 2001). These models of community regulation are all based on the idea that species are imbedded in a complex web of interacting species, and that the strength of these biotic interactions may be influenced by environmental conditions. Species are linked by trophic interactions to species that are above and below them in a food web. Competitive and facilitative interactions may link species within the same trophic level. Changes in temperature or other environmental factors may alter the strength of any of these links, with effects that ripple throughout a food web via species interactions (Hunter and Price 1992).

This community perspective has led some ecologists to caution that climate change may impact communities by altering species interactions (Harte et al. 1992, Ayres 1993, Castilla et al. 1993, Lubchenco et al. 1991, 1993, Root and Schneider 1993, Brown et al. 1997, Davis et al. 1998b, Post et al. 1999). If such alterations change the relative abundance of a given species, effects may then be transmitted through chains of species interactions. These sorts of cascading effects could theoretically arise anywhere in a food web (Hunter and Price 1992).

Evaluating Community-Level Impacts

Is there any empirical evidence that climatic changes can disrupt communities through altered species interactions? In fact, some long-term studies have observed interdecadal variation in community structure and have inferred that altered species interactions played a role. For example, a recent study reported that krill abundance in the Antarctic tended to be much lower during the period from 1984 to 1996 than from 1976 to 1984 (Loeb et al. 1997). Declining krill abundance was associated with a decrease in the frequency of winters with extensive sea-ice cover. In contrast to krill, the abundance of filter-feeding salps (pelagic tunicates) increased with decreasing sea-ice. The authors suggested that open-water conditions support rapid growth of salp populations, and that krill abundance is reduced by strong competition with salps for food (phytoplankton). Krill, in turn, are the major food source for higher trophic levels, such as Adelie penguins, which also appear to be in decline. Thus recent decreases in sea-ice cover may alter competition between salps and krill, leading to changes throughout the Antarctic food web.

In a similar long-term observational study, recent climatic changes appear to have sparked major changes in a Chihuahuan Desert ecosystem during the period from 1977 to 1995 (Brown et al. 1997). In this ecosystem, winter precipitation over this period has increased significantly, apparently triggering large increases in shrub cover. These changes in vegetation may be linked to substantial changes in the abundance of many species of rodents, seed-harvesting ants, and lizards. The authors attribute the wholesale reorganization of this ecosystem to both the direct effects of increased winter moisture and the indirect effects of altered biotic interactions (Brown et al. 1997).

Long-term ecological data sets like those from the Chihuahuan Desert and Antarctica are uncommon, but may be necessary to observe effects that reverberate through food webs (Brown and Heske 1990). These studies suggest a link between climatic changes, species interactions, and community-level changes. However, since these studies were observational, these links can at present only be inferred. Furthermore, in these studies, community changes were observed *first*, and researchers then tried to trace these changes back to specific interactions which may have been altered by climatic changes.

These studies point to important challenges in evaluating community-level impacts of climatic changes. For example, is it possible to obtain direct, experimental evidence of how species interactions are affected by environmental changes? If so, which interactions should be the focus of attention? Can pathways likely to generate community change be identified a priori? To what extent can ecologists hope to predict the likely effects of climate change on natural communities?

Experimentally testing the response of species interactions to environmental change may be tractable in many systems, and is likely to provide valuable mechanistic insight. Plant and ecosystem ecologists have pioneered the approach of using experimental field studies to test how small changes in climatic conditions may affect species interactions. Plastic enclosures or heating lamps have been used to manipulate light, soil temperature, and moisture in the field (Chapin et al. 1995, Harte and Shaw 1995, Henry and Molau

1997). For example, in a recent study of a subalpine meadow in the Rocky Mountains, Colorado, heating lamps were used to reduce soil moisture and increase soil temperature by ~1°C in experimental plots (Harte and Shaw 1995). These small changes in environmental factors altered the competitive ability and the relative abundance of plant species within this community. Similar responses have been documented in other climate warming experiments, such as those by Chapin and colleagues in tussock tundra near Toolik Lake, Alaska (Chapin et al. 1995, Chapin and Shaver 1985, 1996). To date, this experimental approach has been used predominantly to assess changes in plant populations, rather than changes within whole communities or ecosystems.

Addressing changes at the community or ecosystem-level obviously introduces substantial complexity. It is often logistically difficult or even impossible to manipulate environmental factors at large scales in the field. Moreover, even relatively simple communities contain dozens of species and hundreds, if not thousands, of species interactions. Investigating all possible links in the field would be an enormous and impractical task (Davis et al. 1998a). Thus, there is a clear need to identify species interactions that are both highly sensitive to environmental changes, and particularly likely to effect changes in communities.

A large body of experimental work suggests that the composition and diversity of many communities are maintained by a few critical interactions (Paine 1966, 1969, 1992, Power et al. 1996). These interactions frequently involve *keystone species;* that is, a species whose impact on its community or ecosystem is large, and disproportionately large relative to its abundance (Power et al. 1996). Keystone species have been identified in marine, freshwater, and terrestrial ecosystems (Menge et al. 1994, Power et al. 1996). Although these species often exert their effects through consumption, keystone interactions can also include competition, mutualism, dispersal, pollination, and disease (Power et al. 1996).

If key species interactions are sensitive to temperature, they may act as leverage points in natural systems, through which relatively small temperature changes can generate surprisingly large changes in community structure or ecosystem function (Sanford 1999a). Identifying key species interactions in understudied systems remains a significant challenge, but one that can be addressed by experimental and comparative approaches (Power et al. 1996).

Although not all communities contain keystone species, community and ecosystem processes may often be dominated by a few strong interactions against a background of many weak interactions (Paine 1992, Carpenter et al. 1993, Fagan and Hurd 1994).

Quantifying the response of strong interactions to slight changes in temperature is thus a logical starting point in evaluating the potential effects of climate change on communities. The remainder of this chapter demonstrates how experimental and observational approaches can be combined to provide mechanistic insight into how climate change may impact communities. I focus on the interaction between a keystone predator, the sea star *Pisaster ochraceus*, and its principal prey, the rocky intertidal mussels, *Mytilus californianus* and *M. trossulus*. I use my results to suggest how climate change might impact these rocky intertidal communities, and in so doing highlight some of the limitations and challenges of global change research.

Experiments in a Rocky Intertidal Community

The intertidal zone is the area of the shore that lies between the high and low tide marks, and thus is intermittently exposed and covered by the shifting tides. Rocky intertidal communities in the Pacific Northwest are diverse, accessible, and easy to observe and manipulate. As a result, these communities are extremely well described and have long served as model systems to develop and test general ecological theories about disturbance, succession, and biotic interactions (reviewed by Paine 1994).

Study System

I conducted research at wave-exposed rocky intertidal sites within Neptune State Park (44°15'N, 124°07'W), south of Cape Perpetua on the central Oregon coast. This 4 km stretch of coastline is composed of extensive rocky benches, outcrops, pools, and surge channels. The communities here are similar to those of many other wave-exposed regions along the northern Pacific coast of North America (Dayton 1971, Paine 1980, Menge et al. 1994). The high intertidal zone is characterized by fucoid algae and barnacles, the mid zone by dense beds of the California mussel (*Mytilus californianus*), and the low zone by a diverse mixture of algae, seagrass, and invertebrates, including the ochre sea star, *Pisaster ochraceus* (Fig. 4.2).

Figure 4.2. The ochre sea star, *Pisaster ochraceus* feeding at the lower edge of a mussel bed on the Oregon coast. At low tide, sea stars are frequently observed hunched over their prey, including the California mussels (*Mytilus californianus*) shown here. In the absence of sea star predation, this mussel species overgrows and outcompetes other sessile species in the low intertidal zone.

Numerous experimental studies during the past 30 years have identified a subset of strong interactions that maintain the composition and diversity of rocky intertidal communities in the Pacific Northwest (Paine 1966, 1969, 1992, Connell 1970, Dayton 1971, Menge 1992, Menge et al. 1994, Navarrete and Menge 1996). In particular, Paine's classic experiments (Paine 1966, 1969) in Washington State demonstrated that predation by the sea star *Pisaster ochraceus* prevents the mussel *Mytilus californianus* from dominating the low intertidal zone. When sea stars were removed from experimental areas, primary space in the low intertidal zone shifted from a diverse assemblage of invertebrates and algae to a monoculture of *M. californianus*. Paine coined the term *keystone predator* to describe a single predator species (like *Pisaster*) that determines most patterns of community structure. Experimental studies have confirmed that *Pisaster* plays a similar keystone role in wave-swept communities on the central Oregon coast (Menge et al. 1994).

This keystone interaction occurs in a system that is ideally suited to testing the influence of slight temperature change on trophic dynamics. In many terrestrial and aquatic systems, there is a high degree of spatial heterogeneity in the thermal environment (e.g., shaded microhabitats, stratified lakes, above vs. below-ground habitats, etc.). Thus, many ectothermic consumers (e.g., fishes, lizards, insects) may be able to mediate the effects of minor temperature change by selecting preferred microhabitats or limiting activity to certain times (Caulton 1982, Christian et al. 1983, Whicker and Tracy 1987, Gates 1993). These changes in behavior could have important ecological effects, for example by changing where and when consumers are foraging (Rubenstein 1992, Dunham 1993). In addition, changes in behavior complicate attempts to predict the body temperature of consumers, since it is difficult to assess if (or how much) behavioral thermoregulation might offset changes in ambient temperature.

In contrast, most marine benthic consumers (e.g., sea stars, whelks, herbivorous mollusks, sea urchins, etc.) have little opportunity for behavioral thermoregulation when submerged. Sea temperatures along the coast typically vary little over several hundred meters, which is a scale far larger than these organisms' daily movement. Thus, changes in water temperature are experienced as direct changes in the body temperatures of these consumers.

Water temperatures along the Pacific coast vary seasonally, but also vary over shorter time scales in response to coastal upwelling. Episodes of upwelling lasting from several days to three or more weeks are common along the Oregon coast from May to September (Menge et al. 1997). Persistent, strong, southward winds combine with the Coriolis effect to push surface waters offshore (Fig. 4.3), and water temperatures typically drop 3 to 5°C as cold, nutrient-rich water rises from below.

In the Pacific Northwest, densities of *Pisaster* and its impact in the intertidal zone are highest between May and October (Mauzey 1966, Paine 1974, Robles et al. 1995, Sanford 1999b). However, preliminary observations in Oregon suggested that within this season, intertidal sea star activity underwent marked fluctuations associated with oceanographic conditions. During periods of cold-water upwelling, many sea stars appeared to become inactive in low zone channels or shallow subtidal waters (Sanford 1999b). Upwelling patterns along the Pacific coast have changed substantially in recent decades, and some climate models predict further

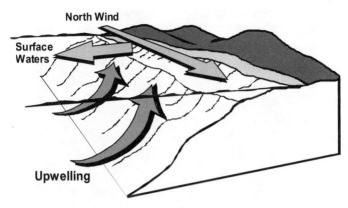

Figure 4.3. Mechanics of coastal upwelling. Strong northerly winds and the Coriolis effect cause surface waters to flow offshore, drawing up cold, nutrient-rich waters from below (after Bakun 1990).

changes associated with global warming (see discussion under "Predicting Changes in Upwelling Intensity"). I therefore conducted experiments to examine how variation in upwelling patterns might alter intertidal communities via effects on this keystone predator.

Design of Field and Laboratory Experiments

Using field experiments, I quantified rates of *Pisaster* predation to test the hypothesis that the strength of the sea star–mussel interaction is reduced during periods of cold-water upwelling. Experiments were conducted at three wave-exposed sites within Neptune State Park. Sites were separated by several hundred meters, and water temperatures varied little (generally < 0.2°C) over these distances. At each site, I identified two large rocky reefs (mean area ± SEM = 132.5 ± 49.7 m²) that were in close proximity to one another yet isolated by surge channels. All sea stars were routinely removed from one reef in each pair and allowed to remain at natural densities on the other.

In April–May 1997, I transplanted 20 clumps of 50 mussels to the low intertidal zone on each of these reefs. *Mytilus californianus* were used in these experiments to gauge the influence of *Pisaster* on the competitively dominant space-holder (Paine 1992). Mussels (shell length = 4.5–5.5 cm) were placed in overlapping rows under

Figure 4.4. Experimental measurement of predation intensity. To the left is a transplanted mussel clump that has been uncaged and is being approached by several sea stars. To the right is a mussel transplant still protected from predation by a mesh cage.

Vexar™ mesh cages that were screwed into the rock (Fig. 4.4). Cages held mussels in place, allowed them to reattach firmly to the rock with byssal threads, and protected them from being eaten by sea stars until cages were removed. This technique has been used successfully by other researchers in Washington and Oregon (Paine 1976, Menge et al. 1994, Navarrete and Menge 1996).

Beginning in mid June, I conducted five consecutive experiments to measure the intensity of sea star predation during periods lasting 14 days each. Starting dates were set a priori as the first day of each spring tide series, so that each experiment consisted of a similar 14-day tidal cycle. At the start of each experiment, I randomly selected four mussel transplants on each reef and removed their cages, thereby exposing these mussels to sources of mortality. I then recorded mussel survivorship daily for the first 6 to 7 days, and again on day 14 (study sites were inaccessible during neap tides, days 8–13). On each of these days, I also recorded the local sea star density (defined as the number of sea stars in a 1 m radius around each clump). Uncaged mussels that remained on day 14 were counted and then removed.

I tested whether variation in rates of predation was associated with changes in environmental factors. At each site I installed a data-logger (Optic StowAway™, Onset Computer Corp., Pocasset, MA) in the low intertidal zone that recorded water temperature (during high tide) or air temperature (during low tide) every 30 minutes. Data-loggers were positioned on horizontal surfaces that were exposed to full sunlight during daylight periods of aerial exposure. From these records at each site, I calculated high tide water temperatures, defined as the mean of all readings during a period from 2 hours before to 2 hours after each high tide. I also extracted the maximum temperature recorded during each low tide period of aerial exposure. Temperatures recorded by data-loggers during periods of aerial exposure can be used to accurately predict sea star body temperatures (and thus potential heat stress) at low tide (Sanford 1999a). The time of low and high tides was estimated using National Oceanic and Atmospheric Administration tide charts.

Since movement and feeding of consumers may be reduced during periods of increased wave stress (see references in Menge and Olson 1990), five wave force meters (Bell and Denny 1994) were deployed in the low zone at each site to record variation in maximum wave forces. Meters were read and reset every 24 hours for the first 5 to 7 days of each period.

I also examined the effects of water temperature on *Pisaster* feeding rates under controlled conditions in the laboratory. Sea stars were maintained under three temperature treatments: (1) constant 12°C, (2) constant 9°C, and (3) a treatment that alternated between two weeks at 12°C, and two weeks at 9°C. The constant 9°C and constant 12°C treatments simulated water temperatures that sea stars would experience during prolonged periods of upwelling or no upwelling, respectively. The alternating treatment simulated the fluctuating water temperatures that are typical of the Oregon coast during upwelling season.

In early June 1996, I collected 48 sea stars (wet weight: range = 118–138 g) from a single site within Neptune State Park. Four individuals were randomly assigned to each of 12 closed 110-liter tanks at Hatfield Marine Science Center in Newport, Oregon. All tanks were held in a cold room at 6 to 8°C, and water temperatures were elevated to set levels and self-regulated to ± 0.1°C using heaters and controllers (Omega Engineering Corp., Controller Model CN76120). Temperatures in treatments were verified independently throughout the experiment by data-loggers submerged in the tanks.

Water was circulated and oxygenated by two pumps in each tank. Water quality was maintained by an in situ wet/dry filter, foam fractionater, and routine water changes with filtered seawater (25% of tank volume/week). Salinity was maintained at 36 ± 1‰, and the experiments were conducted under a 12-hour light:12-hour dark schedule.

All sea stars were initially acclimated for 10 days at 11°C without food. Treatments were then randomly assigned (n = 4 tanks/treatment) and sea stars were provided with mussels ad libitum. *Mytilus trossulus* (shell length = 3.0–4.0 cm) were selected for these experiments since this species is the most common prey item in the diet of *Pisaster* at these sites (Navarrete and Menge 1996, Sanford 1999a), occurring in ephemeral low zone patches below the *M. californianus* beds. The number of mussels consumed per tank was recorded every 14 days.

Results

During the field experiment, mussel mortality was very low on reefs where sea stars were removed (Fig. 4.5A). This suggests that sources of mortality other than predation by sea stars (e.g., dislodgement by waves, consumption by crabs or river otters, etc.) had little effect during these experiments. Thus mortality on reefs where sea stars were at natural densities could be attributed almost entirely to predation by *Pisaster*. Both rates of predation and sea star density were dramatically reduced during a persistent, cold-water upwelling event (Fig. 4.5B, C), during which water temperature dropped about 3°C (Fig. 4.6A).

This drop in predation could not be explained by variation in potential heat stress (Fig. 4.6B) or maximum wave force (Fig. 4.6C), both of which were relatively consistent among periods. Maximum air temperatures and wave forces tended to be low during upwelling—conditions that should tend to *increase* rates of predation (Menge and Olson 1990). I used multiple regression to test whether variation in sea star predation (mean mortality after 14 days in four transplants/site/period) among time periods and sites was associated with (1) water temperature (mean during 27 high tides/period), (2) potential heat stress (mean of maximum low tide air temperature on 5 warmest days/period), or (3) maximum wave force (mean of maximum force/day on 5–7 days/period). Sea star

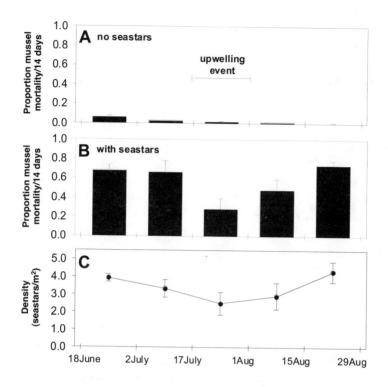

Figure 4.5. Results of field predation experiments. The five consecutive 14-day periods are noted on the x-axis. Persistent upwelling occurred during the third period (July 17–August 1). All data are means ± SEM for three study sites. (A) Mussel mortality at end of each 14-day period on reefs with sea stars removed. (B) Mussel mortality in the presence of sea stars. Predation was significantly reduced during upwelling (analysis of variance (ANOVA), contrast period 3 vs. periods 1, 2, 4, 5; $F_{1,53} = 14.10$, $p < 0.001$). (C) Local sea star density around mussel clumps. Sea star density was also significantly reduced during upwelling (ANOVA, contrast period 3 vs. periods 1, 2, 4, 5; $F_{1,53} = 5.33$, $p = 0.025$).

predation was associated with water temperature ($p < 0.001$), but was unrelated to potential heat stress ($p = 0.93$) or wave forces ($p = 0.60$). Predation was consistently greater at one of the three sites, so site variables were significant in the model. Together water temperature and site explained 86.7% of the variation in mean mussel mortality on reefs with sea stars.

The reduction of local sea star density during upswelling was

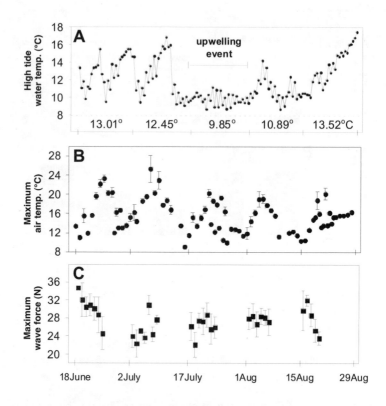

Figure 4.6. Environmental conditions during predation experiments. The five consecutive 14-day periods are noted on the x-axis. All data are means ± SEM for three study sites. (A) High tide water temperature (mean from 2 hours before to 2 hours after each high tide). Temperatures above the x-axis are the overall means for the 27 high tides of each period. (B) Maximum air temperature recorded during each low tide exposure. The periodicity in air temperature reflects the tidal cycle; peak temperatures generally occurred on days when low tide was in the late morning. (C) Maximum wave force recorded during 24-hour intervals, on 5 to 7 days/period.

presumably the result of sea stars remaining inactive in channels or shallow subtidal waters (Sanford 1999b). However, analysis of per capita interaction strength (the difference in rates of mussel mortality on reefs with and without sea stars divided by local sea star density) indicated that the per capita effects of *Pisaster* on mussels were also reduced during upwelling (Sanford 1999a). In other words, in addition to a reduction in the density of foraging sea stars, individual sea stars consumed less during the upwelling period.

Although the influence of salinity on the feeding behavior of *Pisaster ochraceus* is unknown, many sea star species are unable to tolerate salinity below 13 to 23 ‰ (Stickle and Diehl 1987). Therefore a large decrease in salinity might reduce the feeding rate of *Pisaster.* Salinity and temperature were quantified during summer 1999, directly offshore from my study sites, on a mooring located in 15 m of water. The salinity of colder water (< 10°C) typically varied between 33.1 and 34.0 ‰, whereas that of warmer water (>11°C) generally varied between 32.8 and 33.5 ‰ (B. Grantham, unpublished data). Thus, as expected, upwelling events were associated with increased salinity. Salinity changes of this magnitude probably had no measurable effect on sea stars and thus are unlikely to account for the decrease in sea star feeding during upwelling.

In the laboratory experiment, a 3°C temperature difference had a surprisingly large effect on the feeding rate of sea stars. Sea stars maintained in constant 9°C tanks consumed, on average, 29% fewer mussels than did those at a constant 12°C (Fig. 4.7). As in the field experiment, sea stars in the alternating treatment fed intensely at 12°C, reduced feeding at 9°C, and increased feeding again at 12°C.

Discussion: Can We Forecast Community Changes?

The results of these experiments suggest that the strength of the interaction between the keystone predator *Pisaster ochraceus* and its primary prey is directly regulated by slight changes in water temperature. In this section, I use these results to suggest how climate change might impact rocky intertidal communities, and discuss some of the challenges in predicting such climate-related changes.

Linking Predation and Temperature

This sensitivity of *Pisaster* predation to small changes in water temperature is particularly surprising because it occurs over the range of 9 to 13°C, in the middle of this sea star's thermal tolerance range. *Pisaster ochraceus* is found from at least Punta Baja, Baja California, to Prince William Sound, Alaska, and populations living near these

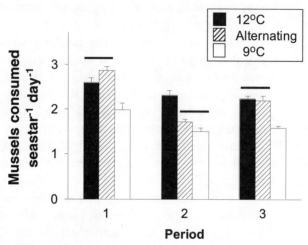

Figure 4.7. Results of laboratory predation experiment. Sea stars were maintained under three temperature treatments: constant 12°C, constant 9°C, and alternating (periods 1–3 = 12°C, 9°C, 12°C, respectively). Bars are mussels consumed per sea star per day (+ SEM) in treatments ($n = 4$ tanks/treatment) during three consecutive 14-day periods. Data were analyzed in a repeated measures analysis of variance. "Treatment" ($F_{2,9}$ = 59.81, $p < 0.001$), "time" ($F_{2,18} = 36.59$, $p < 0.001$), and "time x treatment" ($F_{4,18} = 5.70$, $p = 0.004$) were all significant. Within each time period, shared lines below bars indicate groups whose means did not differ (Tukey-Kramer, $p > 0.05$). Results are presented for the first three periods of the experiment. Thereafter, sea stars apparently became temporarily satiated on the ad libitum diet, and feeding rates declined in all treatments (Sanford 1999b).

range limits experience seasonal water temperatures > 20°C and < 4°C, respectively.

Although the ecological effects of small changes in water temperature have generally been overlooked in marine systems, some decrease in predation rate with declining temperature is expected from a physiological perspective. The direct effects of colder temperatures on rates of movement, metabolism, and digestion should slow feeding rates (Clarke 1987, Cossins and Bowler 1987, Birkeland and Lucas 1990). *Pisaster* shows a lowered metabolic rate at lower temperatures (Paine, pers. comm. in Mauzey 1966) and feeding has been observed to decline or stop completely at low temperatures (< 5°C) in several Atlantic sea star species (Feder and Christensen 1966).

In the field, *Pisaster* predation was sharply reduced during upwelling (Fig. 4.5B), not only because each sea star consumed less, but also because densities of foraging sea stars were locally reduced (Fig. 4.5C). This twofold effect may be a common response of ectothermic consumers to temperature change. Not only should colder temperatures slow feeding rates, but a greater proportion of individuals should become inactive and seek shelter as temperature declines to some species-specific level (e.g., Aleksiuk 1976, Frazer and Gilbert 1976, Hunt 1977, Whicker and Tracy 1987). For example, the rate of predation of ladybird beetles on aphids was found to decrease exponentially as temperature decreased, since both mean walking speed and the proportion of ladybirds that were actively foraging decreased linearly with temperature (Frazer and Gilbert 1976, Kingsolver 1989).

My study indicates a direct link between water temperature and the strength of keystone predation in these rocky intertidal communities. These data provide mechanistic insight into how temperature change impacts an important species interaction. Such information lies at the heart of individual-based and physiologically structured models (Dunham 1993, Kingsolver et al. 1993, Murdoch 1993). These models seek to predict how the influence of environmental change on the physiology of individual organisms translates into population- or community-level effects. Knowledge of how temperature changes influence the *Pisaster-Mytilus* interaction should thus facilitate specific predictions about how anticipated climatic changes would impact these rocky intertidal communities.

There is however a major obstacle to this and similar global change studies; physical models used to predict climatic changes are resolved to a scale far broader than the relatively local scale of ecological studies (Root and Schneider 1993). General circulation models (GCMs) represent the globe as a grid of boxes and use laws of atmospheric physics to predict average conditions in each box. Although improvements have been made, computational limits still restrict GCMs to using boxes that are many kilometers on each side. This scale is too coarse to accurately predict changes in regional climate (Root and Schneider 1993).

Even more uncertain are changes in the frequency, timing, and intensity of seasonal or episodic events such as coastal upwelling, storms, frosts, heat waves, or fire (Root and Schneider 1993). Many researchers have suggested that it is changes in extreme or seasonal events, rather than changes in mean conditions, that may have the greatest effects on natural communities (Dobson et al. 1989, Kingsolver et al. 1993, Root and Schneider 1993, Bhaud et al. 1995). It is possible to generate some estimates of climatic variability from physical models, yet these data are rarely considered by climatic modelers (Root and Schneider 1993). There is thus little information available regarding the sort of variability that may be central to predicting the responses of ecosystems to climate change. For example, in the Pacific Northwest, the density and impact of *Pisaster* in the rocky intertidal zone is highest during the summer (the upwelling season). Thus it is changes in the frequency and intensity of upwelling, rather than changes in annual sea surface temperature, that should play the greater role in regulating sea star predation.

Predicting Changes in Upwelling Intensity

To what extent can we predict how global warming will alter patterns of upwelling along the Pacific coast of North America? Coastal upwelling is a complex process affected by atmospheric and oceanic conditions at local, regional, and hemispheric scales (Barber 1988). The intensity of upwelling (i.e., the temperature of upwelled waters) and its frequency (i.e., the duration of events and number of events per season) varies naturally from year to year and among different sites along the coast. Variation in upwelling is driven primarily by two factors: the strength of alongshore winds and the depth of the thermocline (i.e., the depth below which water temperature is uniformly cold). Understanding how global warming may influence these two factors is the key to understanding how upwelling patterns may be changed.

Alongshore Wind Stress

Unfortunately, current GCMs lack the resolution to accurately estimate regional winds. Present estimates of wind fields generally underestimate historical upwelling indices by a factor of 10 to 20 (Harte et al. 1992). Nevertheless, an understanding of the forces that drive coastal winds suggests that global warming may increase alongshore wind stress. Bakun (1990) argued that global warming may intensify the daytime heating of continental land masses, which should accentuate the atmospheric pressure gradient between the continental low-pressure cell and the higher pressure cell over the cooler ocean. As this gradient strengthens, alongshore wind stress, and consequently coastal upwelling, should intensify. Wind observations from ships at sea suggested that alongshore wind stress off the Northern California coast increased over the period from 1946 to 1988 (Bakun 1990). A more detailed analysis of wind data during this period indicated that a trend of increasing seasonal wind stress was apparent from the U.S.–Mexico border to Cape Mendocino (32–40°N), and also along the central Oregon coast (45°N) (Schwing and Mendelssohn 1997).

Depth of the Thermocline

Coastal upwelling is also strongly influenced by the depth of the thermocline. As the depth of the thermocline increases, a given wind stress draws upwelled waters from shallower, warmer layers, which are typically nutrient depleted (Roemmich and McGowan 1995). In recent decades, there is evidence of a deepened thermocline along at least some regions of the Pacific coast of North America. During the period from 1951 to 1993, surface waters off of Southern California warmed about 1.5°C and the depth of the thermocline increased. During this same period, the biomass of macrozooplankton in this region decreased by an astonishing 70%, presumably because of damped upwelling that reduced nutrient input and primary productivity (Roemmich and McGowan 1995).

The depth of the thermocline is influenced by processes operating at the scale of the entire Pacific basin (Barber 1988). Under normal conditions, the trade winds blow westward across the tropical Pacific and pile up warm surface water in the western Pacific. As a result, the depth of the thermocline is tilted across the Pacific basin, being deep in the western Pacific and shallow in the eastern Pacific, along the margin of North and South America. During El Niño–Southern Oscillation (ENSO) events, the trade winds relax, and the warm-water pool expands eastward. This has the effect of reducing the basin-wide tilt in the thermocline and deepening the thermocline in the eastern Pacific (Barber 1988).

As a result of the deeper thermocline, the frequency and intensity of cold-water upwelling are generally reduced during ENSO years. These changes were apparent along the central Oregon coast during the recent 1997–98 El Niño. During 1997, only 6.3% of May–August high tides fell below 9.5°C, compared to 36.7% of high tides during this same period in 1996 (a non-ENSO year) (Sanford 1999b). Nutrient input into the euphotic zone (and resultant primary productivity) are also typically reduced during El Niño years, leading to well-documented disruptions of marine food webs in upwelling regions (Glynn 1988).

The depth of the thermocline may also be influenced by poorly understood regime shifts in the northeastern Pacific (Lange et al. 1990, Roemmich and McGowan 1995). There appears to be a cyclic warming of the California Current and deepening of the thermocline that occurs on an interdecadal scale, with alternating states lasting roughly 20 to 50 years. A warm phase persisted in the California Current from 1976 until at least the late 1990s and appeared to be associated with a southward shift and intensification of the Aleutian Low. Earlier explanations hypothesizing switches in the relative strength of currents flowing into the California Current versus the Gulf of Alaska do not appear to have physical support (McGowan et al. 1998). Thus it remains unclear what factors trigger these interdecadal regime shifts or whether they are related to ENSO events (Trenberth and Hurrell 1994, McGowan et al. 1998).

Upwelling and Climate Change

Some models indicate that global warming may produce persistent El Niño–like conditions (Meehl and Washington 1996), which would depress the thermocline in the eastern Pacific and weaken coastal upwelling (Peterson et al. 1993). El Niño events have become more frequent and intense in recent decades, perhaps as a result of global climate change (Trenberth and Hoar 1996, 1997). However, since GCMs are currently unable to accurately reproduce ENSO events, it is not possible to definitively link changes in ENSO behavior to global climate change (Trenberth and Hoar 1997).

It is possible that increased daytime heating of the interior valleys along the Pacific coast may increase alongshore wind stress (Bakun 1990), whereas changes in the Pacific Ocean basin may depress the thermocline. The net effect of such changes would be an increase in the *volume* of upwelled water, but the water would be drawn from shallower, warmer layers (Petersen et al. 1993). Available physical data are consistent with this prediction. Seasonal wind stress (April–July) during the period from 1946 to 1990 increased

over the region 32 to 40°N. However, since at least 1970, sea surface temperatures during the upwelling season (April–July) have
generally *increased* over the entire region from 23 to 47°N (Schwing
and Mendelssohn 1997).

Although upwelling patterns in the California Current appear to
be undergoing dramatic changes, it is uncertain whether these
changes are the result of a natural regime shift, changing patterns of
ENSO events, anthropogenic global warming, or some combination
of these phenomena. At present, it is difficult to predict how climate
change will alter patterns of coastal upwelling. However, since the
physical mechanics of upwelling are sensitive to atmospheric and
oceanic conditions, it is likely that climatic disruptions will generate
significant changes in upwelling patterns. These changes will probably be complex and vary among regions.

Potential Changes in Rocky Intertidal Communities

How might rocky intertidal communities along the central Oregon
coast be impacted by changes in the frequency and intensity of
upwelling? My results suggest that a long-term reduction in cold-
water upwelling would cause *Pisaster* feeding rates to remain consistently high rather than fluctuating. In this case, sustained sea star
predation might diminish the vertical extent of the mid-zone mussel beds by shifting the lower limit of these beds to a higher level on
the shore. Since mussel beds provide habitat and shelter for a diversity of species (e.g., crabs, worms, sea cucumbers, and other invertebrates; Suchanek 1979), reducing the areal extent of mussel beds
might also reduce the abundance of these infaunal species.

In addition, increased rates of *Pisaster* predation in the low zone
at these study sites would decrease the abundance of preferred mussel prey, which might increase sea star effects on prey that are currently a minor part of *Pisaster*'s diet (such as acorn barnacles,
limpets, chitons, and whelks; see Navarrete and Menge 1996, Sanford 1999a for diet information). These changes might, in turn, trigger other indirect effects in the community. For example, decreases
in the abundance of herbivorous grazers like limpets and chitons,
along with a reduction in the density of sessile space-holders such
as barnacles, might increase the abundance of low zone algae.

Alternatively, if cold-water upwelling increased, *Pisaster* predation should decrease, potentially allowing the competitively domi-

nant mussel to move down the shore, overgrowing and smothering sessile organisms in the low intertidal zone. As demonstrated by Paine (1966, 1969, 1974), this process might dramatically decrease the diversity of species occupying low zone rock surfaces.

Although these changes are plausible responses to altered upwelling patterns, it is telling that even in this well-studied community, it is difficult to make detailed predictions about community-level changes. This coarse level of resolution may be typical of attempts to predict community-level responses to climate change. The surprising sensitivity of *Pisaster* predation to water temperature suggests that substantial community changes will occur, but provides only general ideas about the specifics of these changes. Moreover, the effects of altered species interactions will be superimposed on the direct impacts of temperature change on intertidal organisms. These direct effects are poorly understood but are likely to vary among species as a result of underlying physiological differences (see earlier section, "Temperature and Species' Geographic Ranges"). An additional unknown factor is how physiological acclimatization and local adaptation to persistent changes in temperature regime might influence the behavior of sea stars and other organisms (Fields et al. 1993).

Despite such complexities, examining the response of key species interactions to temperature change should help identify ecosystems that are particularly vulnerable to climate change. Mechanistic field and laboratory studies, such as those presented in this chapter, are one approach to evaluating the short-term effects of environmental changes on species interactions. Studies quantifying biotic responses to perturbations such as El Niño events may also provide valuable insight (Lubchenco et al. 1993, Paine 1993). Systems with key species interactions that are relatively insensitive to temperature change may have patterns of species composition and diversity that are stable in the face of climatic changes. In contrast, in systems where key interactions are sensitive to temperature, even small climatic changes may effect surprisingly large community changes.

Conclusion

There is a strong need to incorporate the potential impacts of altered species interactions into evaluations of how climate change may affect wildlife and natural systems. From a conservation standpoint, management efforts are unlikely to succeed if they focus solely on ameliorating the direct effects of climate change on an isolated target species. Although ecologists have acknowledged the potential importance of altered species interactions, little progress has been made in addressing these complex effects.

A reasonable strategy is to focus on key species interactions as pathways with a high capacity to produce changes in natural systems. Community ecology offers experimental approaches that can identify key interactions and quantify their strength and sensitivity to changes in environmental factors. In this study, variation in the strength of the sea star–mussel interaction was linked to the frequency and intensity of seasonal upwelling, a process that is highly variable among years and extremely sensitive to atmospheric and oceanographic conditions. The coupling of a temperature-sensitive keystone interaction and a climatically sensitive seasonal process may be a strong indicator of future community changes.

Despite this insight, it is difficult to predict specific changes that may arise in these intertidal communities as a result of global warming. In part, this is because of uncertainty in modeling changes in upwelling patterns. However, predicting exactly how changes in one species may be transmitted directly and indirectly through species networks may ultimately be the greater obstacle. Nevertheless, testing the response of key species interactions to environmental change should help identify ecosystems that are particularly vulnerable to climate change. The results of this study suggest that impacts of climate change on wildlife and ecosystems may be more immediate and complex than is generally realized. Conservation efforts in an era of climate change will require a community-level approach that considers target species within the context of changing species interactions.

Acknowledgments

I am grateful to my advisers, Bruce Menge and Jane Lubchenco, for their invaluable advice during all phases of this work. I thank Laura C. Ryan for field assistance and enthusiastic support, and Philip Sanford for the design and manufacture of the temperature controllers. This chapter was greatly improved by constructive comments from M. Bracken, J. Burnaford, M. Denny, B. Grantham, G. Leonard, J. Lubchenco, B. Menge, F. Saavedra, R. Sagarin, G. Somero, and an anonymous reviewer. L. Weber generously provided laboratory space at Hatfield Marine Science Center. I thank P. Glick for coordinating the National Wildlife Federation Climate Change Program, and S. Schneider and T. Root for editing this volume. In addition to funding from the National Wildlife Federation, this research was supported by a National Science Foundation Predoctoral Fellowship, and the Lerner-Gray Fund for Marine Research, as well as NSF grants to B. Menge, and funds provided to J. Lubchenco and B. Menge by the Andrew W. Mellon and Wayne and Gladys Valley Foundations.

Literature Cited

Aleksiuk, M. 1976. Metabolic and behavioural adjustments to temperature change in the red-sided garter snake (*Thamnophis sirtalis parietalis*): An integrated approach. *Journal of Thermal Biology* 1: 153–156.

Andrewartha, H. G., and L. C. Birch. 1954. *The Distribution and Abundance of Animals*. Chicago: University of Chicago Press.

Ayres, M. P. 1993. Plant defense, herbivory, and climate change. Ch. 6 in P. M. Kareiva, J. G. Kingsolver, and R. B. Huey, eds., *Biotic Interactions and Global Change*. Sunderland, Mass.: Sinauer.

Bakun, A. 1990. Global climatic change and the intensification of coastal upwelling. *Science* 247: 198–201.

Barber, R. T. 1988. Open basin ecosystems. Ch. 9 in L. R. Pomeroy and J. L. Alberts, eds., *Concepts of Ecosystem Ecology: A Comparative View*. Ecological Studies 67. New York: Springer-Verlag.

Barry, J. P., C. H. Baxter, R. D. Sagarin, S. E. Gilman. 1995. Climate-related, long-term faunal changes in a California rocky intertidal community. *Science* 267: 672–675.

Bell, E. C., and M. W. Denny. 1994. Quantifying "wave exposure": A simple device for recording maximum velocity and results of its use at several field sites. *Journal of Experimental Marine Biology and Ecology* 181: 19–29.

Bertness, M. D., and R. Callaway. 1994. Positive interactions in communities. *Trends in Ecology and Evolution* 9: 191–193.

Bhaud, M., J. H. Cha, J. C. Duchêne, and C. Nozais. 1995. Influence of temperature on the marine fauna: what can be expected from a climatic change? *Journal of Thermal Biology* 20: 91–104.

Birkland, C., and J. S. Lucas. 1990. *Acanthaster planci: Major Management Problem of Coral Reefs*. Boca Raton, Fla.: CRC Press.

Breeman, A. M. 1990. Expected effects of changing seawater temperatures on the geographic distribution of seaweed species. Pages 69–76 in J. J. Beukema, W. J. Wolff, and J. Brouns, eds., *Expected Effects of Climatic Change on Marine Coastal Ecosystems.* Dordrecht: Kluwer.

Brown, J. H., and E. J. Heske. 1990. Control of desert-grassland transition by a keystone rodent guild. *Science* 250: 1705–1707.

Brown, J. H., T. J. Valone, and C. G. Curtin. 1997. Reorganization of an arid ecosystem in response to recent climate change. *Proceedings of the National Academy of Sciences of the United States of America* 94: 9729–9733.

Bruno, J. F., and M. D. Bertness. 2001. Habitat modification and facilitation in benthic marine communities. Ch. 8 in M. D. Bertness, S. D. Gaines, and M. E. Hay. *Marine Community Ecology.* Sunderland, Mass.: Sinauer.

Carpenter, S. R., T. M. Frost, J. F. Kitchell, and Timothy K. Kratz. 1993. Species dynamics and global environmental change: A perspective from ecosystem experiments. Ch. 16 in P. M. Kareiva, J. G. Kingsolver, and R. B. Huey, eds., *Biotic Interactions and Global Change.* Sunderland, Mass.: Sinauer.

Castilla, J. C., S. A. Navarrete, and J. Lubchenco. 1993. Southeastern Pacific coastal environments: Main features, large-scale perturbations, and global climate change. Ch. 13 in H. A. Mooney, E. R. Fuentes, B. I. Kronberg, eds., *Earth Systems Responses to Global Change: Contrast between North and South America.* New York: Academic Press.

Caulton, M. S. 1982. Feeding, metabolism and growth of tilapias: Some quantitative considerations. Pages 157–180 in R. V. S. Pullian and R. Lowe-McConnell, eds., *The Biology and Culture of Tilapias.* ICLARM Conference Proceeding 7. Manila: ICLARM.

Chapin, F. S., and G. R. Shaver. 1985. Individualistic growth responses of tundra plant species to environmental manipulations in the field. *Ecology* 66(2): 564–576.

————. 1996. Physiological and growth responses of arctic plants to a field experiment simulating climatic change. *Ecology* 77(3): 822–840.

Chapin, F. S., G. R. Shaver, A. E. Giblin, K. J. Nadelhoffer, and J. A. Laundre. 1995. Responses of arctic tundra to experimental and observed changes in climate. *Ecology* 76(3): 694–711.

Christian, K., C. R. Tracy, and W. P. Porter. 1983. Seasonal shifts in body temperature and use of microhabitats by Galapagos land iguanas (*Conolophus pallidus*). *Ecology* 64(3): 463–468.

Clark, A. 1987. The adaptation of aquatic animals to low temperatures. Pages 315–348 in B. W. W. Grout and G. J. Morris, eds., *The Effects of Low Temperatures on Biological Systems.* London: Edward Arnold.

Connell, J. H. 1970. A predator–prey system in the marine intertidal habitat. I. *Balanus glandula* and several species of *Thais. Ecological Monographs* 40: 49–87.

Cossins, A. R., and K. Bowler. 1987. *Temperature Biology of Animals.* New York: Chapman and Hall.

Davis, A. J., L. S. Jenkinson, J. H. Lawton, B. Shorrocks, and S. Wood. 1998a. Global warming, population dynamics and community structure in a model insect assemblage. Ch. 18 in R. Harrington and N. E. Stork, eds., *Insects in a Changing Environment.* San Diego: Academic Press.

Davis, A. J., J. H. Lawton, B. Shorrocks, and L. S. Jenkinson. 1998b. Individualistic species responses invalidate simple physiological models of community dynamics under global environmental change. *Journal of Animal Ecology* 67: 600–612.

Davis, M. B., and C. Zabinski. 1992. Changes in geographical range resulting from greenhouse warming: Effects on biodiversity in forests. Ch. 22 in R. L. Peters and T. E. Lovejoy, eds., *Global Warming and Biological Diversity.* New Haven: Yale University Press.

Dawson, W. R. 1992. Physiological responses of animals to higher temperatures. Ch. 12 in R. L. Peters and T. E. Lovejoy, eds., *Global Warming and Biological Diversity.* New Haven: Yale University Press.

Dayton, P. K. 1971. Competition, disturbance, and community organization: The provision and subsequent utilization of space in a rocky intertidal community. *Ecological Monographs* 41: 351–389.

Dobson, A., A. Jolly, and D. Rubenstein. 1989. The greenhouse effect and biological diversity. *Trends in Ecology and Evolution* 4: 64–68.

Dunham, A. E. 1993. Population responses to environmental change: Operative environments, physiologically structured models, and population dynamics. Ch. 7 in P. M. Kareiva, J. G. Kingsolver, and R. B. Huey, eds., *Biotic Interactions and Global Change.* Sunderland, Mass.: Sinauer.

Dunson, W. A., and J. Travis. 1991. The role of abiotic factors in community organization. *American Naturalist* 138(5): 1067–1091.

Fagan, W. F., and L. E. Hurd. Hatch density variation of a generalist arthropod predator: Population consequences and community impact. *Ecology* 75: 2022–2032.

Feder, H., and A. M. Christensen. 1966. Aspects of asteroid biology. Ch. 5 in R. A. Boolootian, ed., *Physiology of Echinodermata.* New York: Interscience.

Fields, P. A., J. B. Graham, R. H. Rosenblatt, and G. N. Somero. 1993. Effects of expected global climate change on marine faunas. *Trends in Ecology and Evolution* 8: 361–367.

Foster, D. R., Schoonmaker, P. K., and S. T. A. Pickett. 1990. Insights from paleoecology to community ecology. *Trends in Ecology and Evolution* 5: 119–122.

Frazer, B. D., and N. Gilbert. 1976. Coccinellids and aphids: A quantitative study of the impact of adult ladybirds preying on field populations of pea aphids. *Journal of the Entomological Society of British Columbia* 73: 33–56.

Gates, D. M. 1993. *Climate Change and Its Biological Consequences*. Sunderland, Mass.: Sinauer.

Glynn, P. W. 1988. El Niño–Southern Oscillation 1982–1983: Nearshore population, community, and ecosystem responses. *Annual Review of Ecology and Systematics* 19: 309–345.

Graham, R. W. 1992. Late Pleistocene faunal changes as a guide to understanding effects of greenhouse warming on the mammalian fauna of North America. Ch. 6 in R. L. Peters and T. E. Lovejoy, eds., *Global Warming and Biological Diversity*. New Haven: Yale University Press.

Harte, J., and R. Shaw. 1995. Shifting dominance within a montane vegetation community: results of a climate-warming experiment. *Science* 267: 867–880.

Harte, J., M. Torn, and D. Jensen. 1992. The nature and consequences of indirect linkages between climate change and biological diversity. Ch. 24 in R. L. Peters and T. E. Lovejoy, eds., *Global Warming and Biological Diversity*. New Haven: Yale University Press.

Henry, G. H. R., and U. Molau. 1997. Tundra plants and climate change: The International Tundra Experiment (ITEX). *Global Change Biology* 3 (Suppl. 1): 1–9.

Holbrook, S. J., R. J. Schmitt, and J. S. Stephens, Jr. 1997. Changes in an assemblage of temperate reef fishes associated with a climate shift. *Ecological Applications* 7(4): 1299–1310.

Huey, R. B., and J. G. Kingsolver. 1989. Evolution of thermal sensitivity of ectotherm performance. *Trends in Ecology and Evolution* 4: 131–135.

Huey, R. B., and Stevenson, R. D. 1979. Integrating thermal physiology and ecology of ectotherms: A discussion of approaches. *American Zoologist* 19: 357–366.

Hunt, G. L. 1977. Foraging in a desert harvester ant. *American Naturalist* 111: 589–591.

Hunter, M. D., and P. W. Price. 1992. Playing chutes and ladders: Heterogeneity and the relative roles of bottom-up and top-down forces in natural communities. *Ecology* 73(3): 724–732.

Hutchins, L. W. 1947. The bases for temperature zonation in geographical distribution. *Ecological Monographs* 17(3): 325–335.

Intergovernmental Panel on Climate Change (IPCC). 2001. *Climate Change 2001: The Scientific Basis*, eds. J. T. Houghton, Y. Ding, D. J. Griggs, M. Noguer, P. J. van der Linden, and D. Xiaosu. Contribution of working group I to the third assessment report of the Intergovernmental Panel on Climate Change. Cambridge: Cambridge University Press.

Johnston, K. M., and O. J. Schmitz. 1997. Wildlife and climate change: Assessing the sensitivity of selected species to simulated doubling of atmospheric CO_2. *Global Change Biology* 3: 531–544.

Kareiva, P. M., J. G. Kingsolver, and R. B. Huey, eds. 1993. *Biotic Interactions and Global Change*. Sunderland, Mass.: Sinauer.

Kingsolver, J. G. 1989. Weather and population dynamics of insects: Integrating physiological and population ecology. *Physiological Zoology* 62(2): 314–334.

Kingsolver, J. G., R. B. Huey, and P. M. Kareiva. 1993. An agenda for population and community research on global change. Ch.29 in P. M. Kareiva, J. G. Kingsolver, and R. B. Huey, eds., *Biotic Interactions and Global Change*. Sunderland, Mass.: Sinauer.

Lange, C. B., S. K. Burke, and W. H. Berger. 1990. Biological production off Southern California is linked to climatic change. *Climatic Change* 16: 319–329.

Lévêque, C. 1997. *Biodiversity Dynamics and Conservation: The Freshwater Fish of Tropical Africa*. Cambridge: Cambridge University Press.

Loeb, V., V. Siegel, O. Holm-Hansen, R. Hewitt, W. Fraser, W. Trivelpiece, and S. Trivelpiece. 1997. Effects of sea-ice extent and krill or salp dominance on the Antarctic food web. *Nature* 387: 897–900.

Lubchenco, J., A. Olson, L. Brubaker, S. Carpenter, M. Holland, S. Hubbell, S. Levin, J. MacMahon. P. Matson, J. Melillo, H. Mooney, C. Petersen, R. Pulliam, L. Real, P. Regal, P. Risser. 1991. The sustainable biosphere initiative: An ecological research agenda. *Ecology* 72(2): 371–412.

Lubchenco, J., S. A. Navarrete, B. Tissot, J. C. Castilla. 1993. Possible ecological responses to global climatic change: Nearshore benthic biota of Northeastern Pacific coastal ecosystems. Ch. 12 in H. A. Mooney, E. R. Fuentes, B. I. Kronberg, eds., *Earth Systems Responses to Global Change: Contrast between North and South America*. New York: Academic Press.

Mann, M. E., R. S. Bradley, and M. K. Hughes. 1998. Global-scale temperature patterns and climate forcing over the past six centuries. *Nature* 392: 779–787.

Mauzey, K.P. 1966. Feeding behavior and reproductive cycles in *Pisaster ochraceus*. *Biological Bulletin* 131: 127–144.

McGowan, J. A., D. R. Cayan, and L. M. Dorman. 1998. Climate-ocean variability and ecosystem response in the Northeast Pacific. *Science* 281: 210–217.

Meehl, G. A., and W. M. Washington. 1996. El Niño–like climate change in a model with increased atmospheric CO_2 concentrations. *Nature* 382: 56–60.

Menge, B. A. 1992. Community regulation: Under what conditions are bottom-up factors important on rocky shores? *Ecology* 73: 755–765.

Menge, B. A., and A. M. Olson. 1990. Role of scale and environmental factors in regulation of community structure. *Trends in Ecology and Evolution* 5: 52–57.

Menge, B. A., and J. P. Sutherland. 1976. Species diversity gradients: Synthesis of the roles of predation, competition, and temporal heterogeneity. *American Naturalist* 110: 351–369.

———. 1987. Community regulation: Variation in disturbance, competition, and predation in relation to environmental stress and recruitment. *American Naturalist* 130: 730–757.

Menge, B. A., E. L. Berlow, C. A. Blanchette, S. A. Navarrete, and S. B. Yamada. 1994. The keystone species concept: Variation in interaction strength in a rocky intertidal habitat. *Ecological Monographs* 64(3): 249–286.

Menge, B. A., B. A. Daley, P. A. Wheeler, E. Dahlhoff, E. Sanford, and P. T. Strub. 1997. Benthic–pelagic links in rocky intertidal communities: Bottom-up effects on top-down control? *Proceedings of the National Academy of Sciences of the United States of America* 94: 14530–14535.

Murdoch, W. M. 1993. Individual-based models for predicting effects of global change. Ch. 9 in P. M. Kareiva, J. G. Kingsolver, and R. B. Huey, eds., *Biotic Interactions and Global Change*. Sunderland, Mass.: Sinauer.

Navarrete, S. A., and B. A. Menge. 1996. Keystone predation and interaction strength: Interactive effects of predators on their main prey. *Ecological Monographs* 66(4): 409–429.

Nobel, P. S. 1980. Morphology, surface temperatures, and northern limits of columnar cacti in the Sonoran Desert. *Ecology* 61: 1–7.

Orton, J. H. 1920. Sea temperature, breeding, and distribution in marine animals. *Journal of the Marine Biological Association of the United Kingdom* 12: 339–366.

Pacala, S. W., and G. C. Hurtt. 1993. Terrestrial vegetation and climate change: Integrating models and experiments. Ch.5 in P. M. Kareiva, J. G. Kingsolver, and R. B. Huey, eds., *Biotic Interactions and Global Change*. Sunderland, Mass.: Sinauer.

Paine, R. T. 1966. Food web complexity and species diversity. *American Naturalist* 100: 65–76.

———. 1969. A note on trophic complexity and community stability. *American Naturalist* 103: 91–93.

———. 1974. Intertidal community structure: Experimental studies on the relationship between a dominant competitor and its principal predator. *Oecologia* 15: 93–120.

———. 1976. Size-limited predation: An observational and experimental approach with the *Mytilus–Pisaster* interaction. *Ecology* 57: 858–873.

———. 1980. Food webs: Linkage, interaction strength, and community infrastructure. *Journal of Animal Ecology* 49: 667–685.

———. 1992. Food-web analysis through field measurement of per-capita interaction strength. *Nature* 355: 73–75.

———. 1993. A salty and salutary perspective on global change. Ch. 21 in

P. M. Kareiva, J. G. Kingsolver, and R. B. Huey, eds., *Biotic Interactions and Global Change*. Sunderland, Mass.: Sinauer.

———. 1994. *Marine Rocky Shores and Community Ecology: An Experimentalist's Perspective*. Oldendorf/Lueh, Germany: Ecology Institute.

Park, T. 1954. Experimental studies of interspecies competition. II. Temperature, humidity and competition in two species of *Tribolium*. *Physiological Zoology* 27: 177–238.

Parmesan, C. 1996. Climate and species' range. *Nature* 382: 765–766.

Parmesan, C., N. Ryrholm, C. Stefanescu. J. K. Hill, C. D. Thomas, H. Descimon, B. Huntley, L. Kaila, J. Kullberg, T. Tammaru, W. J. Tennent, J. A. Thomas, and M. Warren. 1999. Poleward shifts in geographic ranges of butterfly species associated with regional warming. *Nature* 399: 579–583.

Peters, R. L., and T. E. Lovejoy, eds. 1992. *Global Warming and Biological Diversity*. New Haven: Yale University Press.

Peterson, C. H., R. T. Barber, G. A. Skilleter. 1993. Global warming and coastal ecosystem response: How Northern and Southern hemispheres may differ in the eastern Pacific ocean. Ch. 2 in H. A. Mooney, E. R. Fuentes, B. I. Kronberg, eds., *Earth Systems Responses to Global Change: Contrast between North and South America*. New York: Academic Press.

Pollack, H. N., Huang, S., and P. Shen. 1998. Climate change record in subsurface temperatures: A global perspective. *Science* 282: 279– 281.

Post, E., R. O. Peterson, N. C. Stenseth, B. E. McLaren. 1999. Ecosystem consequences of wolf behavioural response to climate. *Nature* 401: 905–907.

Pounds, J. A., M. P. L. Fogden, and J. H. Campbell. 1999. Biological response to climate change on a tropical mountain. *Nature* 398: 611– 615.

Power, M. E., D. Tilman, J. Estes, B. A. Menge, W. J. Bond, L. S. Mills, G. Daily, J. C. Castilla, J. Lubchenco, R. T. Paine. 1996. Challenges in the quest for keystones. *Bioscience* 46(8): 609–620.

Robles, C., R. Sherwood-Stephens, and M. Alvarado. 1995. Responses of a key intertidal predator to varying recruitment of its prey. *Ecology* 76(2): 565–579.

Roemmich, D., and J. McGowan. 1995. Climatic warming and the decline of zooplankton in the California Current. *Science* 267: 1324–1326.

Rogers, D. J., and S. E. Randolph. 1993. Distribution of tsetse and ticks in Africa: Past, present, and future. *Parisitology Today* 9: 266–271.

Root, T. L. 1988. Environmental factors associated with avian distributional boundaries. *Journal of Biogeography* 15: 489–505.

Root, T. L., and S. H. Schneider. 1993. Can large-scale climatic models be linked with multiscale ecological studies? *Conservation Biology* 7(2): 256–270.

Rubenstein, D. I. 1992. The greenhouse effect and changes in animal behavior: Effects on social structure and life-history strategies. Ch. 14 in R. L. Peters and T. E. Lovejoy, eds., *Global Warming and Biological Diversity*. New Haven: Yale University Press.

Sagarin, R. D., Barry, J. P., Gilman, S. E., and C. H. Baxter. 1999. Climate-related change in an intertidal community over short and long time scales. *Ecological Monographs* 69: 465–490.

Sanford, E. 1999a. Regulation of keystone predation by small changes in ocean temperature. *Science* 283: 2095–2097.

———. 1999b. Oceanographic Influences on Rocky Intertidal Communities: Coastal Upwelling, Invertebrate Growth Rates, and Keystone Predation. Ph.D. Dissertation, Oregon State University.

Sætersdal, M., and H. J. B. Birks. 1997. A comparative ecological study of Norwegian mountain plants in relation to possible future climatic change. *Journal of Biogeography* 24: 127–152.

Schwing, F. B., and R. Mendelssohn. 1997. Increased coastal upwelling in the California Current System. *Journal of Geophysical Research* 102: 3421–3438.

Scott, D., and M. Poynter. 1991. Upper temperature limits for trout in New Zealand and climate change. *Hydrobiologia* 222: 147–151.

Shelford, V. E. 1913. The reactions of certain animals to gradients of evaporating power and air: A study in experimental ecology. *Biological Bulletin* 25: 79–120.

Stickle, W. B., and W. J. Diehl. 1987. Effects of salinity on echinoderms. Pages 235–285 in M. Jangoux and J. M. Lawrence, eds., *Echinoderm Studies II*, Rotterdam: A. A. Balkema.

Suchanek, T. H. 1979. The *Mytilus californianus* Community: Studies on the Composition, Structure, Organization, and Dynamics of a Mussel Bed. Ph.D. Dissertation, University of Washington.

Sundene, O. 1962. The implications of transplant and culture experiments on the growth and distribution of *Alaria esculenta*. *Nytt Magasin for Botanikk* 9: 155–174.

Tansley, A. G. 1917. On competition between *Galium saxatile* and *Galium sylvestre* on different types of soil. *Journal of Ecology* 5: 173–179.

Thomas, C. D., and J. J. Lennon. 1999. Birds extend their ranges northward. *Nature* 399: 213.

Tilman, D. 1980. Resources: A graphical-mechanistic approach to competition and predation. *American Naturalist* 116: 362–93.

Tomanek, L., and G. N. Somero. 1999. Evolutionary and acclimation-induced variation in the heat-shock responses of congeneric marine snails (genus *Tegula*) from different thermal habitats: Implications for limits of thermotolerance and biogeography. *Journal of Experimental Biology* 202: 2925–2936.

Trenberth, K. E., and T. J. Hoar. 1996. The 1990–1995 El Niño–Southern Oscillation event: Longest on record. *Geophysical Research Letters* 23(1): 57–60.

———. 1997. El Niño and climate change. *Geophysical Research Letters* 24(23): 3057–3060.

Trenberth, K. E., and J. W. Hurrell. 1994. Decadal atmosphere–ocean variations in the Pacific. *Climate Dynamics* 9: 303–319.

Webb, T. 1992. Past changes in vegetation and climate: Lessons for the future. Ch. 5 in R. L. Peters and T. E. Lovejoy, eds., *Global Warming and Biological Diversity*. New Haven: Yale University Press.

Whicker, A. D., and C. R. Tracy. 1987. Tenebrionid beetles in the shortgrass prairie: Daily and seasonal patterns of activity and temperature. *Ecological Entomology* 12: 97–108.

Woodward, F. I. 1987. *Climate and Plant Distribution*. Cambridge: Cambridge University Press.

CHAPTER 5

Testing Climate Change Predictions with the Subalpine Species *Delphinium nuttallianum*

FRANCISCA SAAVEDRA

Changes in climate will have an impact on community structure and composition by affecting species physiology, distribution, and phenology, which will further lead to changes in species interactions. Some species, especially those with short generation times and rapid population growth, might be able to adapt. However, since climate change is expected to occur at a rate to which many species might not be able to adapt, or shift, we might expect extinction of some species (Root 1993, Hughes 2000) and consequently a rearrangement of communities. A recent review of long-term data by Hughes (2000) showed that indeed species in different habitats are responding to climate change. These changes might lead, among other things, to a progressive decoupling of species interactions (e.g., plants and pollinators) and to an increase in invasion by more opportunistic, weedy, or highly mobile species to sites where local populations are going extinct. Ultimately, these changes can lead to communities with low biodiversity and consequently less resistance to perturbation (e.g., droughts, Tilman and Downing 1994, Tilman 1996), according to the diversity–stability hypothesis (Elton 1958).

Studies of species' responses to climate change provide the necessary knowledge to determine which species and communities are most sensitive to climate change. Forecasting the possible consequences of climate change should help to implement measures to mitigate some of the undesirable consequences (e.g., species extinction). For example, it has been argued that conserving areas of high biodiversity/endemism (hotspots) should help to decrease projected rates of species extinctions (Myers et al. 2000). However, if those strategies do not take into account changes in climate, then saving those hotspots might not be enough to limit species extinction in the long-term.

The anthropogenic increase of atmospheric CO_2 will affect plant species directly and indirectly. Plant responses to the direct effects of an increase of CO_2 per se have been widely studied (Bazzaz 1990), but the indirect effects associated with an increase in CO_2, including changes in temperature, precipitation, and snow cover, have been explored less frequently. My purpose in this chapter is to explore an indirect effect of climate change, in the form of changes in snow cover, on the reproductive biology of *Delphinium nuttallianum*, a subalpine perennial species of the Rocky Mountains. Snow was removed experimentally to determine the effects of snow accumulation on the timing of flowering (phenology) of *D. nuttallianum* and to show whether that change has any fitness consequences. In addition, observations of natural populations and of individuals that flower at different times broadened our understanding of the fitness consequences of natural variation in flowering time.

This research will help to predict the fates of some herbaceous subalpine plant species under present and future environmental conditions. If shifts in the growing season occur, as predicted by global climate change models, we may be able to infer the fitness consequences of those changes for native subalpine species. Knowledge of the fitness consequences of a shift in phenology could be used to predict the sensitivity of specific native species to changes in growing season due to global climate change (Galen and Stanton 1991).

I will present here a general review of climate change and its predicted effects on alpine and arctic systems around the world. I also present a short history of climate change models to show the progress made in understanding global climate processes. Three biological questions will be addressed in this chapter. First, does snowmelt date affect the flowering time of *D. nuttallianum?* Second, will some components of fitness in *D. nuttallianum* be affected by an early snowmelt? Finally, how does time of flowering affect fitness of *D. nuttallianum* in natural populations?

Climate Change Predictions

The greenhouse effect is a natural phenomenon. Trace gases in the atmosphere, such as water vapor (H_2O), methane (CH_4), and carbon dioxide (CO_2), can trap infrared (IR) radiation. As a consequence of these gases, the temperature of Earth's lower atmosphere is about 30°C above that which is expected without them (Rodhe et al. 1997). This natural greenhouse effect is not under debate. However, humans have increased this natural greenhouse effect by adding substantially to some of these trace gases (e.g., CO_2, CH_4, but not H_2O). The present debate revolves around the magnitude, timing, and especially the consequences (namely global warming) of the anthropogenic increase of these greenhouse gases (Schneider 1993).

Arrhenius (1896) was the first to predict an increase in levels of atmospheric CO_2 and to predict a corresponding effect on ground temperature. He proposed that the increase in CO_2, resulting from industrial coal burning, was leading to global warming. Indirect measures of atmospheric CO_2 concentrations from preindustrial times (1400 to the present, using ice core records) show that CO_2 levels have indeed increased by approximately 31% since the beginning of the industrial revolution (IPCC 2001a). Direct measurements of the CO_2 concentration since the late 1950s at Mauna Loa (Hawaii) and at the South Pole confirm this trend (Keeling et al. 1995). A parallel increase in mean global temperature of 0.6°C since the beginning of the century has occurred (IPCC 2001a). The observed pattern in temperature change is similar to that predicted by climate models and is very unlikely to be entirely a result of natural climate variation (IPCC 2001a, Santer et al. 1996, Tett et al. 1999). Atmospheric CO_2 levels will probably double from their preindustrial values within less than 100 years (Körner et al. 1996). A doubling of CO_2 and other greenhouse gases should trigger an average temperature increase of 1.4 to 5.6°C, and also should affect precipitation patterns, soil moisture, and snow and ice cover (IPCC 2001a). Specific predictions about changes in precipitation are uncertain. However, long-term studies have shown that mean precipitation has not increased in the last century, while fluctuations about this mean have increased significantly. This fact agrees with the global warming climate scenario, which predicts that higher temperature might result in intensification of the extremes of the hydrological cycle (Tsonis 1996, IPCC 2001a).

Forecasting the possible biological consequences of climate change could provide the guidance needed to prepare for disruptions in communities and ecosystems (Root 1993). The specific biological impacts of climate change are likely to be complex (Root and Schneider 1995), and a great challenge remains to reach some level of predictive power for these impacts. Biological responses to climate change are hard to model, in part, because ecologists tend to work at a much smaller scale than that of climatologists. This makes it hard to generate reliable predictions of broad-scale ecological responses to climate change (Root and Schneider 1995). Most recent climate projections are at a scale of 300×300 km grids. In contrast, most ecological studies are conducted on a local scale (e.g., 50×50 m). Consequently, climate models might provide predictions on a scale that is too coarse to be used to predict ecological phenomena (Schneider 1993).

A second difficulty in predicting biotic responses to climate change involves the anticipated complexity behind the responses of individual species to climate factors. A single component of climate change, global warming, could impact plant species in diverse ways by changing, for instance, rates of growth and the timing of growth (IPCC 2001b). These impacts can in turn affect interactions between plants and animals as well as among plant species, thereby affecting plant fitness, abundance, and distribution and leading to communitywide changes in the relative abundance of species (Bazzaz 1990, Chapin et al. 1995, Galen and Stanton 1995, Harte and Shaw 1995, Shaver et al. 1998, Price and Waser 1998, 2000).

Most projections suggest that the climate will change substantially in a few decades while the response of many species to climate might occur over a time scale of centuries. Accordingly, as the climate changes, many species will be maladapted, species' ranges may shift, and some species may go extinct (Root and Schneider 1993). However, evidence for the rapid evolution of insecticide resistance and heavy metal tolerance in some plant species, especially those with short generation times, suggests that plants could adapt rather quickly to environmental changes. This adaptation will perhaps be helped by reproductive isolation in plant populations. For example, McNeilly and Antonovics (1968) observed a divergence in flowering time between metal-tolerant plant populations on mine tailings and adjacent intolerant populations on normal soils. They interpreted this as a case of divergent selection driving reproductive isolation. It is possible that plant responses to climate change will be substantially slower. Adaptation to particular pollutants generally involves one or a few traits whereas adaptation to climate change is likely to involve many traits. Rates of evolution in response to selection on multiple codependent traits are likely to be lower than when a single trait is involved (Geber and Dawson 1993). On the other hand, existing genetic variation for responses to climate change might be larger than to pollutants since environmental variation is a common phenomenon. Another difficulty in making more reliable predictions about how species may respond to climate change is that heritabilities for many characters are probably low in nature (Stratton 1992, Kelly 1993, but see Mousseau and Roff 1987). Consequently, even if selection pressure is high, evolutionary changes might be small (Galloway 1995). However, even this argument is not clear. Mousseau and Roff (1987) reviewed heritabilities and reported that these generally are not very small in nature. Detailed species-level studies can help clarify whether the negative fitness impacts of climate change are sufficient to lead to the extinction or local displacement of plant and animal species.

Plant–pollinator interactions could also be negatively affected by climate. Bond (1995) hypothesized that the climate cues that trigger plant phenology and insect pollinator development could differ. If climate change affected these cues differently, plants and their pollinators might lose synchrony with each other, impacting community processes and the diversity of plants with more specialized pollinators. Inouye et al. (2000) suggested that a growing disjunction between low and high altitude phenologies might create problems for migratory species that reproduce in high altitude sites. They reported data for migratory robins (*Turdus migratorius*) that have been arriving earlier at high altitude sites in Colorado, whereas snowmelt and plant phenology events in these sites have not changed in the past 25 years. Hummingbirds, which are important pollinators in this high altitude site and are also migratory species, appear to have experienced a similar disjunction (Inouye, pers. comm. 2000).

Experimental Work Testing Plant Responses to Climate Change

There is a large body of literature on the responses of crops, forests, and natural vegetation to the direct effect of elevated CO_2 (Bazzaz 1990). However, as CO_2 increases, temperature, precipitation, snow cover, soil moisture, and mineralization can also change (Schneider 1993). This brief review addresses field experiments that have tested for possible responses in alpine and tundra plant communities or populations to the indirect effects of CO_2 increase, mainly changes in temperature, nutrients, snow cover, and light intensity.

Community-Level Responses

This section reviews field experiments that have tested the indirect effects of CO_2 increase at the community level in alpine and tundra habitats.

Alpine environment, Rocky Mountain Biological Laboratory (RMBL), Colorado (38°57′N, 106°59′W, elevation 2900 m): The effects of climate warming on subalpine meadows have been experimentally simulated by Harte and collaborators since 1990 (Harte and Shaw 1995, Harte et al. 1995, Loik and Harte 1996, Torn and Harte 1996, Saleska et al. 1999, Price and Waser 1998, 2000). Through the use of overhead electric heaters suspended above a subalpine meadow, IR flux has been increased to simulate surface soil warming values similar to that expected if atmospheric CO_2 doubles. Harte and collaborators observed a shift of approximately 10 days in snowmelt date (Harte et al. 1995, Price and Waser 1998), an increase in summer soil temperature of up to 3°C, a reduction of summer soil moisture levels by up to 25% compared to control plots (Harte et al. 1995), and earlier flowering by a representative suite of herbaceous plant species (Price and Waser 1998). Harte and Shaw (1995) suggested that species dominance may have shifted in response to experimental warming. They reported that aboveground biomass, shoot growth rate, and seedling survival of sagebrush (*Artemisia tridentata*) and biomass of shrubby cinquefoil (*Pentaphylloides floribunda*) was greater and aboveground biomass of forbs was lower in heated than control plots. Harte and Shaw (1995) concluded that warming may shift species dominance in this alpine meadow. Further, Saleska et al. (1999) suggested that these changes in plant aboveground biomass and the direct physiological plant response to a decrease in soil moisture could be driving a net loss in soil carbon storage in the warmed plots. However, more recent work (Price and Waser 2000) showed no evidence of community changes during a 5-year study in the same warming experiment. Price and Waser included a year of prewarming data and used a different statistical analysis than Harte and Shaw (1995); the inclusion of the year of prewarming makes the apparent treatment effect on vegetation disappear because the differences between warmed and control plots reported by Harte and Shaw (1995) were present before the heaters were activated. They concluded that vegetation cover at arid high altitude sites may change only slowly under global warming.

Tundra environment, Alaska (68°38'N, 149°34'W, 760 m elevation): Chapin and Shaver (1985) started a long-term study at this site in 1978. They manipulated light intensity, air temperature, and nutrient availability to determine which environmental factors limit growth. They concluded that the abundance of each plant species was limited by a unique combination of environmental conditions (i.e., there is no single factor limiting growth for all species). In 1981, Chapin and collaborators started a new three-factorial experiment to examine the responses of a moist tussock tundra community to manipulations of light, temperature, and nutrients and to determine how global changes in these parameters might affect community and ecosystem processes (Chapin et al. 1995). After a decade of manipulation they found that nitrogen and phosphorus availability tended to increase in response to both elevated temperature (reflecting increased mineralization) and light attenuation (reflecting reduced nutrient uptake by vegetation). Second, a major effect of elevated temperature was to speed plant responses to changes in soil resources, leading to increased nutrient availability through changes in nitrogen mineralization. Finally, species richness was reduced by temperature and nutrient treatment due to the loss of less abundant species. Their controls (unmanipulated plots) changed as well over the decade in the same direction as predicted by their experiments and by pollen records from the warm periods of the Holocene (Ritchie and Cwynar 1982).

Because the study site has been exposed to the highest temperature on record in the region, the change in the unmanipulated plots suggested that climate change is already affecting tundra vegetation. This experiment confirmed the earlier finding (Chapin and Shaver 1985) that the responses of plants to climatic changes are highly individualistic. Pollen records and model simulations of climate change of the last 18,000 years (which could be considered a natural experiment) also indicate that species' responses to climate change are highly individualistic, and that there are discrepancies between observed and simulated patterns; most of these discrepancies appear to reflect an absence of matching habitats, and other factors not included in the model, such as different terrestrial cooling rates in the tropics (COHMAP 1988).

Shaver et al. (1998) found similar responses on wet sedge tundra to those of the Chapin et al. (1995) study. However, they found that wet sedge tundra communities are phosphorus-limited and not nitrogen-limited as was the tussock tundra.

Tundra environment, Alaska (68°38′N, 149°34′W, 780 m elevation): Oberbauer and collaborators extended the growing season of a tundra plant community by shifting the duration of snow cover in summer 1995 and 1996 (Oberbauer et al. 1998). The experiment consisted of one treatment in which experimenters removed snow in the early season and prevented snow accumulation in the late season (extended season), a second treatment in which experimenters extended the season and increased the temperature, and a natural control. The results indicate that total carbon dioxide and methane exchange did not differ among treatments over the season, carbon being lost in all of them. Plant phenology differed between treatments; the extended season treatments triggered an early initiation of bud formation in *Salix pulchra* and *Betula nana.* However, in general, flowering responses were very weak for most of the species (Oberbauer, pers. comm. 1998).

Population-Level Responses

This section reviews field experiments that have tested the indirect effects of CO_2 increase at the population level in alpine and tundra habitats.

Alpine environment, Pennsylvania Mountain, Colorado (39°15′N, 106°03′W, 3550 m elevation): Galen and Stanton (1991), through observations and experiments on a snow bowl population of the alpine buttercup, *Ranunculus adoneus,* showed that the phenology of this species was structured by snow depth. Plants that melted out and flowered late in the season tended to have smaller seeds. They concluded that different parental phenologies affected the fitness of the seedlings; seedlings that germinated from large seeds had a six-fold higher probability of survival than those from small seeds coming from late blooming parents. In this study the most important factor influencing seed size was the time available for growth and seed maturation (late emergence indirectly decreases seed size). A snowpack-duration experiment, in which the growing season length was altered by advancing or delaying snowmelt schedule (Galen and Stanton 1993), demonstrated a direct link between growing season length and plant performance. In the event of changes in season length due to global climate change, Galen and Stanton predicted a strong effect on plant population dynamics. An increase in temperature may promote the production of larger seeds, which may increase the probability of emergence of *R. adoneus* seedlings but may have a negative impact on seed dispersal potential. Further, Galen and Stanton (1999) showed that seedling establishment for this species depends on both growing season length and site fertility. Further, emerging in a given microsite does not assure survival since much of establishment for this species is stochastic.

Arctic environment, the International Tundra Experiment (ITEX): The ITEX was established in the late 1990s. Its initial purpose was to monitor the effects of climate change on major focal plants around the Arctic. Henry and Molau (1997) summarize the short-term responses of some species to climate variation and environmental manipulations (passive warming with open-top chambers (OTCs) and snow depth manipulation) at ITEX sites. In general, responses were species-specific, yet general patterns were found to be similar in ecologically similar sites for some species (e.g., *Dryas*). Plant responses to climate manipulation (e.g., *Saxifraga oppositofolia* and *Cassiope tetragna*) tended to be more pronounced in high latitude arctic sites. In general, both warming and snow depth manipulation induced an earlier phenology and increased seed set, seed weight, and seed germination rates. Further, the responses of species were similar in direction and magnitude across different sites. Reproductive responses to warming also depended on nutrient levels. Mølgaard and Christensen (1997) manipulated temperature in a natural population of *Papaver radicatum* in Disko Island, West Greenland. Temperature was increased by means of "ITEX corners" (angular Plexiglas screens around individual plants), which opened toward one of four different compass directions, therefore creating four different microclimates. The screens also increased snow accumulation. After 4 years, plant biomass in corners facing west, south, and east was significantly higher than in north-facing corners (lower temperature and more snow accumulation) and the controls. The authors also observed an earlier onset of flowering of the treatments compared to the control, and more flowers in plants of the south facing treatment. In another example, Stenström and Jónsdóttir (1997) simulated for two seasons a temperature enhancement by OTCs on the clonal sedge *Carex bigelowii* in Latnjajaure Field Station, Swedish Lapland. Their results showed that the OTCs accelerated the flowering phenology in both years, and sexual reproduction was higher in the OTCs than in the controls. The ITEX report from the 1999 meeting (Hollister 1999) showed that responses to treatment decreased through time, most likely due to nutrient and/or water limitations. These results suggest that nutrient availability might become limiting in the long term. Arft et al. (1999) compiled data from 13 ITEX sites from 1 to 4 years of experimental data. Their results indicate that leaf bud burst and flowering occur earlier in the warmed plots. Vegetative growth was higher in the warmed plots in early years, whereas reproductive effort and success increased in the fourth treatment years. Vegetative and reproductive response to the warming treatment was stronger in herbaceous than woody plants. Finally, sites that had been heated experimentally in the low Arctic had the strongest vegetative response, whereas heated sites in the high Arctic tended to produce more seeds than controls. They concluded that response to warming might vary across climatic zones and functional groups and through time.

Arctic environment: Two sites in the subarctic near Abisko, Swedish Lapland (68°19′N, 18°51′E, 450 m elevation and 68°20′N, 18°41′E, 1150 m elevation), and one in the high Arctic (78°56′N, 11°50′E, 10 m elevation): Havström and collaborators (1993) selected three populations of *Cassiope tetragona* from different parts of its range. The populations were subjected to nutrient, light, and temperature manipulations to simulate the impact of climate change. *C. tetragona* responded differently at the different sites. At the subarctic site (450 m elevation site), nutrient addition caused an increase in growth, whereas in the other two sites (1150 m elevation subarctic and high Arctic site) temperature enhancement treatments had the most significant effect. Havström and collaborators (1995) reported, for the same species in the 450 and 1150 m subarctic sites, a small but significant effect of nutrient addition on flower production, but temperature manipulation had no effect. However shading caused a significant and large decrease of flower production.

Arctic studies in Europe: A semi-polar desert site (78°56'N, 11°50'E) and an arctic dwarf shrub heath site (68°21'N, 18°49'E): Wookey and collaborators (1993) manipulated temperature, soil moisture, and nutrients to simulate the impact of climate change. The dominant species at each site responded differently to the manipulations; temperature enhancement affected phenology and seed set in the semi-polar desert site (dominated by *Dryas octopetala* ssp. *octopetala*), but had no effect in the arctic dwarf shrub heath site (dominated by *Empetrum hermaphroditum*). Water and/or nutrients affected fruit production significantly at the latter site. Vegetative and reproductive development of *Polygonum viviparum* at the semi-polar site responded positively to nutrient addition, while temperature enhancement affected only reproductive development. Increasing soil moisture had no significant effect (Wookey et al. 1994).

Modeling Climate Change

A model developed by Arrhenius (1896) quantified for the first time the radiation budget of the atmosphere and the surface. Although Arrhenius lacked modern technology and meteorological data, he was able to create a climate model that agrees remarkably well with modern-day models (Ramanathan and Vogelmann 1997). Arrhenius's interest in the greenhouse effect was motivated by his desire to understand temperature variation during the Quaternary, but he applied his results to predict the consequences of industrial emission of CO_2 on future climate change (Rodhe et al. 1997).

Arrhenius's pioneering work was followed by a series of new models. Simple surface, energy-balanced models (one-dimensional) in the 1950s gave rise to the more complex one-dimensional models in the 1960s. Manabe and Whetherald (1967), with their one-dimensional radiative-convective equilibrium model, gave rise to the modern era of equilibrium global climate models applied to atmospheric CO_2 simulations (Schneider and Thompson 1981). Three-dimensional general circulation models (GCMs) were the next stage in simulating atmospheric responses to increases of CO_2 (Manabe and Stouffer 1979, 1980, Manabe and Wetherald 1975, 1980).

GCMs have proven to be effective simulators of some of the physical mechanisms that control global warming (Manabe 1997).

Despite the enormous progress in modeling greenhouse effects, there remains a great amount of uncertainty about the magnitude and the distribution of climate change. A clear reflection of this uncertainty is the range in estimates for the increase in temperature if atmospheric CO_2 were to be doubled (between 7.4°C and 5.6°C, IPCC 2001a).

Ecosystem responses to climate change have been evaluated by several models. The Vegetation and Ecosystem Modeling and Analysis Project (VEMAP) addressed the response of multiple resource limitation to environmental variability at the scale of the United States (0.5° grid). VEMAP researchers (Kittel et al. 1995, VEMAP 1995) calibrated six different models to the same initial conditions and then used the models to simulate vegetation responses (at equilibrium) using simulated climate change associated with a doubling of CO_2. The group was interested in the comparison of different models (biogeochemistry models and vegetation type distribution models) and in the determination of their equilibrium sensitivity to climate change and atmospheric CO_2 concentrations. VEMAP (1995) concluded that models' differences in sensitivity to changes in physical variables versus internal nutrient limitations lead to differences in sensitivity to climate change as well. Later Schimel et al. (1997) extended the initial work by comparing the three VEMAP biogeochemistry models (TEM, Century, and Biome-BGC). The three models simulate the cycles of carbon, nitrogen, and water for terrestrial environments but have differences between them. For example, the three models represent different ecosystems with different degrees of aggregation and have different sensitivities to different biogeochemical and physiological processes (e.g., Leaf Area Index (LAI) was calculated in Biome-BGC and Century but not in TEM). The models were compared using current climate and CO_2 concentrations. The authors concluded that the models responded differently to climate change along an environmental gradient, most likely because they differ in sensitivity to changes in physical variables versus internal nutrient limitations. In other words, differences in sensitivity and spatial variability within vegetation types are reflected in three models' different responses to simulations of a doubling in CO_2 (Schimel et al. 1997).

Another group of scientists used individual-based models to

predict ecosystem responses to climate change. For example, these models simulate the dynamic of a forest by following the dynamic of individual trees (growth, birth, and death). This approach, exemplified by the JABOWA (Botkin et al. 1972) and the FORET (Shugart and West 1977) models, has been applied to a variety of forest systems around the world. Shugart (1990), in a review on the use of ecosystem models to assess the consequences of climate change, concluded that although these models have been successful in predicting large dynamic processes, they are limited by their spatial scale. For that reason he proposed that ecosystem models be used in conjunction with individual-based models, such as gap-models, in ecosystem studies.

The next step will be to test these models by experimentation. As an intermediate between mathematical models and field experiments, model laboratory systems have been used to understand and predict population and ecosystem responses to climate change (Lawton 1995). These model systems avoid some of the difficulties and limitations inherent to field studies (Bazzaz's opinion in Thébaud and Johnston 1997). As one example, Davis and collaborators (1998) used a microcosm experiment to test the effects of global warming on species distribution. They concluded that climate change predictions based on changes in climatic zones alone could be misleading. Their experiment suggested that dispersal and species interactions must be included in order to predict biotic responses to climate change. However laboratory systems also have limitations; there is always the uncertainty that the results do not reflect what happens or might happen in "natural communities."

In summary, experimental studies exploring the indirect effects of an increase of CO_2 (e.g., temperature increase, snowpack decrease, etc.) on alpine and arctic plants have shown that responses to climate change manipulation are often (1) species-specific (Chapin et al. 1995, Harte and Shaw 1995, Henry and Molau 1997, Price and Waser 1998, Shaver et al. 1998), (2) variable between sites (Havström et al. 1993, Henry and Molau 1997, Arft et al. 1999, Hollister 1999), and (3) change across time (Chapin et al. 1995, Chapin and Shaver 1996, Arft et al. 1999, Hollister 1999). Overall, climate change simulations suggest that response to global warming in arctic and alpine systems will be mediated by nutrient and water availability. In xeric-subalpine environment, climate change will likely be mediated by summer drought, and long-term studies may be needed to detect vegetation responses to climate change in these habitats (Price and Waser 2000).

Study Site and System

The study was carried out in subalpine meadows of the Colorado Rocky Mountains, near the RMBL (38°57′N, 106°59′W, 2900 m elevation), in southwestern Colorado. I chose subalpine meadow plants for two reasons. First, mountain environments are expected to be very sensitive to global warming (IPCC 1996a). Second, there is little field experimental work on the impacts of global warming on nonforest environments.

I chose *Delphinium nuttallianum* Pritzel (formerly *D. nelsonii* Greene) (Ranunculaceae), a perennial herb, to test the potential impact of climate change. *D. nuttallianum* is one of the first species to flower in the meadows near the RMBL. It takes from 3 to 7 years for *D. nuttallianum* to flower for the first time and longer to reach full flower production (Waser and Price 1991). Each plant produces only one inflorescence, which has 1 to 15 flower buds (Waser and Price 1981). The species is found in the dry meadows of the Rockies throughout the western United States. It ranges from the northern states of South Dakota and Idaho to the southern states of Colorado, Utah, and northern Arizona (Harrington 1964, Waser and Price 1990). The biology of this species has been studied in great detail and the timing of flowering has been shown to be highly sensitive to climate manipulation (Price and Waser 1998). Also, Inouye and McGuire (1991) found a positive correlation between snowpack accumulation and both *D. nuttallianum* floral abundance and flowering phenology (high snow accumulation results in more flowers and later flowering). Finally, there is evidence that *D. nuttallianum* is an important species in the subalpine community, as an early nectar source for both hummingbirds (Waser 1976, Inouye et al. 1991) and queen bumblebees (Inouye and McGuire 1991).

Does Snowmelt Date Affect the Flowering Time of *Delphinium Nuttallianum?*

Price and Waser's (1998) observations on the global warming experiment at the RMBL (Harte et al. 1995, Harte and Shaw 1995) suggested that if winter precipitation patterns (in the form of snow) don't change, flowering date will shift with global warming by shifting the snowmelt date. If precipitation changes, these predictions could fail. Current models of climate change, however, have no clear predictions about whether precipitation will increase or decrease at a regional level. The most likely scenario is that more snow will fall in the winter, snow will melt earlier, and flowering time will be earlier (Harte, pers. comm. 2000). For example recent work in the Sierra Nevada, California, shows that snow accumulation from 1937 to 1997 in sites above 2,400 m elevation is higher yet still melts earlier (Johnson 1998). At the RMBL, there has been a trend toward higher winter snow accumulation since 1973, and no change in date of snowmelt, although more years are needed to determine if this trend is significant (Inouye et al. 2000). Experimental warming led to both an earlier snowmelt date (Harte et al. 1995, Price and Waser 1998) and corresponding changes in the mean date of flowering (Price and Waser 1998). There was a significant and positive relationship between the mean timing of plant reproduction and snowmelt in eight out of ten species measured: *Claytonia lanceolata, Erythronium grandiflorum, Mertensia brevistyla* (early-flowering species); *Delphinium nuttallianum, Lathyrus leucanthus, Potentilla pulcherrima* (mid-flowering species); *Eriogonum subalpinum,* and *Campanula rotundifolia* (late-flowering species). There was no significant positive relationship between the mean time of reproduction and snowmelt date for two of the late flowering species: *Ipomopsis aggregata* (no relationship) and *Seriphidium vaseyanum* (significant negative relationship). Four species, among them *D. nuttallianum,* flowered significantly earlier in the heated plots. Early flowering species tended to respond more strongly to experimental warming, suggesting that snowmelt is the primary cue for timing of reproduction in these species. In contrast, species response was less clear when durations of flowering and fruiting and fruit set between treatments were compared (Price and Waser 1998). However, six species showed significant decreases in duration of flowering with later snowmelt, and eight out of nine species showed higher frequencies of fruit maturation (not individually significant, but significant by a binomial test) in the heated plots.

The evidence that snowmelt triggers an early flowering time was only indirect, since the warming experiment changed not only snowmelt time but many other parameters (e.g., soil moisture, soil temperature). A snow removal experiment allows a more direct test of the effect of snowmelt time on *D. nuttallianum* flowering time. Removing snow experimentally should affect the phenology of *D. nuttallianum*. When snowcover is decreased in this way, snowmelt date should occur earlier and reproduction of *D. nuttallianum* should start earlier in the season. Snowmelt date is associated with other environmental factors, including soil temperature, water availability, nutrient availability, exposure to cold, and length of the growing season. Indeed, manipulating snow levels to speed snowmelt increased both spring soil temperature and spring soil moisture.

Methods

In 1996 and 1997, I used two of the three sites (one at 2920 and the other at 3170 m elevation) from an ongoing snow removal experiment (Dunne 2000) to test my hypothesis. Snow was removed so as to induce a shift in snowmelt matching that was observed in the global warming experiment (about 1 week) (Harte and Shaw 1995, Harte et al. 1995). At each site five pairs (blocks) of 4 × 4 m plots were established. One plot from each pair was assigned as a control (C), and the other as a snow removal treatment (T). Snow was partially removed by shoveling in late March and April 1996 and 1997. Shoveling speeded melting of the remaining snow within each treatment plot by an average of 4 to 8 days. Each pair of plots was located at a slightly different altitude within each site. I collected data on the phenology of *D. nuttallianum* at the upper site (3170 m elevation) in 1996 and 1997 and at the middle site (2920 m elevation) in 1997.

Analysis

In 1996 and 1997, I counted the number of flowers per plot every other day (if possible) to determine whether snow removal shifted the peak of flowering. Number of open flowers per day was scaled as a percentage of the maximum number of open flowers at the peak of flowering at each plot (relative flower number). I used the MIXED procedure in SAS (Littell et al. 1996) for repeated measures analysis to compare the phenologies of the control (C) versus the treatment (T) plots, with treatment and date as the fixed effects, block as the random effect, and date in the repeated statement. I used the repeated statement to model the variance–covariance structure of the random effect. I used Compound Symmetry (CS) to model the random effect and the option "ddfm = betwithin" to calculate the degrees of freedom.

In 1997 I also counted the total number of flowers per plant per plot at the upper site. The number of days, following the first census, until maximum number of flowers was also analyzed. Assumptions for analysis of variance (ANOVA) were checked, and there was no need to transform the data. I used MIXED procedure in SAS (Littell et al. 1996) for one-way ANOVA, with treatment as a fixed effect and block as a random effect.

Results

In 1996, for the upper site (3170 m), analysis of the time of flowering showed that the effect of the snowmelt date (date) is significant ($p < 0.001$), snow removal treatment (treatment) alone is not significant ($p > 0.5$), and the interaction date × treatment is significant ($p < 0.01$) (Table 5.1, Fig. 5.1).

In 1997, similar results were found for the middle (2920 m) and the upper sites (3170 m) (Tables 5.2 and 5.3, Figs. 5.2 and 5.3).

Table 5.1. Repeated Measures Analysis. Effect of Snow Removal on Flowering Date (1996, 3170 m).

Source	Effect	LSdf	Ddf	F	P
Treatment	Fixed	1	8	0.16	0.6973
Block	Random	4			
Block × Treatment	Random	4			
Date	**Fixed**	**12**	**92**	**28.50**	**0.0001**
Date × Treatment	**Fixed**	**12**	**92**	**4.46**	**0.0001**
Error	Random	92			

Two treatments, five blocks, and 13 repeated measures. Treatment and date were treated as fixed effects and block as a random effect. Repeated = date; LSdf = least squares degrees of freedom, Ddf = denominator degree of freedom. Significant effects are shown in boldface type.

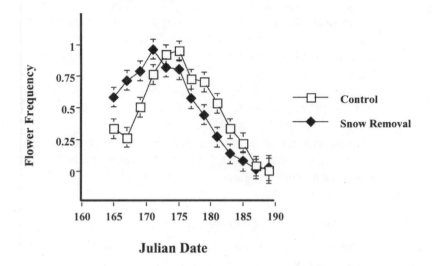

Julian Date

Figure 5.1. Snow removal experiment 1996, 3170 m. Plot of the average frequency of flowers and date (total number of flowers were transformed to a relative frequency from the maximum number of flowers per plot). Julian date 165 corresponds to June 13. Values (± standard error) are based on the mean of five plots.

In 1997 individual plants were followed during the flowering season in the upper site and number of flowers per individual plant was counted every other day. I analyzed the number of days to reach maximum number of flowers, following the first observation day. One-way ANOVA showed that the snow-removal treatment had a significant effect on phenology ($p < 0.05$) when numbers of days from the first census to the peak of flowering is analyzed. Treatment plants reached their maximum number of flowers on average $1^{1}/_{2}$ days earlier than the controls.

In summary, 1996 and 1997 experimental data indicate that the time of flowering was earlier on snow removal plots. In addition, flowering duration was longer in the snow removal plots (see middle site, 1997; Fig. 5.2). The snow removal effect on flowering varied significantly across time; difference in the proportion of open flowers between controls and snow removal treatment was significant only at the beginning of the season. In 1997, individual-plant

Table 5.2. Repeated Measures Analysis. Effect of Snow Removal on Flowering Date (1997, 2920 m).

Source	Effect	LSdf	Ddf	F	P
Treatment	**Fixed**	**1**	**8**	**5.86**	**0.0419**
Block	Random	4			
Block × Treatment	Random	4			
Date	**Fixed**	**12**	**96**	**49.85**	**0.0001**
Date × Treatment	**Fixed**	**12**	**96**	**11.12**	**0.0001**
Error	Random	96			

Two treatments, 5 blocks and 13 repeated measures. Treatment and date were treated as a fixed effects and block as a random effect. Repeated = date; LSdf = least squares degrees freedom, Ddf = denominator degree of freedom. Significant effects are shown in boldface type.

Table 5.3. Repeated Measures Analysis. Effect of Snow Removal on Flowering Date (1997, 3170 m).

Source	Effect	LSdf	Ddf	F	P
Treatment	Fixed	1	8	0.17	0.6892
Block	Random	4			
Block × Treatment	Random	4			
Date	**Fixed**	**9**	**72**	**98.73**	**0.0001**
Date × Treatment	**Fixed**	**9**	**72**	**4.40**	**0.0001**
Error	Random	72			

Two treatments, five blocks and 10 repeated measures. Treatment and date were treated as a fixed effects and block as a random effect. Repeated = date; LSdf = least squares degrees freedom, Ddf = denominator degree of freedom. Significant effects are shown in boldface type.

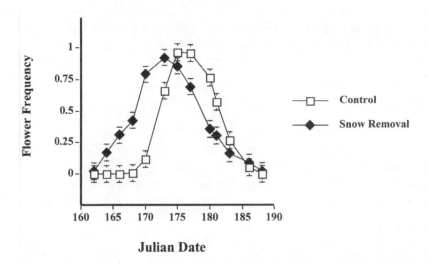

Figure 5.2. Snow removal experiment 1997, 2920 m. Plot of the average frequency of flowers versus Julian date (total number of flowers were transformed to a relative frequency from the maximum number of flowers per plot). Julian date 162 corresponds to June 11. Values (± standard error) are based on the mean of five plots.

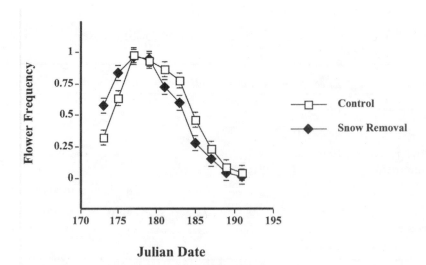

Figure 5.3. Snow removal experiment 1997, 3170 m. Plot of the average frequency of flowers versus Julian date (total number of flowers were transformed to a relative frequency from the maximum number of flowers per plot). Julian date 173 corresponds to June 22. Values (± standard error) are based on the mean of five plots.

analysis in the upper site showed that plants in the treatment plots reach their peak of flowering earlier than the controls.

Will Some Components of Fitness in *Delphinium Nuttallianum* Be Affected by an Early Snowmelt?

Inouye and McGuire (1991) found, by relating phenological data from 1973 to 1989 to weather data and snowpack data for the study area, a positive correlation between snowpack accumulation and both *D. nuttallianum* floral abundance and flowering phenology (high snow accumulation results in more flowers and later flowering). However, the net effect of snowmelt date on plant fitness is not well known, nor is the temporal allocation of plants to flower production.

Observations at the global warming experiment (Saavedra 2000) showed that between 1994 and 1998, *D. nuttallianum* had fewer flowers in heated plots than in controls. Since heating led to an earlier snowmelt these results support the idea that earlier snowmelt may have a negative effect on flower production (Inouye and McGuire 1991). Also, the density of flowering *D. nuttallianum* plants was lower in the heated plots than in the controls.

If plants that flower earlier have higher probabilities of frost damage, then earlier flowering could affect the reproduction of *D. nuttallianum* by decreasing the number of flowers and indirectly reducing seed production. Flowering date could also have a direct effect on seed number and seed quality; plants that flower earlier may be at a disadvantage if few pollinators are present, resulting in fewer visits per plant. For an outcrossing species like *D. nuttallianum* this may result in decreased seed number. It may be possible that the rate of outcrossing for *D. nuttallianum* is lower for individuals that flower earlier in the season, as Schmitt (1983) suggested for Linanthus bicolor, if early populations have significantly fewer pollinators than late flowering populations.

Shifting the flowering date could have a negative impact on the resilience of the community if shifting were to affect plant–plant and plant–pollinator interactions. Hummingbirds rely heavily on the sequence in time of flowering of three flowers for nectar at the RMBL: *Delphinium nuttallianum, Ipomopsis aggregata* and *Delphinium barbeyi* (Waser 1978). Waser and Real (1979) suggested that the early flowering species (*D. nuttallianum*) might act as a mutualistic partner with the late flowering species. They showed evidence that the years of lower density of *D. nuttallianum* were also the years with lower activity of broad-tailed hummingbirds and lower seed set for *I. aggregata*. The broad-tailed hummingbird is a summer resident species whose members colonize and nest in the RMBL in response to *D. nuttallianum* (Waser 1976). If early snowmelt leads to poor flowering of *D. nuttallianum*, this can affect the colonization and fecundity of this pollinator, and, consequently, other species that depend on it for pollination services, including *I. aggregata*.

I have used the snow removal experiment to test whether a change in snowpack duration and thus date of snowmelt may affect the fitness of *D. nuttallianum* plants. If snowcover is shorter, snowmelt date will occur earlier and reproduction of *D. nuttallianum* should start earlier in the season. As a consequence, flower production, seed number, and seed quality could be lower due to an absolute shift in flowering time. Perhaps most important, unless pollinators shift their activity in synchrony with the plants, plants that flower early might miss the period of highest bumblebee activity, consequently lowering seed quality through a reduction in pollinator visits.

Methods

In the upper and middle sites in 1996 and 1997, I collected (1) the mature closed fruits to determine number of seeds and (2) seeds from mature closed or open fruits to be used to determine seed weight and seedling size in a common garden. The seeds that were weighed were always selected from the first or second flower from the inflorescence (counting from the bottom) to correct for flower-level variation within inflorescences. It might have been better to include seeds coming from flowers in higher positions within the inflorescence, to take into account within-plant variance. This would be a good idea for future work, but my interest was in plant to plant and plot to plot variation and not so much in variation within plant; thus I normalized the within-plant variation by selecting only first and second fruits. I tried when possible to collect seeds from 10 individuals per plot and

10 seeds per individual. I also counted the total number of flowers per individual in 1997.

In 1996 a total of 900 seeds from the upper site (3170 m) was weighed and planted in a common garden close to the lab (2900 m). Plastic trays with potting soil were used to create 0.5 × 0.5 m plots (10 replicates). In each plot 90 seeds from different plants were assigned randomly to each of the 90 cells. The seeds from the natural and the removal treatments were planted randomly in the same plot. There was approximately an equal number of seeds per treatment. Seedling size in terms of leaf area was estimated by multiplying the number, length, and width of the cotyledons that emerged in 1997 (Waser and Price 1994).

Analysis

Assumptions for ANOVA (normality and homogeneity of the variance) were checked. *Average flower number per plant* was square root transformed to correct for lack of normality. *Seed number* was square root transformed before statistical analysis to correct for lack of normality. Since I did not have data for the middle site in 1996, I ran two analyses of variance for seed number, one for the upper site for year 1996–97; and a separate analysis for 1997 for the upper and middle site. For *seed weight* I calculated an average weight per plant; there was no need to transform the data. There was no need to transform leaf area.

I used the MIXED procedure in SAS (Littell et al. 1996) for ANOVA, with treatment, year, and site as a fixed effect and block as a random effect for most of the analysis. I used the Satterthwaite option to estimate degrees of freedom except for average seed number (Table 5.5), because the estimation made by SAS was not correct.

Results

Number of Flowers (1997)

Average flower number per plant did not respond significantly to snow removal manipulation for either of the two sites in 1997 ($p > 0.05$) (Table 5.4), nor was a trend evident. In the upper site, plants from the snow removal treatment tended to produce more flowers

Table 5.4. Two-way ANOVA. Effect of Snow Removal (1997, 2920 m and 3170 m) on the Average Number of Flowers per Plant.

Source	Effect	LSdf	Ddf	Variance Component	F	P
Treatment	Fixed	1	12.5		0.46	0.5087
Site	Fixed	1	12.5		1.87	0.1953
Treatment × Site	Fixed	1	12.5		1.80	0.2040
Block (Site)	Random	4		0.0		
Treatment × Block (Site)	Random	4		0.037		
Error	Random	297		0.328		

Treatment was treated as a fixed effect and blocks within site as a random effect. LSdf = least squares degrees freedom, Ddf = denominator degree of freedom. Significant effects are shown in boldface type. Variance component values are reported for random effect and F ratio and probabilities are reported for fixed effect.

Table 5.5. Two-way ANOVA. Effect of Snow Removal on Seed Number per Fruit, Upper Site 1996, 1997.

Source	Effect	LSdf	Ddf	Variance Component	F	P
Treatment	Fixed	1	4		0.36	0.5789
Year	Fixed	1	105		1.40	0.2390
Block	Random	4		0.0289		
Treatment × Year	**Fixed**	**1**	**105**		**4.00**	**0.0482**
Treatment × Block	Random	4		0.0		
Error	Random	105		1.7325		

Treatment and year were treated as a fixed effect and block as a random effect. LSdf = least squares degrees freedom, Ddf = denominator degree of freedom. Significant effects are shown in boldface type. Variance component values are reported for random effect and F ratio and probabilities are reported for fixed effect.

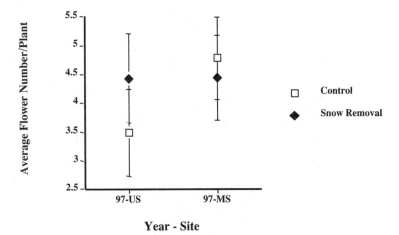

Figure 5.4. Average number of flowers per plant 1997, 2920 and 3170 sites. US, upper site; MS, middle site. Error bars represent 95% comparison limits between control and treatments in the same year–site combination.

than the controls, whereas in the middle site this trend was reversed (Fig. 5.4). These results contrast with observations in the warming experiment (average flower number per plant is lower in the heated plots) indicating that flower number is not dependent on snow date alone (Saavedra 2000).

Seed Number (1996, 1997)

An ANOVA of seed number for the upper site (1996 and 1997) shows a significant interaction effect of treatment × year (Table 5.5). Neither of these effects was statistically significant by itself. Plants from the controls produced more seeds per fruit than those from the snow removal in 1996, but the trend is reversed in 1997 (Fig. 5.5). Analyses of seed number for 1997 in the upper and middle site show a significant effect of treatment × site (Table 5.6). In the upper site, plants from the snow removal treatment produced more seeds than the controls, whereas in the middle site this trend was reversed (Fig. 5.5).

In both 1996 and 1997, in the upper site, the average seed number for the snow removal treatments did not differ significantly from the controls (t-tests, $p > 0.05$). In 1997, however, the middle site

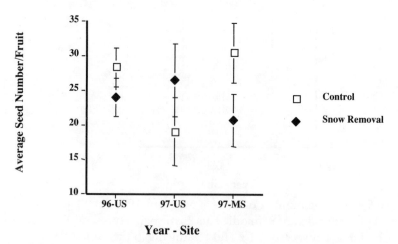

Year - Site

Figure 5.5. Average seed number per fruit 1996 and 1997, 3170 m. US, upper site; MS, middle site. Error bars represent 95% comparison limits between control and treatments in the same year–site conditions.

Table 5.6. Two-way ANOVA. Effect of Snow Removal on Seed Number per Fruit, Upper Site and Middle Site in 1997.

Source	Effect	LSdf	Ddf	Variance Component	F	P
Treatment	Fixed	1	25.2		0.18	0.6789
Site	Fixed	1	25.2		0.61	0.4434
Block (Site)	Random	4		0.0		
Treatment × Site	**Fixed**	1	25.2		**7.19**	**0.0127**
Treatment × Block (Site)	Random	4		0.0110		
Error	Random	69		1.4429		

Treatment and site were treated as a fixed effect and block within site as a random effect. LSdf = least squares degrees freedom, Ddf = denominator degree of freedom. Significant effects are shown in boldface type. Variance component values are reported for random effect and F ratio and probabilities are reported for fixed effect.

showed a response. The controls in the middle site produced fruits with significantly more seeds than the snow-removal treatments (p = 0.0204) (Fig. 5.5).

In summary, the effect of snow removal treatment on seed set per fruit varied between sites and years. Only in 1997 and in one site could one infer a negative effect of an early snowmelt date on seed number. It is possible that there is a strong relationship between snowmelt date and seed number, but the treatment did not shift snowmelt date sufficiently to detect this relationship. A regression analysis might be more sensitive to the responses of plants to snowmelt date across sites and could be used in the future to complement the ANOVA.

Seed Weight (1996, 1997)

Seed weight was used as an estimator of seed quality (Stanton 1984). There is a significant year effect and a treatment × year interaction. Most of the weight variation can be explained by plant to plant variation (86% of the variance comes from the error term) (Table 5.7). Overall, seeds from the snow removal experiment tend to weigh more than the controls, but the difference is not statistically significant. Seeds in 1997 weighed significantly less than seeds from 1996.

Within-year and site comparisons indicate that in 1996, at both

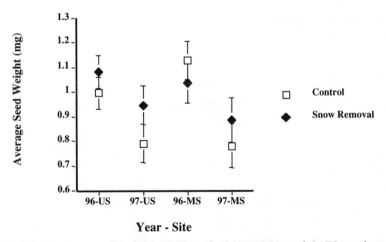

Figure 5.6. Average seed weight 1996 and 1997, 2920 and 3170 m sites. US, upper site; MS, middle site. Error bars represent 95% comparison limits between control and treatments in the same year–site combination.

Table 5.7. Three-way ANOVA. Effect of Snow Removal on Average Seed Weight, Upper and Middle site 1996, 1997 (2599 seeds from 249 individuals).

Source	Effect	LSdf	Ddf	Variance Component	F	P
Treatment	Fixed	1	15.8		1.85	0.1929
Site	Fixed	1	15.8		0.0	0.9446
Year	**Fixed**	**1**	**229**		**51.58**	**0.0001**
Block (Site)	Random	4		0.0		
Treatment × Site	Fixed	1	15.8		1.54	0.2325
Treatment × Year	**Fixed**	**1**	**229**		**5.19**	**0.0236**
Site × Year	Fixed	1	229		1.75	0.1870
Treatment × Site × Year	Fixed	1	229		1.16	0.2835
Treatment × Block (Site)	Random	4		0.00634		
Year x Block (Site)	Random	4		0.0		
Treatment × Year × Block (Site)	Random	4		0.0		
Error	Random	225		0.03853		

Treatment, site, and year were treated as a fixed effect and block within site as a random effect. LSdf = least squares degrees freedom, Ddf = denominator degree of freedom. Significant effects are shown in boldface type. Variance component values are reported for random effect and F ratio and probabilities are reported for fixed effect.

altitudes, seed weight did not respond significantly to treatment ($p > 0.05$). In 1997, seed weight responded significantly to the snow removal treatment only in the upper site ($p = 0.0505$); in the upper site, seeds from the snow removal treatment weighed significantly more than those in the control plots, while no significant difference was found in the middle site that year (Fig. 5.6). It is possible that an increase in the growing season's length is more important at higher elevations. Galen and Stanton (1991, 1993) working at 3,550 m elevation in a similar environment, found a significant effect of snowpack-duration on seed weight for another member of the Ranunculaceae family. As with seed number, it will be useful to conduct regression analyses comparing the effect of snowmelt date on seed weight across sites.

Seedling Size in a Common Garden (1996 cohort)

After 1 year there was a significant difference in growth (cotyledon area estimate) between controls and treatments for the seeds coming from the upper site ($p = 0.033$) (Table 5.8). The seedlings from the snow removal treatment had larger leaves (cotyledons). The mean leaf area for the controls was 20.32 mm^2 (SE = 0.78, $n = 370$) compared to 22.14 mm^2 (SE = 0.77, $n = 394$) for seedlings from the snow removal treatment. Most of the variance in growth came from individual seed variation (86%).

Table 5.8. One-way ANOVA. Effect of Snow Removal on Leaf Area Estimate.

Source	Effect	LSdf	Ddf	Variance Component	F	P
Treatment	**Fixed**	**1**	**8.06**		**6.6**	**0.033**
BlockSe	Random	9		3.5036		
Treatment × BlockSe	Random	9		0.0		
BlockMa	Random	4		0.0		
Treatment × BlockMa	Random	4		0.9101		
Treatment × BlockSe × BlockMa	Random	36				
Error	Random	699		26.1127		

Seeds were collected in snow removal experiment, 1996, upper site and grown in common gardens at 2920 m ($n = 763$). LSdf = least squares degrees freedom, Ddf = denominator degree of freedom. BlockSe = block of seeds, BlockMa = block maternal plants. Significant effects are shown in boldface type. Variance component values are reported for random effect and F ratio and probabilities are reported for fixed effect.

In summary, only some of the measures of fitness measured seem to respond to snow removal experiments. Flower number was not affected in the expected way (Table 5.4). Seed number per fruit was significantly less in the snow removal treatments than in the controls only in the middle site for the year 1997 (Fig. 5.5). Seed weight response seems to be year-specific (Table 5.7), seeds from the treatment weighed significantly more than the controls only in the upper site, 1997. Finally, the leaf area of seedlings (1996-cohort) from the upper site was significantly larger for the treatments than for the controls.

How Does Timing of Flowering Affect Reproduction in Natural Populations?

Shifting the date of snowmelt experimentally with a warming experiment is one approach for investigating the possible effect of climate change. In order to circumvent the limitations of the experimental approach and to have a better understanding of the fitness consequences of flowering at different dates, I complemented the experimental observations in 1997 with observations of two populations that naturally flower at different times.

Flowering time has been a topic of interest for many years. Questions related to phenology are diverse, ranging from descriptions of the temporal pattern of flower presentation to attempts to understand the selective forces that determine the shape of the flowering curves (Thomson 1980) and their timing (e.g., Waser 1978, 1979, O'Neil 1997 at the population level; Poole and Rathcke 1979, Rathcke 1988 at the community level). Within a region, not all populations flower at the same time (Bliss 1956, Billings and Bliss 1959, Jackson 1966, Galen and Stanton 1991). Campbell (1987) showed that, among six populations of *Veronica cusickii* in the Olympic Mountains, Washington, that differ in the duration of snow cover and concordantly in flowering time, only three were pollen limited. Campbell and Halama's (1993) experimental work showed that seed production in *Ipomopsis aggregata* was a complex result of the interaction of pollen and resource levels; in addition, their results showed that the effect of pollen and resources on seed production changes through the season. An increase of resources lengthens flower production, producing significantly more flowers only in the late season, whereas pollen addition increased significantly seed set per flower primarily in the first 2 weeks.

The flowering phenology of *D. nuttallianum* has already been

studied in the meadows of the RMBL, and those studies have suggested that competition for pollination is a selective force acting to
maintain divergent flowering times of *D. nuttallianum* and sympatric
Ipomopsis aggregata (Waser 1978). My preliminary work in 1997
showed that the pollinator guild changes over the season. The early
population flowered when no *I. aggregatta* were in flower and few
other *D. nuttallianum* populations were in flower. Bumblebees and
hummingbirds pollinated the early population. In contrast, the late
population flowered when most *D. nuttallianum* populations were in
flower and *I. aggregata* was in flower as well. Solitary bees and bumblebees pollinated the late population.

Many studies in the literature have related time of flowering and
seed set per flower or per plant. Few of these studies have looked at
the quality of the seeds produced through the season. To my knowledge only Galen and Stanton (1991, 1993) and Lacey and Pace
(1983) have done so. They used, respectively, seed weight and germination rates, and germination, survivorship, and growth to assess
the quality of the offspring from individuals flowering at different
times. Another indirect way to estimate quality of the seeds is to
measure outcrossing rates. There are few papers that study the relationship between time of blooming and outcrossing rates and
among them none does so experimentally. Stephenson (1982) conducted a nonexperimental study of *Catalpa speciosa* and concluded
that plants that flower later in the season have higher outcrossing
rates. Only outcrossed flowers form fruits in this species, so fruit
production became his estimate for outcrossing. Zimmerman
(1987), working with *Polemonium foliosissimum,* also concluded that
late plants have higher outcrossing potential. He based his conclusion on pollinators' movements through the season. For *D. nuttallianum,* experimental crossings indicated that self versus outcrossed
seeds differed in fitness. Outcrossed individuals outperformed the
self-pollinated progeny (outbred progeny, grew larger, survived
longer, and reached sexual maturity faster) (Waser and Price 1994).

Methods

In 1997 I selected, near the RMBL, two populations of *D. nuttallianum.* These two populations are located at the same altitude and
are separated by only a few hundred meters, but differ in snowmelt
date. The early population (EP) is on a south-facing slope that melts
out earlier than the other site (LP). I established a 4 × 4 m study

plot in the EP site and a 3 × 4 m plot in the LP site. I tagged individual flowers and plants as they began flowering.

I recorded (1) the number of flowers every other day to determine the *peak of flowering*. I marked individual flowers with colored strings as they emerged to look at the effect of date of flowering on seed quantity and seed quality. I also determined (2) *the number of seeds* per flower and (3) *the average seed weight* through the season.

I compared seed number and seed weight between early and mid cohorts across population. Early cohort refers to flowers that open at the beginning of the season when flowers are scarce for pollinators and mid cohort refers to flowers that open during peak flowering time. In the 1997 EP population, the early cohort is defined by flowers that opened between June 6 and 8, while the mid cohort includes flowers that opened between June 14 and 18. In the 1997 LP population, early cohort corresponds to flowers that opened between June 23 and 25 and mid cohort to flowers that opened between June 30 and July 4.

Analysis

Data were tested for normality and homogeneity of variance before analysis. There was no need to transform the data for number of seeds and seed weight.

Analysis of covariance (ANCOVA) was done using the MIXED procedure in SAS (Littell et al. 1996) to test for the effect of flowering time on seed number and seed weight. Time and site were used as fixed effects and fruit position as the covariate for both seed number and seed weight analysis.

Results

Phenology

The flowering curve of the EP extended from June 2 to June 26 and for the LP from June 23 to July 14 (Fig. 5.7). The early population reached the peak of blooming June 14 and the late population on July 2.

Seed Number

After adjusting for fruit position, ANCOVA indicates that flowering time affects seed production ($p < 0.0001$); overall, early flowers in

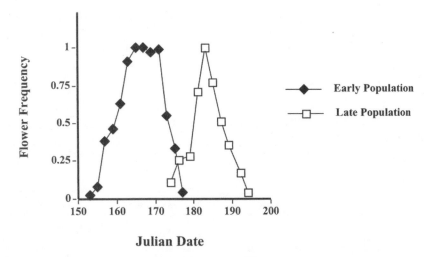

Julian Date

Figure 5.7. Flowering phenology for the early population (EP) and the late population (LP). Natural observations, 1997. Average frequency of flowers are plotted against Julian date (total number of flowers were transformed to a relative frequency from the maximum number of flowers per plot). Julian date 153 corresponds to June 2.

1997 produced more seeds than mid-season flowers. Site and the interaction time × site were not significant.

Average seed number per flower for the EP decreased through the season ($r_s = -0.373, p < 0.001$). Early flowers (opened June 6–8) produced significantly more seeds than flowers from the peak of flowering (opened June 14–18) ($p = 0.0093$). Early flowers produced an average of 32.01 seeds in contrast to 23.43 seeds for the mid- season flowers (Fig. 5.8A).

Similar to the EP, in the LP, the average seed number per flower decreased through the season ($r_s = -0.370, p < 0.001$). Early flowers (opened June 23–25) produced significantly more seeds than flowers from the peak of flowering (opened June 30–July 4) ($p = 0.0001$). Early flowers produced an average of 36.22 seeds in contrast to 22.91 seeds for the mid season (Fig. 5.8B).

Seed Weight

After adjusting for fruit position, ANCOVA indicates that flowering time ($p < 0.01$) and site ($p < 0.01$) affects significantly average seed weight; overall seeds from early flowers in 1997 weighed more than seeds from mid-season flowers. Seeds from EP weighed more than seeds from LP, but this might be compensation, since plants from

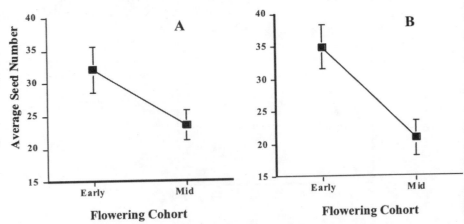

Figures 5.8. Average seed number per fruit versus flowering cohort, two populations. For the early population: Early, flowers that opened between June 6 and 8; Mid, flowers that opened between June 14 and 18. For the late population: Early, flowers that opened between June 23 and 25; Mid, flowers that opened between June 30 and July 4. Error bars represent one standard error. (a) Early population 1997. (b) Late population 1997.

LP populations produced on average more seeds than the EP population. The interaction time × site was not significant.

Seeds from early flowers were significantly heavier than those from the peak of flowering for both sites ($p < 0.05$). In EP the average seed weight for early seeds was 1.065 mg and for the mid-season seeds was 0.904 mg (Fig. 5.9A). In LP the average seed weight for early seeds was 0.908 mg and for the mid-season seeds was 0.741 (Fig. 5.9B).

Discussion of Empirical Studies

Results from the warming experiment (Price and Waser 1998) suggested that warming will trigger an early flowering of *D. nuttallianum* by shifting the date of snowmelt; however, the evidence was only indirect since warming altered many other microclimate parameters besides snowmelt (Harte et al. 1995). This snow removal experiment corroborated their findings and provided direct evidence that a change in snowmelt date will trigger changes in flower phenology. Moreover, the results suggest that shifting snowmelt date by, on average, 10 days, as could occur under a $2 \times CO_2$ climate scenario, might affect some aspects of the reproductive biology of *D. nuttal-*

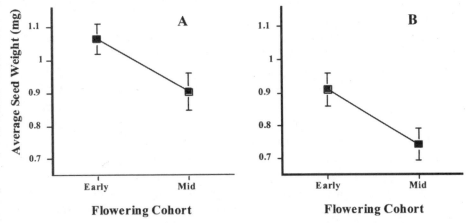

Figures 5.9. Average seed weight per fruit versus flowering cohort, two populations. For the early population: Early, flowers that opened between June 6 and 8; Mid, flowers that opened between June 14 and 18. For the late population: Early, flowers that opened between June 23 and 25; Mid, flowers that opened between June 30 and July 4. Error bars represent one standard error. (a) Early population 1997. (b) Late population 1997.

lianum, such as seed weight and seed set, and not others, such as flower production. Furthermore the results suggest that responses are year and site specific. Finally, this result indicates that other factors (e.g., moisture, pollinators, air temperature, etc.) modulate the responses to snowmelt date.

Contrary to what was expected from Inouye and McGuire's long-term studies (1991), in 1997 flower number did not respond significantly to a difference in snowpack (T versus C) (Table 5.4). A more detailed analysis of Inouye and McGuire's data revealed that the number of flowers per plant did not vary with snow accumulation, but the number of flowering plants per plot did. So apparently plant density seems to respond positively to snowpack across years. Observations from the warming experiment support this idea. The density of flowering *D. nuttallianum* was significantly lower in the heated plots (earlier snowmelt and drier conditions) than the control plots (Saavedra 2000). However in neither of the two sites from the snow removal experiment (1997 data) could I detect a significant difference in flowering plant density between snow removal treatments (lower snowpack) and controls, and it is possible that more years might be needed to see a density effect.

Decreasing snowpack had a significant effect on seed set and seed weight, but the direction of the change depended upon both site and year. In the upper site, seed number per fruit showed no significant response to treatment (1996, 1997), yet the trend was different for different years. In the upper site (1997) snow removal had a small, but significant, positive effect on seed weight. Seeds from the snow removal plots weighed on average 16% more than seeds coming from the control plots (the effect on weight was not compensated by seed number, as both were increased in the treatment plots). In 1996 the trend was the same but the difference is not statistically significant. The increase in seed weight from snow removal in 1997 is similar to that observed by Galen and Stanton (1991, 1993) when they extended the growing season for *Ranunculus adoneus* by 12 days. Advancing snowmelt led to an average increase in seed mass of 33% in their experiment. Although the trend is the same as that found in my study, the magnitude of the change is different. The sample size I used for the seed weight analysis was large and allowed detection of even small statistical differences between the treatments. Are these differences in seed weight biologically significant? Growing the seeds from the 1996 cohort in a common garden showed that after a year seeds that weighed more (from the snow removal treatment) had significantly larger seedlings. If seedling size is correlated to survival, then one could infer that having a larger snow-free season could induce an increase in *D. nuttallianum* seedling survival in a high altitude site. However, other effects may be important. Seedling establishment is dependent on the biotic and abiotic environment. One way to test this hypothesis would be to follow the fate of the seedlings through many years in controls and snow removal plots, instead of in common gardens. In doing so, both plants and their environment will change simultaneously. Even though such a plot is only a small proportion of the landscape, at least it would allow estimates of the seedling's response in a matrix that is closer to that expected under natural conditions. Experiments conducted by Stanton and Galen (1997) and Stanton et al. (1997) show that the problem is more complex. Sites that flowered earlier produced bigger seeds that have an advantage in all sites in the snowbed. Further, a snow removal experiment showed that seedling establishment is not a simple function of snowmelt regime; it varies significantly between early, more fertile sites and later, less fertile sites (Galen and Stanton 1999).

In the middle site, in 1997, seed number per fruit was negatively and significantly affected by the snow removal treatment. Fruits from the snow removal plots had on average 32% fewer seeds than the controls (this could be compensation, since seeds from the snow removal treatment weighed 12% more than the controls). In 1996 and 1997, snow removal did not have a statistically significant effect on seed weight. In 1996, seeds from the snow removal plots weighed on average 8% less than the seeds coming from the controls, whereas in 1997 the trend was reversed (seeds from the snow removal treatment weighed 12% more than the controls).

The experimental approach in my study was complemented by observations of the phenology and the reproduction of two natural populations that naturally flower at different times (early and late snowmelt sites). The observations showed that in nature the phenology of different populations in a given area diverges in time more than the range produced by the experimental snow manipulation. A 10-day shift in snowmelt date is within the range of variation between close sites in an area, and certainly within the range of yearly variation (Barr, pers. comm. 2000. Moreover, if different flowering cohorts within each of the populations are considered analogous to the snow removal plot (early flowers) and control (mid/late flowers) from the snow removal experiment, one could reason that flowering earlier is advantageous. In both sites, early flowers produced more seeds and heavier seeds than middle flowers. This reasoning is contradictory to our findings in the middle site where flowering earlier (snow removal) had a negative and significant effect on seed number in 1997, but agrees with the results for the upper site, 1997, where snow removal had a positive effect on seed weight and number.

Although plant–pollinator interactions are in most cases not obligate (Waser et al. 1996), they could be negatively affected if global warming were to affect their synchrony. For instance, a lack of synchrony between hummingbirds and their host plant might not cause a total failure in sexual reproduction, yet it might decrease seed set (Waser 1979). The snow removal manipulation shifted the flowering time of *D. nuttallianum* but did not affect the time of emergence of bumblebees nor the arrival time of the hummingbirds (different spatial scales), and as such it offered the possibility of testing for hypothetical effect of changes in the synchrony between pollinator and their host plants.

Nesting patterns of broad-tailed hummingbirds (*Selasphorus platycercus*) at the RMBL are correlated with the temporal and spatial abundance of *D. nuttallianum* flowers (Waser 1976). The population size of broad-tailed hummingbirds seems to be correlated to floral abundance during the breeding season (Inouye et al. 1991) and nesting time parallels shifts in flower phenology between years (Waser 1976). Waser and Real (1979) reported that the years of lower density of *D. nuttallianum* were the years with lower activity of broad-tailed hummingbirds and lower fecundity of *I. aggregata*. If the distribution of *D. nuttallianum* were to be affected by climate change, we could expect changes in the abundance and fecundity of plants and pollinators that depend on it.

Indirectly, the fecundity of many plant species could be negatively affected if climate change affects the local and long-distance migratory pattern of hummingbirds. Rufous hummingbirds (*Selaphorus rufus*) breed in northwestern North America and arrive (on southward migration) after *D. nuttallianum* has flowered in the study area. Scarlet gilia (*Ipomopsis aggregata*) is the primary source of nectar for this hummingbird species (Calder 1987). There is evidence that the migratory pattern of rufous hummingbirds has changed in recent decades. Hill and collaborators (1998) reported that the number of wintering rufous hummingbirds has increased exponentially in the southeastern United States in the last 20 years. This evidence plus the capacity of hummingbirds to track changing resources (Bond 1995), suggests that if time or abundance of flower resources were to change with climate change, these hummingbirds might be able to shift their nesting sites and migration routes and move out of this habitat to forage elsewhere. Decreases in the local diversity of pollinators could probably have an impact on the resilience of the community (Aizen and Feinsinger 1994). It also could cause a decline of bird pollinators over the long term. For instance, Martin (1997) reported that a high elevation site in Arizona had in the past 5 years experienced both the wettest and the driest summers on record, and breeding birds at the site have moved up and down in elevation following the climate gradient. The changes in the preferred habitats have caused a significant decrease in their breeding success and will likely cause a population decline in the long term. Given that the local distribution of the birds along the altitudinal gradient is similar to their latitudinal distribution, Martin proposed that the reduction in nesting success might reflect larger shifts along latitudes related to climate change (Price 1997, Root 1997).

The reproductive responses of *D. nuttallianum* to snow manipulation are small and conflicting. One possibility is that snowmelt differences between controls and snow removal plots were only 4 to 8 days on average. Stronger treatment effects might be needed to simulate $2 \times CO_2$ effects more accurately and to observe a response on seed parameters. Another possibility is that reproduction is not the most sensitive stage that may be affected by global warming. Observations in the warming experiment suggested that year to year flowering-plant emergence (density of flowering plants) could be more sensitive to warming than is reproduction (number of flowers per plant) (Saavedra 2000).

In light of the positive effect that a lower snowpack had on seed weight in the upper site (1997), or the negative effect on seed set in the middle site (1997), the long-term output is hard to predict. If timing of snowmelt were to change in a warmer climate, the effect might depend on the site as well as year, and on the correlated moisture patterns. These observations illustrate the importance of testing the impact of global warming at a larger temporal and spatial scale.

A warmer climate will likely induce a series of changes that will affect ecosystems through changes in the mean air and soil temperature, nutrient cycling, length of growing season, and phenology and relative fitness of species. In high latitude and altitude sites (like my study site) higher temperature should induce an early snowmelt and, as a consequence, induce a shift in plant phenology. The warming experiment (Harte et al. 1995) and the snow removal experiment (Dunne 2000) simulate only partially the effects of global warming. The warming experiment induces an early snowmelt and drier, warmer soil through a downward IR flux, and although it may be a reasonable mimic of some aspects global warming it has limitations (it doesn't warm the air over the unenclosed plots, it doesn't change CO_2 concentrations, and it doesn't elicit potential covarying changes in precipitation; Price and Waser 1998). The snow removal experiment mimics the early snowmelt, which is a good simulation of the primary effect of warming expected in ecosystems with winter snowpack. But by changing the snow cover (snowmelt date), other variables were also changed (for instance decreases in water availability, drier soils in early summer), and other variables like total IR will be the same after the snowmelt for treatments and controls. These limitations and the small spatial time scale of the experiments limit our capacity to predict in the long term the direction of the changes, but they give us a starting point to explore possible consequences of climate change in alpine ecosystems.

Conclusion

There is an increasing awareness among members of the scientific community, including ecologists and evolutionary biologists, of the environmental changes occurring at a planetary scale. Most believe now that climate change is really taking place, and would like to understand the mechanism of climate change and to evaluate the consequences of those changes. Observations, experiments, and models are all helping to provide a better understanding of the magnitude of the environmental change, its consequences, and ways to cope with it.

There has been growing emphasis among ecologists to help direct scientific research to be useful for environmental decision making. For example, Lubchenco et al. (1991) and the initiatives proposed by the ESA (Ecological Society of America) and AERC (Association of Ecosystem Research Centers) (1993) called for syntheses of ecological information to help decision makers who are addressing modern environmental problems. The ESA proposed three research priorities in its Sustainable Biosphere Initiative (SBI): Global Change, Biological Diversity, and Sustainable Ecological Systems. The SBI propositions also assume that basic research toward understanding ecological principles is required in order to resolve our environmental problems. Another example of these ecological initiatives is presented by Vitousek et al. (1997). They provided both an overview of human effects on Earth's ecosystems (terrestrial and marine) and a policy recommendation. Vitousek et al. (1997) suggested that the extent of our impact on Earth should also affect the way we think about Earth. They recommended three complementary actions: first to reduce the rate at which we alter the environment, second to accelerate our effort to understand ecosystems and how they interact with many of the components of global change, and third, to manage Earth's ecosystems for sustainability. Finally, Ehrlich (1997, p. 15) encouraged ecologists to communicate pertinent information to decision makers and to the general public, and especially young scientists "to help bind up the world's wounds wherever possible."

There is substantial evidence that the levels of CO_2 and other greenhouse gases have indeed increased as a result of human activity, and that, as a consequence, global temperatures have increased. Nonetheless, we, as ecologists, have mainly proposed solutions that do not imply a radical change in our lifestyles. It is impossible to go back and fix the past to avoid all the undesirable consequences of our modern lifestyle. The production of heat and electrical power by burning fossil fuels is so intimately tied to our standard of living in our modern societies that there is little public debate over the merits of decreasing energy consumption. We prefer for now to understand, to predict, and to mitigate; we do not want radical change, at least not yet. What is radical change? A radical change would be to simplify desires, and in the process, to minimize luxury consumption and the unnecessary use of energy. What is it that we really need?

The limitation of extravagant desires, overconsumption, and wastefulness provides only a partial solution for our environmental problems. Ultimately, more profound changes need to be made in our way of thinking and relating. Specifically, we need to strengthen the realization held by some that everything is interconnected, and that everything we do here and now has a consequence in the future. As Bohm and Edwards emphasized in their book *Changing Consciousness*, we need to further develop our sense of wholeness and to search for ways to forgo more fragmentary approaches to reality. These changes in consciousness can encompass our understanding of both ecological dynamics and the cultural and spiritual connections between the Earth's inhabitants.

Acknowledgments

I am enormously grateful for the unconditional support from my adviser David Inouye, who has dealt with me so patiently and whose previous investigations inspire my work. This work was enriched by conversations with Peter Abrams, Allison Brody, Larry Douglass, Jay Evans, and Rick Williams. I am especially thankful to Mary Price and Nick Waser for sharing ideas and the joy of doing science. Thanks also to several Earthwatch volunteers, Tim Brown, Betsy Heartfield, and Jennifer Reithel for helping me in the field. I am very thankful to Jen Dunne and her collaborators for letting me use the snow removal experiment: John Harte, Kevin Taylor, and others. Thanks to the University of Maryland and the Rocky Mountain Biological Laboratory and staff for helping me to realize this project. Special thanks to billy barr, Kevin Donovan, Sonda Eastlack, and Gary Entsminger. Thanks to the National Wildlife Federation Climate Change Fellowship, NSF grants (DEB-94-08382 and IBN-98-14509) to David Inouye, NSF grant (DEB- 96-28819) to John Harte, and Earthwatch for their financial support.

Literature Cited

Aizen, M. A., and P. Feinsinger. 1994. Forest fragmentation, pollination, and plant reproduction in a Chaco dry forest, Argentina. *Ecology* 75: 330–351.

Arft, A. M., M. D. Walker, J. Gurevitch, J. M. Alatalo, M. S. Bret-Harte, M. Dale, M. Diemer, F. Gugerli, G. H. R. Henry, M. H. Jones, R. D. Hollister, I. S. Jónsdóttir, K. Laine, E. Lévesque, G. M. Marion, U. Molau, P. Mølgaard, U. Nordenhäll, V. Raszhivin, C. H. Robinson, G. Starr, A. Stenström, M. Stenström, Ø. Totland, P. L. Turner, L. J. Walker, P. J. Webber, J. M. Welker, and P. A. Wookey. 1999. Responses of tundra plants to experimental warming: Meta-analysis of the International Tundra Experiment. *Ecological Monographs* 69: 491–511.

Arrhenius, S. 1896. On the influence of carbonic acid in the air upon the temperature on the ground. *Philosophical Magazine* 41: 237–276.

Bazzaz, F. A. 1990. The response of natural ecosystems to the rising of global CO_2 levels. *Annual Review of Ecology and Systematics* 21: 167–196.

Billings, W. D., and L. C. Bliss. 1959. An alpine snowbank environment and its effects on vegetation, plant development, and productivity. *Ecology* 40: 388–397.

Bliss, L. C. 1956. A comparison of plant development in microenvironments of arctic and alpine tundras. *Ecological Monographs* 26: 303–337.

Bohm, D., and M. Edwards. 1991. *Changing Consciousness: Exploring the Hidden Source of the Social, Political, and Environmental Crises Facing Our World.* San Francisco: Harper.

Bond, W. J. 1995. Effects of global change on plant–animal synchrony: Implications for pollination and seed dispersal in Mediterranean habitats. Pages 181–202 in J. M. Moreno and W. C. Oechel, eds., *Global Change and Mediterranean-Type Ecosystems.* New York: Springer-Verlag.

Botkin, D. B., J. F. Janak, and J. R. Wallis. 1972. Some ecological consequences of a computer model of forest growth. *Journal of Ecology* 60: 849–872.

Calder, W. A., III. 1987. Southbound through Colorado: Migration of rufous hummingbirds. *National Geographic Research* 3: 40–51.

Campbell, D. R. 1987. Interpopulation variation in fruit production: The role of pollination-limitation in the Olympic Mountains. *American Journal of Botany* 74: 269–273.

Campbell, D. R., and K. J. Halama. 1993. Resource and pollen limitations to lifetime seed production in a natural plant population. *Ecology* 74: 1043–1051.

Chapin, F. S., III, and G. R. Shaver. 1985. Individualistic growth response of tundra plant species to environmental manipulation in the field. *Ecology* 66: 564–576.

————. 1996. Physiological and growth responses of arctic plants to a field experiment simulating climatic change. *Ecology* 77: 822–840.

Chapin, F. S., III, G. R. Shaver, A. E. Giblin, K. J. Nadelhoffer, and J. A. Laundre. 1995. Responses of Arctic tundra to experimental and observed changes in climate. *Ecology* 76: 694–711.

COHMAP. 1988. Climate change in the last 18,000 years: Observation and model simulations. *Science* 241: 1043–1052.

Davis, A. J., L. S. Jenkinson, J. H. Lawton, B. Shorrocks, and S. Wood. 1998. Making mistakes when predicting shifts in the species range in response to global warming. *Nature* 391: 783–786.

Dunne, J. A. 2000. Climate Change Impacts on Community and Ecosystem Properties: Integrating Manipulations and Gradient Studies in Montane Meadows. Ph.D. Dissertation, University of California, Berkeley.

Ehrlich, P. R. 1997. *A World of Wounds: Ecologists and the Human Dilemma.* Oldendorf/Luhe, Germany: Ecology Institute.

Elton, C. S. 1958. *The Ecology of Invasions by Animals and Plants.* London: Methuen.

ESA, and AERC. 1993. *National Center for Ecological Synthesis, Scientific Objectives, Structure, and Implementation.* Report from the joint committee of the Ecological Society of America and the Association of Ecosystem Research Centers.

Galen, C., and M. L. Stanton. 1991. Consequences of emergence phenology for reproductive success in *Ranunculus adoneus* (Ranunculaceae). *American Journal of Botany* 78: 978–988.

————. 1993. Short-term responses of alpine buttercups to experimental manipulations of growing season length. *Ecology* 74: 1052–1058.

————. 1995. Responses of snowbed plant species to changes in growing-season length. *Ecology* 76: 1546–1557.

————. 1999. Seedling establishment in alpine buttercups under experimental manipulations of growing season. *Ecology* 80: 2033–2044.

Galloway, L. F. 1995. Response to natural environmental heterogeneity: Maternal effects and selection of life-history characters and plasticities in *Mimulus guttatus*. *Evolution* 49: 1095–1107.

Geber, M. A., and T. E. Dawson. 1993. Evolutionary responses of plants to global change. Pages 179–197 in P. Kareiva, J. Kingsolver, and R. Huey, eds., *Biotic Interactions and Global Change.* Sunderland, Mass.: Sinauer.

Harrington, H. D. 1964. *Manual of the Plants of Colorado*. Denver: Sage.

Harte, J., and R. Shaw. 1995. Shifting dominance within a montane vegetation community: Results of a climate-warming experiment. *Science* 267: 876–880.

Harte, J., M. S. Torn, F.-R. Chang, B. Feifarek, A. P. Kinzig, R. Shaw, and K. Shen. 1995. Global warming and soil microclimate: Results from a meadow-warming experiment. *Ecological Applications* 5: 132–150.

Havström, M., T. V. Callaghan, and S. Jonasson. 1993. Differential growth responses of *Cassiope tetragona*, in arctic dwarf-shrub, to environmental perturbations among three contrasting high- and subarctic sites. *Oikos* 66: 389–402.

Havström, M., T. V. Callaghan, and S. Jonasson. 1995. Effects of simulated climate change on the sexual reproductive effort of *Cassiope tetragona*. Proceedings of contributed and poster papers from the international conference. Pages 109–114 in T. V. Callaghan, W. C. Oechel, T. Gilmanov, J. I. Holten, B. Maxwell, U. Molau, B. Sveinbjörnsen, and M. Tyson, eds., *Global Change and Arctic Terrestrial Ecosystems*. Oppdal, Norway: Commission of the European Communities Ecosystems Research Report, Brussels.

Henry, G. H. R., and U. Molau. 1997. Tundra plants and climate change: The International Tundra Experiment (ITEX). *Global Change Biology* 3 (Suppl. 1): 1–9.

Hill, G. E., R. R. Sargent, and M. B. Sargent. 1998. Recent changes in the winter distribution of rufous hummingbirds. *Auk* 115: 240–245.

Hollister, R. D. 1999. *Plant Responses to Climate Change: Integration of ITEX Discoveries*. East Lansing: Michigan State University.

Hughes, L. 2000. Biological consequences of global warming: Is the signal already apparent? *Trends in Ecology and Evolution* 15: 56–61.

Inouye, D. W., and A. D. McGuire. 1991. Effects of snowpack on timing and abundance of flowering in *Delphinium nelsonii* (Ranunculaceae): Implication for climatic change. *American Journal of Botany* 78: 997–1001.

Inouye, D. W., W. A. Calder, and N. M. Waser. 1991. The effect of floral abundance on feeder censuses of hummingbird populations. *Condor* 93: 279–285.

Inouye, D. W., B. Barr, K. B. Armitage, and B. D. Inouye. 2000. Climate change is affecting altitudinal migrants and hibernating species. *Proceedings of the National Academy of Sciences* 97: 1630–1633.

Intergovernmental Panel on Climate Change (IPCC). 2001a. *Climate Change 2001: The Scientific Basis*, eds., J. T. Houghton, Y. Ding, D. J. Griggs, M. Noguer, P. J. van der Linden, and D. Xiaosu. Contribution of working group I to the third assessment report of the IPCC. Cambridge: Cambridge University Press.

————. 2001b. *Climate Change 2001: Impacts, Adaptation, and Vulnerability*, eds., J. J. McCarthy, O. F. Canziani, N. A. Leary, D. J. Dokken, and K. S. White. Contribution of working group II to the third assessment report of IPCC. Cambridge: Cambridge University Press.

Jackson, M. T. 1966. Effects of microclimate on spring flowering phenology. *Ecology* 47: 407–415.

Johnson, T. R. 1998. *Climate Change and Sierra Nevada Snowpack*. M.S. thesis, University of California, Santa Barbara.

Keeling, C. D., T. P. Whorf, M. Wahlen, and J. van der Plicht. 1995. Interannual extremes in the growth of atmospheric CO_2. *Nature* 375: 666–670.

Kelly, C. A. 1993. Quantitative genetics of size and phenology of life-history traits in *Chamaecrista fasciculata*. *Evolution* 47: 88–97.

Kittel, T. G. F., N. A. Rosenbloom, T. H. Painter, D. S. Schimel, and VEMAP Modeling Participants. 1995. The VEMAP integrated database for modeling United States ecosystem/vegetation sensitivity to climate change. *Journal of Biogeography* 22: 857–862.

Körner, C., F. A. Bazzaz, and C. B. Field. 1996. The significance of biological variation, organism interactions, and life histories in CO_2 research. In C. Körner and F. A. Bazzaz, eds., *Carbon Dioxide, Populations, and Communities*. San Diego: Academic Press.

Lacey, E., and R. Pace. 1983. Effect of flowering and dispersal times on offspring fate in Daucus carota (Apiaceae). *Oecologia* 60: 274–278.

Lawton, J. H. 1995. Ecological experiments with model systems. *Science* 269: 328–331.

Littell, R. C., G. A. Milliken, W. W. Stroup, and R. D. Wolfinger. 1996. *SAS System for Mixed Models*. Cary, N.C.: SAS Institute.

Loik, M. E., and J. Harte. 1996. High-temperature tolerance of *Artemisia tridentata* and *Potentilla gracilis* under a climate change manipulation. *Oecologia* 108: 224–231.

Lubchenco, J., A. M. Olson, L. B. Brubaker, S. R. Carpenter, S. P. Hollad, S. P. Hubell, S. A. Levin, J. A. MacMahon, P. A. Matson, J. M. Melillo, H. A. Mooney, C. H. Peterson, H. R. Pulliam, L. A. Real, P. J. Regal, and P. G. Risser. 1991. The sustainable biosphere initiative: An ecological research agenda. *Ecology* 72: 371–412.

Manabe, S. 1997. Early Development in the study of greenhouse warming: The emergence of climate models. *Ambio* 26: 47–51.

Manabe, S., and R. J. Stouffer. 1979. A CO_2 climate sensitivity study with a mathematical model of the global climate. *Nature* 282: 491–493.

————. 1980. Sensitivity of a global climate model to the increase of CO_2 concentration in the atmosphere. *Journal of Geophysical Research* 85: 5529.

Manabe, S., and R. T. Wetherald. 1967. Thermal equilibrium for the atmosphere with a given distribution of relative humidity. *Journal of Atmospheric Sciences* 24: 241–259.

——. 1975. The effect of doubling the CO_2 concentration on the climate of a general circulation model. *Journal of Atmospheric Sciences* 32: 3–15.

——. 1980. On the distribution of climate change resulting from the increase in CO_2 content of the atmosphere. *Journal of Atmospheric Sciences* 37: 99–118.

Martin, T. E. 1997. Demographic cost of shifting microhabitat use by birds in response to changing climate. *Workshop on the Impacts of Climate Change on Flora and Fauna*. Boulder, Co.: Birdlife International, WWF.

McNeilly, T., and J. Antonovics. 1968. Evolution of closely adjacent plant populations. IV. Barriers to gene flow. *Heredity* 23: 205–218.

Mølgaard, P., and K. Christensen. 1997. Response to experimental warming in a population of *Papaver radicatum* in Greenland. *Global Change Biology* 3: 116–124.

Mousseau, T. A., and D. A. Roff. 1987. Natural selection and the heritability of fitness components. *Heredity* 59: 181–197.

Myers, N., R. A. Mittermeier, C. G. Mittermeier, G. A. B. da Fonseca, and J. Kent. 2000. Biodiversity hotspots for conservation priorities. *Nature* 403: 853–858.

Oberbauer, S., G. Starr, and E. W. Pop. 1998. Effect of extended growing season and soil warming on carbon dioxide and methane exchange of tussock tundra in Alaska. *Journal of Geophysical Research* 103: 29075–29082.

O'Neil, P. 1997. Natural selection on genetically correlated phenological characters in *Lythrum salicaria* L. (Lythraceae). *Evolution* 51: 267–274.

Poole, R., and B. Rathcke. 1979. Regularity, randomness, and aggregation in flowering phenologies. *Science* 203: 470–471.

Price, J. 1997. Potential impacts of global climate change on summer distributions of some North American birds. *Workshop on the Impacts of Climate Change on Flora and Fauna*. Boulder, Co.: Birdlife International, WWF.

Price, M. V., and N. M. Waser. 1998. Effects of experimental warming on plant reproductive phenology in a subalpine meadow. *Ecology* 79: 1261–1271.

——. 2000. Responses of subalpine meadow vegetation to four years of experimental warming. *Ecological Applications* 10: 811–823.

Ramanathan, V., and A. M. Vogelmann. 1997. Greenhouse effect, atmospheric solar absorption and the earth's radiation budget: From the Arrhenius-Langley era to the 1990s. *Ambio* 26: 38–46.

Rathcke, B. 1988. Flowering phenologies in a shrub community: Competition and constraints. *Journal of Ecology* 76: 975–994.

Ritchie, J. C., and L. C. Cwynar. 1982. The Late Quaternary vegetation of the North Yukon. Pages 113–126 in D. M. Hopkins, J. V. Matthews Jr.,

C. E. Schweger, and S. B. Young, eds., *Paleoecology of Beringia*. New York: Academic Press.

Rodhe, H., R. Charlson, and E. Crawford. 1997. Svante Arrhenius and the greenhouse effect. *Ambio* 26: 2–5.

Root, T. L. 1993. Effect of global climate change on North American birds and their communities. Pages 280–292 in P. Kareiva, J. Kingsolver, and R. Huey, eds., *Biotic Interactions and Global Change*. Sunderland, Mass.: Sinauer.

———. 1997. Changes over 30 years in spring arrival dates of migrating birds: Is spring arriving three weeks earlier? *Workshop on the Impacts of Climate Change on Flora and Fauna*. Boulder, Co.: Birdlife International, WWF.

Root, T. L., and S. H. Schneider. 1993. Can large-scale climatic models be linked with multiscale ecological studies? *Conservation Biology* 7: 256–270.

———. 1995. Ecology and climate: Research strategies and implications. *Science* 269: 334–339.

Saavedra, F. 2000. Potential impact of climate change on the phenology and reproduction of *Delphinium nuttallianum* (Ranunculaccae). Ph.D. dissertation. University of Maryland, College Park, Maryland.

Saleska, S. R., J. Harte, and M. S. Torn. 1999. The effect of experimental ecosystem warming on CO_2 fluxes in a montane meadow. *Global Change Biology* 5: 125–141.

Santer, B. D., K. E. Taylor, T. M. L. Wigley, T. C. Johns, P. D. Jones, D. J. Karoly, J. F. B. Mitchell, A. H. Oort, J. E. Penner, V. Ramaswamy, M. D. Schwarzkopf, R. J. Stouffer, and S. Tett. 1996. A search for human influences on the thermal structure of the atmosphere. *Nature* 382: 39–46.

Schimel, D. S., VEMAP Participants, and B. H. Braswell. 1997. Continental scale variability in ecosystem processes: Models, data, and the role of disturbance. *Ecological Monographs* 67: 251–271.

Schmitt, J. 1983. Density-dependent pollinator foraging, flowering phenology, and temporal pollen dispersal patterns in *Linanthus bicolor*. *Evolution* 37: 1247–1257.

Schneider, S. H. 1993. Scenarios of global warming. Pages 9–23 in P. Kareiva, J. Kingsolver, and R. Huey, eds., *Biotic Interactions and Global Change*. Sunderland, Mass.: Sinauer.

Schneider, S. H., and S. L. Thompson. 1981. Atmospheric CO_2 and climate: Importance of the transient response. *Journal of Geophysical Research* 86: 3135–3147.

Shaver, G. R., L. C. Johnson, D. H. Cades, G. Murray, J. A. Laundre, E. B. Rastetter, K. J. Nadelhoffer, and A. E. Giblin. 1998. Biomass and CO_2 flux in wet sedge tundras: Responses to nutrients, temperature, and light. *Ecological Monographs* 68: 75–97.

Shugart, H. H. 1990. Using ecosystem models to assess potential consequences of global climatic change. *Trends in Ecology and Evolution* 5: 303–307.

Shugart, H. H., and D. C. West. 1977. Development of an Appalachian deciduous forest succession model and its application to the assessment of the impact of the chestnut blight. *Journal of Environmental Management* 5: 161–170.

Stanton, M. L. 1984. Seed variation in wild radish: Effect of seed size on components of seedling and adult fitness. *Ecology* 65: 1105–1112.

Stanton, M. L., and C. Galen. 1997. Life on the edge: Adaptation versus environmentally mediated gene flow in the snow buttercup, *Ranunculus adoneous*. *American Naturalist* 150: 143–178.

Stanton, M. L., C. Galen, and J. Shore. 1997. Population structure along a steep environmental gradient: Consequences of flowering time and habitat variation in the snow buttercup, *Ranunculus adoneous*. *Evolution* 51: 79–94.

Stenström, A., and I. S. Jónsdóttir. 1997. Responses of the clonal sedge, *Carex bigelowii*, to two seasons of simulated climate change. *Global Change Biology* 3: 89–96.

Stephenson, A. 1982. When does outcrossing occur in a mass-flowering plant? *Evolution* 36: 762–767.

Stratton, D. A. 1992. Life-cycle components of selection in *Erigeron annuus*. II. Genetic variation. *Evolution* 46: 107–120.

Tett, S., P. A. Stott, M. R. Allen, W. J. Ingram, and J. F. B. Mitchell. 1999. Causes of twentieth-century temperature change near the earth's surface. *Nature* 399: 569–572.

Thébaud, C., and A. Johnston. 1997. Plant responses to global changes in CO_2: Unfinished business? *Trends in Ecology and Evolution* 12: 425–426.

Thomson, J. D. 1980. Skewed flowering distributions and pollinator attraction. *Ecology* 61: 572–579.

Tilman, D. 1996. Biodiversity: Population versus ecosystem stability. *Ecology* 77: 350–363.

Tilman, D., and J. A. Downing. 1994. Biodiversity and stability in grasslands. *Nature* 367: 363–365.

Torn, M. S., and J. Harte. 1996. Methane consumption by montane soils: Implications for positive and negative feedback with climate change. *Biogeochemistry (Dordrecht)* 32: 53–67.

Tsonis, A. A. 1996. Widespread increases in low-frequency variability of precipitation over the past century. *Nature* 382: 700–702.

VEMAP Participants. 1995. Vegetation/Ecosystem modeling and analysis project (VEMAP): Comparing biogeography and biogeochemistry models in a continental-scale study of terrestrial ecosystems to climate change and CO_2 doubling. *Global Biogeochemical Cycles* 9: 407–438.

Vitousek, P. M., H. A. Mooney, J. Lubchenco, and J. M. Melillo. 1997. Human domination of Earth's ecosystems. *Science* 277: 494–499.

Waser, N. M. 1976. Food supply and nest timing of broad-tailed hummingbird in the Rocky Mountains. *Condor* 78: 133–135.

———. 1978. Competition for hummingbird pollination and sequential flowering in two Colorado wildflowers. *Ecology* 59: 934–944.

———. 1979. Pollinator availability as a determinant of flowering time in Ocotillo (*Fouquieria splendens*). *Oecologia* 39: 107–121.

Waser, N. M., and M. V. Price. 1981. Pollination choice and stabilizing selection for flower color in *Delphinium nelsonii*. *Evolution* 35: 376–390.

———. 1990. Pollination efficiency and effectiveness of bumble bees and hummingbirds visiting *Delphinium nelsonii*. *Collectanea Botanica* 19: 9–20.

———. 1991. Outcrossing distance effects in *Delphinium nelsonii*: Pollen loads, pollen tubes, and seed set. *Ecology* 72: 171–179.

———. 1994. Crossing-distance effects in *Delphinium nelsonii*: Outbreeding and inbreeding depression in progeny fitness. *Evolution* 48: 842–852.

Waser, N. M., and L. Real. 1979. Effective mutualism between sequentially flowering plant species. *Nature* 281: 670–672.

Waser, N. M., L. Chittka, M. V. Price, N. M. Williams, and J. Ollerton. 1996. Generalization in pollination systems, and why it matters. *Ecology* 77: 1043–1060.

Wookey, P. A., A. N. Parsons, J. M. Welker, J. A. Potter, T. V. Callaghan, J. A. Lee, and M. C. Press. 1993. Comparative responses of phenology and reproductive development to simulated environmental change in subarctic and high arctic plants. *Oikos* 67: 490–502.

Wookey, P. A., J. M. Welker, A. N. Parsons, M. C. Press, T. V. Callaghan, and J. A. Lee. 1994. Differential growth, allocation and photosynthetic responses of *Polygonum viviparum* to simulated environmental change at a high arctic polar semi-desert. *Oikos* 70: 131–139.

Zimmerman, M. 1987. Reproduction in *Polemonium*: Factors influencing outbreeding potential. *Oecologia* 72: 624–632.

CHAPTER 6

Modeling Potential Impacts of Climate Change on the Spatial Distribution of Vegetation in the United States with a Probabilistic Biogeography Approach

ELENA SHEVLIAKOVA

Anthropogenic activities are leading to rapid changes in land cover and are responsible for emissions of greenhouse gases into the atmosphere, particularly carbon dioxide (CO_2). These activities may bring about climatic changes characterized by an increase in average global temperature of 1 to 5°C over the next century and changes in precipitation regime (IPCC 1996a). Climatic changes of such magnitude are likely to alter the distribution of terrestrial ecosystems on a large scale. The motivations for halting or slowing down climate change arise entirely from our moral considerations and concerns about various adverse impacts. Ecological impacts deserve special attention.

There is a growing recognition of the significance of a healthy biosphere for the well-being of society. In the last two decades, the paradigms of sustainable development and Earth stewardship penetrated and influenced public debate around the globe. People have become more concerned with the well-being of ecosystems and the role they might play in maintaining and promoting environmental quality. There are two general reasons to be concerned about ecosystems' health and functioning. First, ecosystems have an instrumental value; that is, society can use them for economic and scientific purposes. Loss of ecosystems and species will limit humans' ability to understand the determinants of diversity, population dynamics, ecosystem regulation, and resource-sharing strategies. Second, ecosystems are the object of aesthetic appreciation and moral attention and possess intrinsic value. That is, we care about their existence in and of itself, rather than just about the benefits they confer on us.

Simulations of future regional and global vegetation distributions and results from experiments on individual plants indicate that climate change is expected to have significant effects on the functioning and distribution of major ecosystems (Walker and Steffen 1997). Ongoing loss and fragmentation of habitats from human land practices, combined with the anticipated impacts of climate change, have led to a widely voiced concern about preservation of biodiversity (Pimm et al. 1995). Expansion of agricultural and urban areas will accelerate in the future and will bring about even more dramatic, possibly catastrophic, losses of biodiversity.

Modeling vegetation distribution structure is an important step in understanding fauna's response to climate change because vegetation provides shelter, foraging and breeding grounds, and migration and wintering sites. Insights into how these grounds and sites can potentially change contribute to our understanding of potential impacts of climate change on terrestrial biodiversity. Animals may be able to adapt to new habitats if the assemblages of vegetation are structurally similar to the current conditions, even if the mix of vegetation species may change (Graham 1992). However, if rapid climate change and habitat fragmentation are combined, a markedly different structure of ecosystems (the interactive web of flora and fauna) will limit animals' ability to adapt to and occupy the new habitats in as diverse and robust a form (Webb 1992). According to Myers and Lester (1992), vegetation structure is a key variable in choice of breeding sites for many northern shorebirds. For example, two shorebird species (sanderlings and knot) breed almost exclusively in tundra above 75°N. Expected reduction in the extent of tundra or its structure may lead to significant declines in population of these species. Other studies indicate that migration abilities of North American wintering birds may depend on availability of specific vegetation along their migration paths (Sorenson et al. 1998). Changes in vegetation structure can also affect animals' behavioral characteristics and their competitive abilities. Riechert (1986) demonstrated that spiders are much less aggressive in wet riparian areas than in dry grasslands.

The continuing difficulty of assessing ecological impacts of climate change is due in part to the complexity in the coupling of the flow of water, energy, and carbon between the biosphere and the atmosphere (Bonan 1995). Numerous studies demonstrate the importance of the biosphere for the functioning of global biogeochemical cycles and of the climate system itself (Cox et al. 2000, Pacala et al. 2001). Recent results of CO_2 fertilization experiments indicate that the beneficial effect of elevated CO_2 on plants may be significant, but less than originally expected in the earlier studies (Walker and Steffen 1997, DeLucia et al. 1999, Schlesinger and Lichter 2001). The magnitude of CO_2 fertilization will depend on a variety of other factors, such as changes in ambient environment and rates of nitrogen deposition. A number of modeling studies suggest that drastic changes could occur in ecosystem distribution and functioning, resulting in the loss of biodiversity and the release of large amounts of CO_2 into the atmosphere (IPCC 1996b, Morgan et al. 2001).

Modeling Global-Scale Vegetation–Climate Interactions

In recent years, the climate change community has taken a number of different approaches toward understanding physical, ecological, and biogeochemical aspects of vegetation–climate interactions. Studies have been performed on different spatial and temporal scales, from an individual leaf to the entire globe (IPCC 1996b). Global impact studies have been performed with biogeography and biochemistry models (e.g., VEMAP 1995). The newly emergent approach is the integrated study of biosphere and climate interactions (Foley et al. 1996, Betts et al. 1997).

Biogeography models, also known as global equilibrium vegetation transfer or bioclimatic models, attempt to simulate global equilibrium distributions of vegetation under current and future climates. Most of these models are correlative in nature and rely on observed associations between climate and vegetation (Prentice et al. 1992; Siegel et al. 1995). Biogeochemistry models focus on the simulation of nutrient flows among vegetation, soil, and the atmosphere. Typically, these models deal only with changes in carbon and nitrogen flows and net primary productivity (NPP) due to climate change. These models assume constant distributions of vegetation and soil types. The Vegetation/Ecosystem Modeling and Analysis Project (VEMAP) study (VEMAP 1995) compared the sensitivity of different biogeochemistry models to assumptions about different vegetation distributions, carbon fertilization effects, and climate change.

Development of dynamic global vegetation models (DGVMs) remains a high-priority task in predicting realistic impacts of future climate changes on vegetation. The approach of Woodward et al. (1995) aims to model global vegetation in terms of Leaf Area Index (LAI), NPP, and stomatal conductance. These are calculated with a plant-productivity and phytogeography model. This model is able to react to different disturbance regimes. A different strategy is advanced by Haxeltine and Prentice (1996), Ni et al. (2000) and Foley et al. (1996). Their approach is based on the concept that vegetation system dynamics are shaped to maximize NPP. Thus, for a given climate, a plant functional type can be identified that can survive the local conditions (soil and climatic) while maximizing NPP. The third strategy is incorporated in the process-based model Hybrid of Friend et al. (Friend et al. 1997). It is based on the stand modeling approach and is able to project spatially, temporally, and biologically detailed responses to climate change. The dynamic models are more sophisticated than earlier equilibrium models in representing physiological processes, describing vegetation in terms of plant functional types or species types, and in simulating abilities of different types to compete for resources. The dynamic models are highly complex, computationally intensive, and require a large number of parameters. The pattern of simulated vegetation distributions from the current climate conditions compares favorably to actual vegetation distributions in very few models (e.g., Moorcroft et al., in press).

While DGVMs continue to be developed and improved (Cramer et al. 2001), our understanding of the likelihood and magnitude of potential changes in vegetation structure can be improved by analyzing current vegetation distribution using the equilibrium biogeography approaches. For example, a probabilistic biogeography model has been used to evaluate ecological risk of possible changes in the potential forest distribution over areas of Switzerland under different climate change scenarios (Kienast et al. 1996). The quantification of probabilities of different changes in vegetation composition (i.e., risk) can be used in a number of ways: by policy makers in climate change policy development; in ecosystem impacts studies to identify high risk areas and vegetation types for further more detailed research; to educate the general public about risks to various ecosystems given a particular climate scenario; and to identify high-risk ecosystems in terms of species loss for monitoring or conservation practices. Because biogeography models are much simpler in structure and computational and parameter requirements than DGVMs, these equilibrium models or transient models derived from them continue to be used in integrated assessments (e.g., ICAM [Dowlatabadi and Morgan, in preparation] and IMAGE-2 [Alcamo 1994]). For example, Kirilenko and Solomon (1998) further explored the possibility of incorporating dynamic vegetation responses, such as migration, withdrawal, and immigration, into the equilibrium bioclimatic schemes of BIOME1 (Prentice et al. 1992).

Modeling Vegetation–Climate Relationships with Equilibrium Models

Typically, equilibrium biogeography models are correlative in nature, deterministic, and use heuristics in the form of process-based rules to classify vegetation types for a given set of climatic and soil variables (VEMAP 1995). Most climatic constraints are estimated with rigid environmental-envelope boundaries and ignore interactions among climatic variables (Box 1981). These models are often criticized for lacking causal relationships between climate and plant physiology, and thus their presumed ability to deal only with the current distribution of biomes. However, at the global scale, climate largely controls the distribution of vegetation, particularly at the aggregate level. Biogeography models such as Mapped Atmosphere-Plant-Soil system (MAPSS) (Neilson 1995) and BIOME1 (Prentice et al. 1992) have combined elements of correlative and mechanistic approaches and are now focusing on modeling trends in distribution of potential vegetation. The later mechanistic models, such as BIOME 3.0 (Haxeltine et al. 1996), are able to account for different mechanistic processes and describe vegetation as a combination of plant functional types (PFTs) able to compete for resources.

An alternative to a rule-based bioclimatic classification is a probabilistic approach (Siegel et al. 1995). In this approach, relationships between climatic and geomorphologic variables and distribution of vegetation types are determined empirically from available data using multivariate nonparametric density functions of vegetation occurrence. The probabilistic approach to modeling vegetation distribution was originally proposed in the 1960s by Richard Goodall but had only limited use due to the lack of computational power at that time. This approach has been used on a regional scale to simulate current and future distributions of plant species in the Alpine region (Kienast et al. 1996), and on the west coast of the United States (Prentice, pers. comm. 1995, regarding study of Bartlein). In a probabilistic model a prevalence of a specific vegetation type is determined from a statistical relationship. Parameters of such statistical models are estimated from the climate, terrain, soil, and vegetation distribution data. Siegel et al. (1995) analyzed the relationship between the Olson vegetation types distribution and average climate characteristics using a multinomial generalized logit model. For the three continuous explanatory variables T, P, and Z and the dependent categorical variable vegetation type with 17 levels of response the model is

$$\log\left(\frac{\pi_j(T, P, Z)}{\pi_{17}(T, P, Z)}\right) = \alpha_j + \beta_j T + \gamma_j P + \eta_j Z, \, j = 1,..., 16$$

$$\pi_{17}(T, P, Z) = 1 - \sum_{j=1}^{16} \pi_j(T, P, Z)$$

where the model parameters α_j, β_j, γ_j, and η_j were estimated with the categorical data modeling (CATMOD) procedure of the SAS statistics package. The major drawback of a multinomial logit model is its inability to provide a good approximation for the highly nonlinear relationships between occurrence of vegetation types and the set of explanatory variables. An alternative to multinomial logit model is a method based on nonparametric density estimation. This method addresses the issue of nonlinearity and allows for a larger set of independent variables. Nonparametric density estimation and its results are described in the following text.

Choice of Independent Variables for a Probabilistic Model

It is important to acknowledge that one could select a different set of independent climate and soil variables and types of vegetation classification to describe bioclimatic relationships. Prentice (1990) argues that "any combination of terms has potential, and yet there is no perfect index from a biological point of view." There is widespread acknowledgment that climatic factors such as ambient temperature, incident solar radiation, and water availability play an important role in the distribution of plants (Whittaker 1978). These climatic factors are complex in nature and can therefore be described by different ensembles of variables.

In this study, 13 aggregated types were used to characterize different vegetation communities/land cover (Table 6.1). The distribu-

Table 6.1. Vegetation Types and Their Relative Areas.

Dominant Vegetation/Land Cover	Percentage of the Conterminous U.S. Area
1. Crops	29.63
2. Inland water	2.29
3. Wetlands	0.07
4. Grasses	21.63
5. Shrubs: creosote, mesquite, cactus, and saltbrush	10.86
6. Oak, hickory, deciduous woodlands, few subtropical plants	8.07
7. Beech and maple	3.04
8. Birch, aspen, willow and alder	5.27
9. Spruce, fir, larch, hemlock, and cedar	3.92
10. Pine	11.65
11. Pinion juniper	2.33
12. Tundra	0.16
13. Barren/Ice	1.08

tion of aggregate vegetation types was derived from the U.S. Geological Survey 1 km Seasonal Land Cover Data Set. This data set has more than 200 different land cover categories. However, each of the 200 categories can be represented by a combination of a smaller number of vegetation types. For example, category 165, "Mixed Forest (Pine and Oak)," consists of two vegetation types, pine and oak, with a weighting factor of 0.5 assigned to each of them. The 200 categories of the original data set provide sufficient information to describe each of 1 km grid cells over the conterminous United States in terms of fractions of the 26 distinctive vegetation types. However, some of these vegetation types occur sparsely (e.g., willow or red cedar), and therefore can be aggregated with vegetation types with similar climatic ranges and physiological characteristics. Information about vegetation composition from a 1 km grid was further aggregated into the information on a 1° latitude by 1° longitude grid.

A number of independent variables were investigated in order to explore model specification issues in the simulation of vegetation prevalence. The climate variables were selected to represent seasonal and annual characteristics of climate important for plant growth and development: average annual temperature and precipitation, spring, summer and autumn precipitation, minimum and maximum temperatures, and biotemperature. Although soil characteristics play an important role in vegetation functioning, most soil-related variables were not included due to the gross resolution of the available data on a global scale and the consequently low estimates of statistical significance. Climatic variables were derived from the Cramer database of average monthly temperatures and precipitation for 30-minute resolution.

In correlative bioclimatic models, temperature serves as a surrogate for solar radiation available for vegetation growth and development. Although there might be intraannual variations in temperature, more long-term patterns persist and are most important to plants (Box 1981, Woodward 1987). Temperature levels affect metabolic rates of plants, uptake and loss of water and CO_2, and their growth and development. Temperature extremes determine the limits of potential plant distribution and adaptation. Box (1981) suggested that the extreme temperatures can be represented by the average maximum and minimum (monthly) values and/or by the annual range of monthly mean temperatures. Mean temperature of the warmest month (TMAX) is a proxy for the circumstances under which a plant's respiration exceeds plant photosynthetic production and when metabolic collapse may occur. Mean temperature of the coldest month (TMIN) represents the cold tolerance capacities of vegetation types. TMIN also may represent chilling requirements: successful functioning of seasonal plants may require periods of dormancy and vernalization, which are induced by the low temperatures. Prentice et al. (1992) suggested that some plant types might not need cold periods (e.g., tropical evergreen and raingreen forests). Box (1981) suggested that different values of TMIN are among the key determinants in prevalence of evergreen and summergreen plants.

Plants require a period that is sufficiently warm for growth. Such requirements often have been expressed through biotemperature and growing degree days Biotemperature is a sum of temperatures that exceed a chosen threshold T_0 (e.g., 0°C or 4°C). The integral of biotemperature over an annual period represents growing degree days (GDD). The value of appropriate T_0, and subsequently the computed value of GDD, varies for different types of plants. Plants native to colder environments generally have lower GDDs than those from warmer environments (Prentice et al. 1992). Requirements for a certain level of biotemperature or GDD might be explained by the plants' respiration requirements. According to this hypothesis, under low temperatures, respiration becomes insufficient, followed by inhibition of cellulose synthesis.

The availability of water plays an important role in the water and energy balance of vegetation. The lowest average monthly precipitation (PMIN) represents plants' tolerance to drought. This metric is particularly important for evergreen plants (Box 1981). Occurrence of a vegetation type under low values of PMIN indicates that a particularly severe drought must occur before plants undergo changes in specific processes (e.g., dormancy) during their life cycle. Box (1981) noted that PMIN is particularly useful for making the distinction between raingreen and evergreen vegetation. Highest average monthly precipitation (PMAX) provides a measure of the amount of water required to trigger the beginning and/or continuation of vegetation growth. This constraint on vegetation growth plays a particularly important role in climates with seasonal or irregular precipitation. Low occurrence of a vegetation type under high values of PMAX may reflect an inability of certain plants (e.g., succulents) to function in excessive water conditions or an inability to compete with other plants for water resources. Average precipitation in the warmest month (PTMAX) serves as a metric for classifying the severity of summer drought. In some climates, precipitation reaches a maximum in the warmest month (e.g., boreal, temperate continental, and tropical climates). In others (e.g., Mediterranean climates), precipitation is lowest in the month with the highest temperature. Annual average precipitation (PMEAN) represents the overall moisture available to plants. An alternative to this set of precipitation variables is a set of seasonal precipitation variables (i.e., averages in winter, spring, summer, and autumn). For example, Chapin and Starfield (1997) used summer precipitation as one of the proxies for climate in modeling ecosystem responses to climate change in arctic Alaska.

Many equilibrium models include different measures of evapotranspiration that describe the physical process of water transfer into the atmosphere by evaporation from the soil and transpiration through the vegetation. Evapotranspiration depends both on climatic factors and surface characteristics (e.g., vegetation). Climatic factors include temperature, net radiation, wind speed, and humidity. Vegetation cover defines evapotranspiration through both mechanical (e.g., surface roughness) and physiological (e.g., stomatal aperture) means. The potential evapotranspiration (PET) is the amount of water that would be released to the atmosphere under natural conditions from a bare surface with sufficient, but not excessive, water availability. The relationship between available moisture and solar radiation is often expressed through such variables as moisture index (MI) (Box 1981), the inverse of moisture index—dryness index (DI) (Monserud et al. 1993), or potential evapotranspiration ratio (PETR). The annual moisture index (MI) represents the relationship between potential water loss and total water available:

$MI = PRCP/PET,$
where   *PRCP* is annual precipitation.
        PET is annual potential evapotranspiration.

The difference between MI, DI, and PETR is that each uses a different method for calculating annual potential evapotranspiration. Box (1981) used the Thornthwaite method. Monserud and others (Monserud et al. 1993) based their estimation on the work by Budyko.

Three different criteria were applied for the choice, independent of variables: physiological relevance, explanatory power, and data availability. Stepwise discriminant analysis, multiple comparison of means, and multicolinearity tests were used to identify the minimum set of independent variables that explain variations in the distribution of vegetation prevalence under present climatic and geomorphologic conditions. Similarity between mean values of independent variables for many vegetation types suggests that it is impossible to accurately predict a specific vegetation type occurrence as a function of a single climatic or geomorphologic variable. However, combinations of such variables may provide a means for the determination of climatic conditions under which a specific vegetation type is more likely or less likely to prevail. Furthermore, the analysis indicates that the combination of average annual precipitation and temperature could only explain less than half of the variation in the distribution of vegetation types. In order to explain the remaining variation, seasonal climate characteristics (e.g., TMIN, TMAX) or specific physiologically relevant variables (e.g., biotemperature) need to be included. Although soils play an important role in vegetation functioning, most soil-related variables were not included due to insufficient resolution of available data, leading to low statistical significance. Other variables such as potential evapotranspiration or moisture index, initially considered important, were excluded because they are multicolinear with other variables. The results of stepwise discriminant analysis suggest that a set of seasonal precipitation variables (e.g., average spring precipitation) has more explanatory power than a set of precipitation extremes (e.g., PMIN or PMAX). In this study eight climatic variables were used: TMIN, TMAX, TMEAN, BIOT, PMEAN, Pwinter, Pspring, and Psummer.

Nonparametric Density Estimation Method

Once a set of climatic variables is identified, multivariate nonparametric density estimation is applied to estimate probability density functions of each vegetation type. Nonparametric density estimation methods are well understood and have been increasingly used in practice for both univariate and multivariate analysis in such areas as chemical and electrical engineering and medical biostatistics. A comprehensive review of different density estimation methods can be found in Silverman (1986). The advantages of this approach are: (1) the class conditional distribution estimates and the decision surfaces are more flexible, (2) a more direct and significant role is given to the data, (3) no a priori assumptions about the parametric family of distribution are required, and (4) both categorical and continuous variables can be used as explanatory variables.

In a nonparametric density approach, every observation represents a center of the sampling interval. By placing a kernel density K(x) around each observation one can define a probability density function:

$$\hat{f}(x) = \frac{1}{Nh}\sum_{i=1}^{N} K\left(\frac{x - X_i}{h}\right), \text{ where}$$

$K(x)$ is a kernel density function such as $\int_{-\infty}^{\infty} K(x)\,dx = 1$,
N is a number of observations,
h is a smoothing parameter or bin width,
X_i is the $i-$ th observation of a variable x.

The kernel density estimator can be thought as a sum of "bumps" centered at each available observation (see Fig. 6.1, adapted from Silverman 1986). The type of a kernel density function determines the shape of the bumps. The bin width determines the smoothness of the estimator. Unlike histograms, kernel density estimators are continuous. There are different types of kernel den-

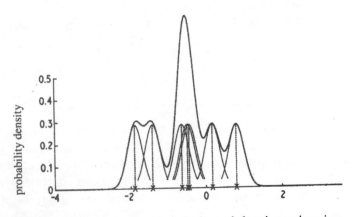

Figure 6.1. Schematic representation of a kernal density estimation.

sity functions: Gaussian, triangular, rectangular, and so forth. The Gaussian kernel was used in this model. The window width affects the smoothness of the estimator. With wider windows, the distributions obtained become less jagged.

Changes in Vegetation Composition under Four Different Climate Change Scenarios

The nonparametric density functions were estimated for the current climate described by the combinations of eight climatic variables. In order to compare results of estimations with the observations, an index V of deviation from an observed state was computed for every grid cell i in the following manner:

$$V_i = \sqrt{\sum_j (\Delta f_j)^2} \Big/ \sqrt{2}$$

where Δf_j represents change in probability of prevalence of a vegetation type j. The V index always lies between 0 and 1, with 1 being complete change in vegetation composition and 0 no change. Figure 6.2 shows distribution of the V index computed from the observed fractions of vegetation and those estimated with probability density function under current climate. Figure 6.2 illustrates that

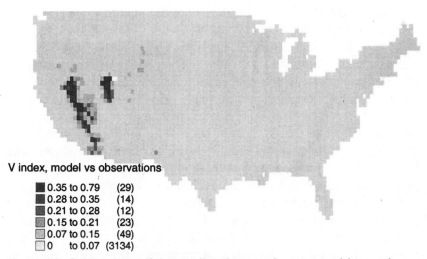

V index, model vs observations

■ 0.35 to 0.79   (29)
■ 0.28 to 0.35   (14)
■ 0.21 to 0.28   (12)
▨ 0.15 to 0.21   (23)
▧ 0.07 to 0.15   (49)
□ 0   to 0.07 (3134)

Figure 6.2. Comparison of the predicted vegetation composition under current climate to the observed vegetation composition.

Table 6.2. Range of changes in the climatic variables under four different climate scenarios.

Variable	GISS	GFDL	CCC	MM4
TMIN, °C	2.5 to 7.2	2.3 to 7.1	2.1 to 7.8	0.5 to 7.9
TMAX, °C	2.2 to 7.0	1.9 to 7.9	2.4 to 4.9	0.5 to 5.1
TMEAN, °C	3.1 to 5.3	3.0 to 6.6	2.2 to 5.9	2.3 to 5.3
BIOT,°C	1.7 to 5.3	1.6 to 5.4	1.0 to 5.4	1.3 to 4.1
PMEAN, mm/mn	-7 to 49	-19 to 63	-17 to 19	-19 to 434
Pwinter, mm/mn	-21 to 77	-72 to 16	-36 to 21	-32 to 231
Pspring, mm/mn	-15 to 48	-41 to 83	-26 to 87	-52 to 176
Psummer, mm/mn	-5 to 48	-57 to 240	-21 to 52	-83 to 471
Pautumn, mm/mn	-35 to 73	-15 to 86	-23 to 41	-50 to 1424

under current climate conditions the composition of vegetation is estimated very accurately in the majority of grid cells.

The estimated density functions then were used to calculate the probability of occurrence for each of the 11 vegetation types for each 1 degree by 1 degree grid cell and each of the four scenarios of climate change. Inland water bodies and the extent of agricultural and urban areas were considered constant under all scenarios. The climate changes were modeled as changes in *each* of the average and seasonal climatic variables. The new sets of climates were derived from VEMAP Climate Change Scenario Databases. The new sets of climate variables for each scenario were recomputed from the monthly temperature and precipitation variables obtained from three global circulation models (GISS, GFDL, and CCC GCMS) and one mesoscale atmospheric model (MM4) under assumptions of $2 \times CO_2$ atmospheric concentration. The ranges of changes in each of the climatic variables are shown in Table 6.2. Additionally, Figures 6.3A, B and 6.4A, B display spatial patterns of changes in average temperature and precipitation. The table and figures indicate that all four scenarios predict a similar magnitude of warming, but the spatial patterns of warming differ markedly. The models' predictions vary in both the magnitude and the spatial distribution of the expected changes in precipitation.

The distributions of the V index for the four different climate change scenarios are presented in Figures 6.5A and 6.5B. Light gray indicates little or no change, and dark gray indicates substantial changes in composition of vegetation. In all four cases the major changes in composition of vegetation occur along the West Coast, in the Rocky and Appalachian Mountains, and in the Northeast of the United States. The Midwest and parts of the Southwest do not exhibit strong changes in vegetation composition. This result could be attributed to the fact that these regions are in agricultural or urban areas and were assumed to be unchanged. Changes also appear to be small in the lower part of the Southwest (e.g., Texas). However, in this case the small V index can be attributed to the fact that all four climate change scenarios show relatively small warming there without drastic reductions in precipitation. The West Coast and Rocky Mountains exhibit significant changes in composition of vegetation under all four scenarios. In the Northeast, Southwest, and Midwest, the most pronounced changes in vegetation composition occur under the GFDL scenario and the least under the CCC scenario. Table 6.3 illustrates the magnitude of possible changes in the total areas dominated by different vegetation types. Although these estimates were computed at two different climate–vegetation equilibria and do not represent the actual transient responses of vegetation, they still show whether the areas occupied by a specific vegetation type will decrease or increase with changing climate. For

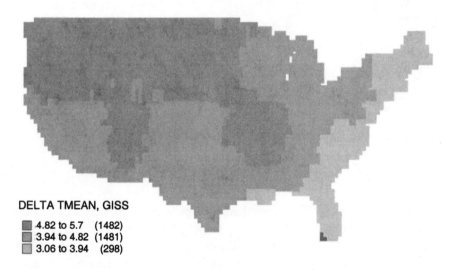

DELTA TMEAN, GISS

■ 4.82 to 5.7   (1482)
▨ 3.94 to 4.82 (1481)
☐ 3.06 to 3.94   (298)

Figure 6.3a. Changes in average temperature.

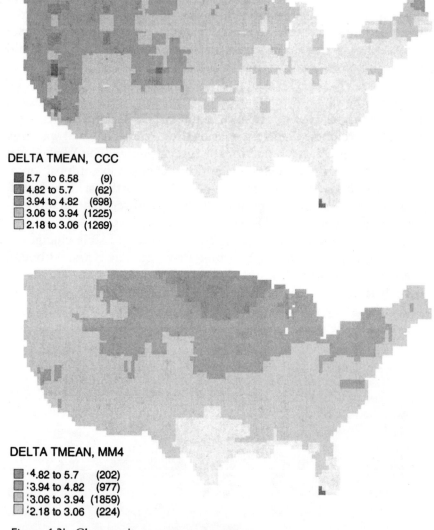

DELTA TMEAN, CCC

- ■ 5.7 to 6.58   (9)
- ▨ 4.82 to 5.7   (62)
- ▥ 3.94 to 4.82   (698)
- ▢ 3.06 to 3.94 (1225)
- ▢ 2.18 to 3.06 (1269)

DELTA TMEAN, MM4

- ▨ 4.82 to 5.7   (202)
- ▨ 3.94 to 4.82   (977)
- ▢ 3.06 to 3.94 (1859)
- ▢ 2.18 to 3.06   (224)

Figure 6.3b. Changes in average temperature.

example, it is likely that under all four scenarios the areas currently occupied by such coniferous species as spruce, fir, larch, and hemlock will decrease. Figure 6.6 shows a possible direction of migrations for some vegetation types under GFDL and GISS climate change scenarios. Deciduous (e.g., oak) and coniferous (e.g., pine,

DELTA PMEAN, GISS

▓ 39.5 to 51.3	(4)
▒ 27.7 to 39.5	(25)
░ 15.9 to 27.7	(79)
░ 4.1 to 15.9	(1108)
□ -7.7 to 4.1	(2045)

Figure 6.4a. Changes in average annual precipitation.

spruce) vegetation types show generally northward movement in their prevalence under the assumed changes in climate. However, the extent and spatial configuration vary depending on the magnitude of warming and precipitation regime.

Table 6.3. Changes in the Area of Different Vegetation Types (sq. km)

Dominant vegetation or land cover	GISS	GFDL	CCC	MM4
1. Crops	0	0	0	0
2. Water	0	0	0	0
3. Wetlands	5,598	37,272	9,244	8,371
4. Grasslands	11,095	155,650	108,262	-2,378
5. Shrubs	189,468	133,091	235,071	153,880
6. Oak, hickory, etc.	-114,757	527	-12,015	-39,096
7. Beech and maple	-114,380	-148,409	-85,055	-102,507
8. Birch, aspen, etc.	-169,568	-225,532	-138,341	-142,558
9. Spruce, fir, etc.	-37,458	-51,881	-54,281	-25,840
10. Pine	233,935	146,065	-28,997	199,386
11. Pinion juniper	48,079	-9,528	-159	-24,550
12. Tundra	-6,329	-4,008	-6,630	-5,279
13. Barren /Ice	-45,616	-32,099	-27,174	-19,014

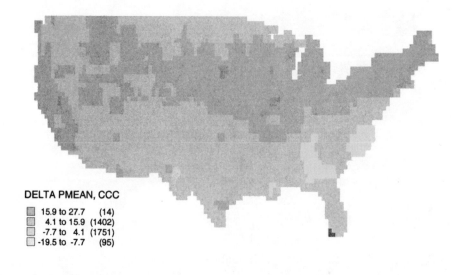

DELTA PMEAN, CCC

- 15.9 to 27.7   (14)
- 4.1 to 15.9 (1402)
- -7.7 to 4.1 (1751)
- -19.5 to -7.7   (95)

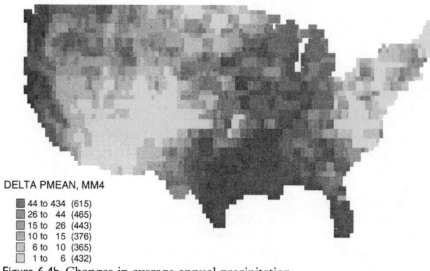

DELTA PMEAN, MM4

- 44 to 434 (615)
- 26 to 44 (465)
- 15 to 26 (443)
- 10 to 15 (376)
- 6 to 10 (365)
- 1 to   6 (432)

Figure 6.4b. Changes in average annual precipitation.

Conclusion

The objective of this study is to formulate and to test simple, physically based, probabilistic models of vegetation–climate relationships for the assessment of the likelihood of changes in regional vegeta-

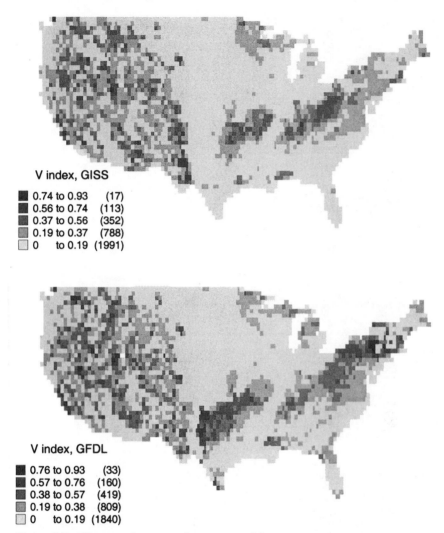

Figure 6.5a. Changes in vegetation composition.

tion composition under several different scenarios of climate change. Nonparametric density estimation approach can serve as an exploratory tool for the analysis of highly nonlinear interactions between vegetation, soil, and climate. It allows one to produce and compare estimates of vegetation prevalence with different ensembles of independent variables and vegetation classification. The approach easily permits one to work with mixtures of vegetation

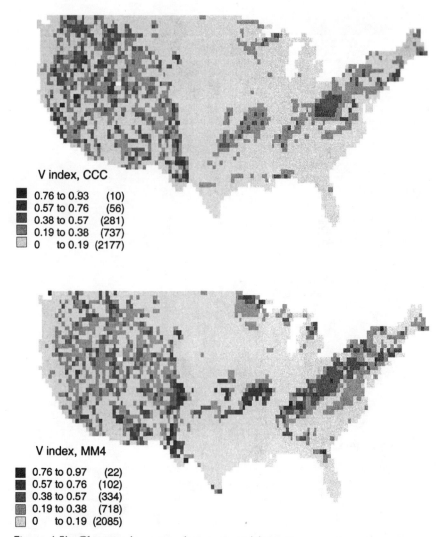

Figure 6.5b. Changes in vegetation composition.

types rather than a single dominant category. The major conclusions of the study are: (1) there are likely to be significant changes in the composition of U.S. vegetation under all four assumed scenarios of climate change; (2) warmer and drier climates are likely to cause reduction in the prevalence of many coniferous species (e.g., spruce, fir, larch, etc.) and deciduous species (maple, birch, etc.); (3)

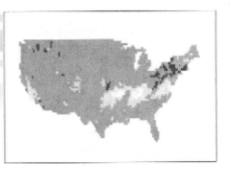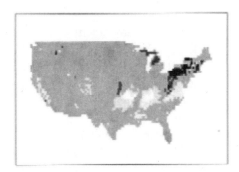

Oak and hickory

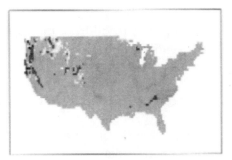

Spruce, fir, larch, hemlock or cedar

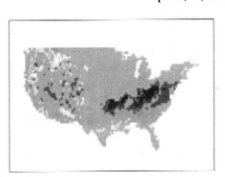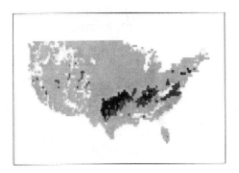

Pine

Figure 6.6. Changes in vegetation prevalence under GISS and GFDL climate change scenarios.

warmer and drier climates may substantially increase prevalence of shrubs and grass, particularly in the Western United States; and (4) wetter climates without significant warming could increase prevalence of pine species in the Southeast.

This probabilistic modeling approach to biogeography has a number of weaknesses. It does not consider factors such as elevated CO_2, nitrogen deposition, and changes in the disturbance regimes (e.g., fires). Although it captures the nonlinear relationship between vegetation prevalence and climate, it does not employ a process-based, mechanistic explanation of this relationship. It uses information from climate models, which operate on a scale much larger than the scale of a plant resource allocation or the scale of population or community dynamics. Several different strategies have been proposed to address the problem of scaling. Root and Schneider (1995) proposed a "strategic cyclical scaling" to model climate–ecosystem interactions: Given a model parameterization being scale dependent, one must cycle back and forth across the scales testing mechanistic constructions at a large scale and then cycle back to see if the processes are adequately represented. Hurtt et al. (1998) advocate development of biological models that are formulated at an intermediate scale of biological detail and are tested at different spatial and temporal scales.

Most important, as in many other biogeography models, the probabilistic model relies on the assumption that the current distribution of vegetation is in equilibrium with the current climate. The dynamics and end points of the changes in vegetation composition will depend on the rate of climate change and the degree of fragmentation in land cover as well as on nonlinear responses of individual components of ecosystems. For example, different taxa—insects, birds, and animals—can have differential transient responses in response to changes in climate and/or disturbance regime. Communities of species may become disaggregated during such transitions and may have responded in unexpected ways to their environment and climate (Root and Schneider 1993). Changes in land practices due to the growing population will further exacerbate impoverishment of the biosphere. While the assumption of equilibrium is clearly not true in an absolute sense, the errors due to current disequilibrium are presumed to be small enough that the predictions can still serve as a useful starting point for assessing where and how vegetation composition can change.

Acknowledgments

This work was supported by National Wildlife Federation Climate Change and Wildlife Fellowship, by Electric Power Research Institute grant (RP 3236), and by National Science Foundation grant (SBR-9521914).

Literature Cited

Alcamo, J., ed. 1994. *IMAGE 2.0: Integrated Modeling of Global Climate Change*. Dordrecht: Kluwer.

Bachelet, D., R. P. Neilson, J. M. Lenihan, and R. J. Drapek. 2001. Climate change effects on vegetation distribution and carbon budget in the United States. *Ecosystems* 4: 164–185.

Betts, R. A., P. M. Cox, S. E. Lee, and F. I. Woodward. 1997. Contrasting physiological and structural vegetation feedbacks in climate change. *Nature* 387: 796–799.

Bonan, G. B. 1995. Land–atmosphere interactions for climate system models: Coupling biophysical, biogeochemical and ecosystem dynamical processes. *Remote Sensing of Environment* 51: 57–73.

Box, E. O. 1981. *Macroclimate and plant form: An introduction to predictive modeling in phytogeography.* The Hague: Junk.

Cannel, M. G. R., and R. I. Smith. 1983. Thermal time, chill days and precipitation of budburst in *Picea sitchenis. Journal of Applied Ecology* 20: 951–963.

Chapin, F. S., III, and A. M. Starfield. 1997. Time lags and novel ecosystem response to transient climatic changes in arctic Alaska. *Climatic Change* 35: 449–461.

Cox, P. M., R. A. Betts, C. D. Jones, S. A. Spall, and I. J. Totterdell. 2000. Acceleration of global warming due to carbon-cycle feedbacks in a coupled climate model. *Nature* 408: 184–187.

Cramer, W., A. Bondeau, F. I. Woodward, I. C. Prentice, R. A. Betts, V. Brovkin, P. M. Cox, V. Fisher, J. A. Foley, A. D. Friend, C. Kucharik, M. R. Lomas, N. Ramankutty, S. Sitch, B. Smith, A. White, and C. Young-Molling. Global response of terrestrial ecoregion structure and function to CO_2 and climate change: Results from six dynamic global vegetation models. *Global Change Biology* 7: 357–373.

DeLucia, E. H., Hamilton J. G., Naidu S. L. Thomas R. B., Andrews J. A., Finzi A., Lavine M., Matamala R., Mohan J. E., Hendrey G. R. and Schlesinger W. H. 1999. Net primary production of a forest ecosystem with experimental CO_2 enrichment. *Science* 284: 1177–1179.

Dowlatabadi, H., and M. G. Morgan (in preparation). The Integrated Climate Assessment Model. Center for Integrated Study of the Human Dimensions of Global Change. Pittsburgh.

Foley, J. A., I. C. Prentice, N. Ramankutty, S. Levis, D. Pollard, S. Sitch, and A. Haxeltine 1996. An integrated biosphere model of land surface processes, terrestrial carbon balance, and vegetation dynamics. Global Biogeochemical Cycles 10(4): 603–628.

Friend, A. D., A. K. Stivens, R. G. Knox, and M. G. R. Cannel 1997. A process-based, terrestrial biosphere model of ecosystem dynamics (Hybrid v3.0). Ecological Modelling 95: 247–287.

Gates, D. M. 1993. Climate Change and Its Biological Consequences. Sunderland, Mass.: Sinauer.

Graham, R. W. 1992. Late Pleistocene Fauna Changes as a Guide to Understanding Effects of Greenhouse Warming on the Mammalian Fauna of North America. Pages 76–90 in R. L. Peters and T. E. Lovejoy, eds., Global Warming and Biological Diversity. New Haven: Yale University Press.

Haxeltine, A., and I. C. Prentice 1996. BIOME3: An equilibrium terrestrial biosphere model based on ecophysiologycal constraints, resource availability, and competition among plant functional types. Global Biogeochemical Cycles 10(4): 693–709.

Hurtt, G. C., P. R. Moorcroft, S. W. Pacala, and S. A. Levin. 1998. Terrestrial models and global change: Challenges for the Future. Global Change Biology 4: 581–590.

The Intergovernmental Panel on Climate Change (IPCC). 1996a. Climate Change 1995: The Science of Climate Change, eds., J. T. Houghton, L. G. Meira Filho, B. A. Callander, N. Harris, A. Kattenberg, and K. Maskell. Contribution of working group I to the second assessment report of the IPCC. Cambridge: Cambridge University Press.

———. 1996b. Climate Change 1995: Impacts, Adaptations and Mitigation of Climate Change: Scientific Technical Analysis, eds., R. T. Watson, M. C. Zinyowera, and R. H. Moss. Contribution of working group II to the second assessment report of the IPCC. Cambridge: Cambridge University Press.

Kienast, F., B. Brzeziecki, and O. Wildi 1996. Long-term adaptation potential of central European mountain forests to climate change: A GIS-assisted sensitivity assessment, Forest Ecology and Management 80: 133–153.

Kirilenko, A. P., and A. M. Solomon 1998. Modeling dynamic vegetation response to rapid climate change using bioclimatic classification. Climatic Change 38: 15–49.

Monserud, R. A., and R. Leemans 1992. Comparing global vegetation maps with the Kappa statistic. Ecological Modelling 62: 275–293.

Monserud, R. A., N. M. Tchebakova, and R. Leemans. 1993. Global vegetation change predicted by the modified Budyko model. Climatic Change 25: 59–83.

Moorcroft, P. R., G. C. Hurtt, and S. W. Pacala. 2001 (in press). Scaling rules for vegetation dynamics: A new terrestrial biosphere model for global change studies, *Ecological Monographs.*

Morgan, M. G., L. F. Pitelka, and E. Shevliakova. 2001. Estimates of climate change impacts on forest ecosystems. *Climatic Change* 49: 279–397.

Myers, J. P., and R. T. Lester. 1992. Double jeopardy for migrating animals: Multiple hits and resource asynchrony. Pages 76–90 in R. L. Peters and T. E. Lovejoy, ed., *Global Warming and Biological Diversity.* New Haven: Yale University Press.

Neilson, R. P. 1995. A model for predicting continental-scale vegetation distribution and water balance. *Ecological Applications* 5: 362–385.

Ni, J., M. T. Sykes, I. C. Prentice, and W. Cramer. 2000. Modelling the vegetation of China using the process-based equilibrium terrestrial biosphere model BIOME3. *Global Ecology and Biogeography* 6: 463–479.

Pimm, S., G. J. Russell, J. Gittleman, and T. M. Brooks. 1995. The future of biodiversity. *Science* 289: 347–350.

Pacala, S. W., G. C. Hurtt, D. Baker, P. Peylin, R. A. Houghton, R. A. Birdsey, L. Heath, E. T. Sundquist, R. F. Stallard, P. Ciais, P. Moorcroft, J. P. Caspersen, E. Chevliakova, B. Moore, G. Kohlmaier, E. Holland, M. Gloor, M. E. Harmon, S. M. Fan, J. L. Sarmiento, C. L. Goodale, D. Schimel, and C. B. Field. 2001. Consistent land- and atmosphere-based U. S. carbon sink estimates. *Science* 292: 2316–2320.

Prentice, I. C., W. Cramer, S. P. Harrison, R. Leemans, R. A. Monserud, and A. Solomon. 1992. A global biome model based on plant physiology and dominance, soil properties and climate. *Journal of Biogeography* 19: 117–134.

Prentice, K. C. 1990. Bioclimatic distribution for general circulation model studies. *Journal of Geophysical Research* 95(D8): 11811–11830.

Riechert, S. 1986. Spider fights as a test of evolutionary game theory. *American Scientist* 74: 604.

Root, T. L., and S. H. Schneider. 1993. Can large-scale climatic models be linked with multiscale ecological studies? *Conservation Biology* 7(2): 256–270.

———. 1995. Ecology and climate: Research strategies and implications. *Science* 269: 331–341.

Schlesinger, W. H. and J. Lichter. 2001. Limited carbon storage in soil and litter of experimental forest plots under increased atmospheric CO_2. *Nature* 411: 466–469.

Siegel, E., H. Dowlatabadi, and M. J. Small. 1995. Modeling of ecosystem prevalence due to the climate change with multinomial multivariate logistic regression. *Journal of Biogeography* 22: 875–879.

Silverman, B. W. 1986. *Density Estimation for Statistics and Data Analysis.* London: Chapman and Hall.

Sorenson, L. G., R. Goldberg, T. L. Root, and M. G. Anderson. 1998. Potential effects of global warming on waterfowl populations breeding in the Northern Great Plains. *Climatic Change* 40: 343–369.

Thornthwaite, C. W. 1948. An approach toward a rational classification of climate. *Geographical Review* 38: 55–94.

VEMAP. 1995. Vegetation/ecosystem modeling and analysis project: Comparing biogeography sand biogeochemistry models in a continental-scale study of terrestrial ecosystem responses to climate change and CO_2 doubling. *Global Biogeochemical Cycles* 9: 407–437.

Walker, B., and W. Steffen 1997. An overview of the implications of global change for natural and managed terrestrial ecosystems. *Conservation Ecology* 1(2). Available online at http://www.consecol.org/vol1/iss2/art2.

Webb, T., III. 1992. Past changes in vegetation and climate: Lessons for the future. Pages 59–75 in R. L. Peters and T. E. Lovejoy, ed., *Global Warming and Biological Diversity*. New Haven: Yale University Press.

Whittaker, R. H., ed. 1978. *Classification of Plant Communities*. The Hague: Junk.

Woodward, F. I. 1997. *Climate and Plant Distribution*. Cambridge: Cambridge University Press.

Woodward, F. I., T. M. Smith, and W. R. Emanuel 1995. A global land primary productivity and phytogeography model. *Global Biogeochemical Cycles* 9(4): 471–490.

CHAPTER 7

Climate Change and the Susceptibility of U.S. Ecosystems to Biological Invasions: Two Cases of Expected Range Expansion

ERIKA S. ZAVALETA AND JENNIFER L. ROYVAL

At least 5000 nonnative species, including over 2100 exotic plant species and 2000 alien insect species, have invaded North America since the arrival of European explorers on the continent (Niemela and Mattson 1996, Vitousek et al. 1997). Many of these accidental colonists have become established in the United States, where they exact a growing toll on the native character and diversity of ecosystems. Several established invasives have achieved widespread, public notoriety for their damaging effects, such as the intentionally introduced kudzu weed, *Pueraria lobata,* that has buried much of the Southeast under thick mats of vines. Others have yet to be widely recognized for the damage they cause to U.S. ecosystems and indigenous species. For example, nonnative fish species, which are generally perceived by the public as benign or even beneficial, have already driven 44 native freshwater fish species to threatened or endangered status (Vitousek et al. 1997).

Climate change has the potential to allow range expansions of invasive species in the United States through at least two pathways.

First, rapid climate change may cause significant dieback of sessile and slow-migrating indigenous organisms unable to tolerate new temperature and precipitation conditions. Diebacks of forest trees associated with species range shifts, in particular, have resulted from climate change events in recent geological history (Woodward 1987, Prentice et al. 1991). Many invasive species cannot establish successfully in the absence of disturbance; their success reflects the growing extent of anthropogenic disturbances such as air pollution, intensive agriculture, livestock grazing, and timber harvest in the United States (USOTA 1993, Rejmanek and Richardson 1996). For disturbance-adapted invaders, climate change–associated diebacks of perennial vegetation communities—forests, woodlands, and shrublands—and accompanying disruption of faunal communities may provide new and widespread opportunities for disturbance-loving exotic species to colonize and spread.

The second way that climate change may worsen invasive species impacts is by enabling poleward and uphill range expansions. While Eurasia is recognized as the source of many North American exotics, harmful invasive species in the United States consist disproportionately of tropical and subtropical organisms with cold-limited current naturalized distributions. A stratified random sample of 10% of the *harmful,* nonindigenous plant and animal species occurring in the United States indicates that 48% (n = 24 species) are likely to benefit from an overall warming trend,[1] while only 4% (n = 2 species) are likely to suffer range contractions.[2] Another 48% (n = 24) of the sampled species have characteristics less clearly predictive of directional shifts (Table 7.1)

Table 7.1. The Likely Responses to Climate Change of 50 Randomly Chosen, Nonindigenous Species in the United States.

Species	Common Name	Expected Response	Evidence
INSECTS AND ARACHNIDS			
Apis mellifera scutellata	African honey bee	+	Tropical native range; NI range very limited to souther: states
Blatella asahinai	Asian cockroach	+	NI range limited to Florida
Parlatoria ziziphi	Black parlatoria scale	+	Citrus pest, tropical and subtropical global range, N range limited to Florida

Species	Common Name	Expected Response	Evidence
Cactoblastis cactorum	Cactus moth	+	Tropical origin; food plant is *Opuntia* cactus, found over-whelmingly in hot, dry regions
Ctenarytaina eucalypti	Eucalyptus psyllid	+	Native to Australia/NZ, host plant is eucalyptus; NI only in California
Poinsettia whitefly	Bemisia tabaci	+	NI range limited to southern California
Ctenarytaina longicauda	Tristania psyllid	0	Origin unclear
Osmia cornuta	Mason bee	0	Native to Spain, southern Europe; NI only in California
Bythotrephes cederstroemi	Spiny water flea	0	NI in Great Lakes
Lymantria monacha	Nun moth	−	Siberian origin; food trees coniferous
Yponomeuta cagnagella	Spindle tree ermine moth	−	NI in Vermont
OTHER INVERTEBRATES			
Achatina fulica	African giant snail	+	Tropical origin; NI in Florida
Penaeus monodon	Giant tiger shrimp	+	Tropical origin (coasts of Thailand, India, Indonesia, the Philippines); NI in North Carolina and south
VERTEBRATES			
Xenopus laevis	African clawed frog	+	Tropical origin; NI in Arizona
Alectoris chukar	Chukar partridge	+	Tropical origin; NI in Hawaii
Herpestes auropunctatus	Indian mongoose	+	Tropical origin; NI in Hawaii
Pycnotus jocosus	Red-vented bulbul	+	NI in Hawaii
Dasypus novemcinctus	Nine-banded armadillo	+	NI in Florida
Boiga irregularis	Brown tree snake	+	Tropical origin; tropical NI range
Tetraogallus himalayensis	Himalayan snowcock	0	NI in central latitudes of continental United States
Sturnus vulgaris	Starling	0	NI throughout United States and Canada
Sus scrofa	European hog	0	NI in central latitudes of United States; not clearly T-limited
Oryctolagus cuniculus	European rabbit	0	NI in central latitudes; not clearly T-limited

continues

Table 7.1. *Continued*

Species	Common Name	Expected Response	Evidence
PLANTS			
Eichornia asurea	Rooted water hyacinth	+	NI in FL, TX, CA, AL only
Psidium cattleianum	Strawberry guava	+	NI in Hawaii
Lycinium ferocissimum	African boxthorn	+	Subtropical origins; NI in subtropics (Australia and New Zealand)
Cortaderia jubata	Andean pampas grass	+	Does not tolerate long freezes; NI in California, Hawaii only
Musa spp.	Banana	+	Tropical origins; NI in Hawaii
Orobanche spp.	Broomrape	+	Tropical origins; NI in Georgia
Mimosa pigra pigra	Catclaw mimosa	+	NI in Florida & Australia; tropical origins
Colubrina asiatica	Lather leaf	+	NI only in Florida
Mangifera indica	Mango	+	Tropical origin
Tamarix spp.	Saltcedar	+	Restricted NI range
Camapanula rapunculoides	Creeping bellflower	0	NI in central latitudes
Hedera helix	English ivy	0	NI in eastern deciduous zone
Lepidium latifolium	Perennial pepperweed	0	NI in Washington state
Sporobolus virginiflorus	Poverty grass	0	No clear evidence of T limitation
Solanum eleagnifolium	Silver-leaf nightshade	0	NI in central U.S. latitudes
Hibiscus trionium	Venice mallow	0	NI in Washington state
Anchusa officinalis	Common bugloss	0	NI in Washington state
Centaurea diffusa	Diffuse knapweed	0	NI at central latitudes
Hieracium pratense	Yellow hawkweed	0	NI in Washington state
Centaurea jacea	Brown knapweed	0	NI in Washington state
Echinochloa crus-galli var. *frumentacea*	Japanese millet	0	Native to Japan
Rosa multiflora	Multiflora rose	0	Central U.S. latitudes
Hieracium aurantiacium	Orange hawkweed	0	NI in Washington state
Hypericum perforatum	St. Johnswort	0	NI in Washington state
Proboscidea louisianica	Unicorn plant	0	NI in Washington state
Salix spp.	Weeping willow	0	NI in central latitudes
Avena fatua	Wild oats	0	NI in central latitudes

Identified as harmful by the U.S. Office of Technology Assessment (1993); + , more likely to increase in areal extent in the United States under warming; 0, unclear or no change likely in U.S. areal extent under warming; -, more likely to decrease in U.S. areal extent under warming; NI, nonindigenous; T, temperature.

(Zavaleta, unpublished data). Several notorious invaders, such as the African honey bee (*Apis mellifera scutellata*), the red imported fire ant (*Solenopsis* spp.), the kudzu vine (*Pueraria lobata*), the avian malaria vector (*Culex quinquefasciatus*), and the brown tree snake (*Boiga irregularis*) fall into the category of species likely to be able to expand north, uphill, and inland if warming occurs.

Invasive species, by and large, pose threats to native wildlife and ecosystems in several ways. Exotic animals may outcompete native wildlife for limited food, space, water, and other resources. In Florida, for example, the endangered Oskaloosa darter (*Etheostoma okaloosae*) has suffered population declines because of competition from the introduced brown darter (*E. edwini*). Exotic animals may also prey on native wildlife: the brown tree snake (*Boiga irregularis*) in the U.S. territory of Guam has caused the extinction of at least five species or subspecies of birds to date through consumption of their eggs and chicks (USOTA 1993). Exotic animals and exotic plants also impact native wildlife by altering habitat: the invasion of the melaleuca tree (*Melaleuca quinquinerva*) is changing the topography, soils, hydrodynamics, and vegetation structure of the Everglades wetlands system with impacts on its diversity and abundance of wildlife (Ewel 1986, Simberloff et al. 1997). Nearby, the invasive water hyacinth (*Eichhornia crassipes*) completely covers surface waters (USOTA 1993), hampering efforts of the endangered snail kite (*Rostrhamus sociabilis*) to find its prey. Finally, introduced diseases and disease vectors often threaten native wildlife. In Hawaii, the introduced mosquito *Culex quinquefasciatus* carries an exotic strain of avian malaria that threatens the remaining native birds in the archipelago (Atkinson et al. 1995). In all, exotic species are believed by the U.S. Fish and Wildlife Service to have contributed to the listing of 160 species as threatened or endangered under the Endangered Species Act and to be the *principal* cause of listing for 41 of these species (USOTA 1993). Species invasions now threaten biodiversity more than any other single human impact except land use change. To understand how climate change will affect native wildlife and ecosystems, it is critical to consider how climate change will affect the nature of the global invasives problem.

This chapter focuses on the possibility that warming and drying will expand range sizes of invasive species. However, warming is just one component of climate change, climate change only one of many co-occurring global environmental changes, and range size only one aspect of the severity of the invasions problem (Lonsdale 1999, Parker et al. 1999). It is beyond the scope of this chapter to predict broad scenarios, such as what regions are most likely to be affected by invasives in the future, because such predictions would require consideration of other features of invasives responses, such as abundance changes, and other features of climate and global changes, such as changes in climate variability and land use. We focus on synthesizing and adding to current evidence that the responses of invasive species range sizes, a single component of invasion severity, are likely to increase overall in response to warming, a single component of global change.

Predicting Future Range Shifts

Quantifying future species range changes can be accomplished with only a limited degree of precision. To begin with, existing models of future climate change lack detailed spatial and temporal resolution. Range shifts may occur on the scale of tens to hundreds of kilometers, while current climate models typically divide the globe into homogeneous 300 km grid cells. Smoothing of topographic features, such as whole mountain ranges, within these grid cells renders model output accuracy uncertain even at this coarse scale. The unknown course of future CO_2 emissions also makes specific temporal predictions impossible. Even models that address, specifically, a doubling of atmospheric CO_2 concentrations vary substantially in their predictions about how much warming will occur. While climate models are increasingly consistent in their large-scale predictions of temperature change, they cannot yet provide precise temperature change estimates at the regional level, or precipitation change estimates at any scale. Models also cannot accurately predict the effects that increased atmospheric CO_2 will have on climate features such as temperature variability on various time scales and the distribution and frequency of extreme events, both of which will undoubtedly modify species responses. Finally, current model predictions do not capture regional- to local-scale feedbacks that will affect species responses, such as albedo changes in response to decreased snow cover that may lead to increased regional warming. Hence, range shift predictions reflecting particular climate changes cannot be attached to particular times in the future, and their accuracy depends on the degree to which actual, regional climate trajectories ultimately match the coarse-scale climate model outputs used to build the range predictions.

Even perfect climate predictions would not enable highly certain predictions of range changes. While climate has been found the principal factor controlling the distribution of vegetation on a regional scale (e.g., Ohmann and Spies 1998) and research on a variety of fauna has clarified the importance of absolute climatic limits to the distributions of animals (e.g., Root 1988a, 1988b, Barry et al. 1995), climate certainly is not the only factor determining species ranges. Competition and a suite of other biotic factors; edaphic conditions; the particulars of a species' history; and other human-caused global and environmental changes, especially transport and land use change, all play significant roles in determining species distributions. In a case where rapidly changing climate is expected to drive rapid range shifts, new complexities arise: the migration rates of individual species, particularly plants, may be unable to keep pace with the rate of warming (Clark et al. 1998). Dieback of plant communities, particularly forests, may alter habitat suitability for animals or shift the competitive balance among species. Changes in disturbance regimes expected to accompany warming, such as increases in wildfire, also affect species interactions (Harte et al. 1992). Local species losses, gains, and shifts in relative abundance may allow some organisms to spread beyond previous biotic range limits, or may constrict the distributions of organisms subject to new biotic pressures (Davis et al. 1998). Because we do not know what all of the ecological outcomes of climate change will be, we cannot account for how they will affect future ranges of particular species.

However, it is possible to evaluate the probable direction and magnitude of range shifts under climate change because climate models can now provide a restricted range of likely warming trajectories, and because climate is commonly accepted as playing the principal role in determining the limits of species distributions (Woodward 1987). Noxious invaders are particularly amenable to this approach, since the most successful invasive species display reduced dependence on biotic conditions in their naturalized ranges. Competitive dominance over a wide range of resident species, rapid dispersal capability, and tolerance of a wide range of habitat conditions such as vegetation and soil types are typical of harmful invaders such as kudzu (USOTA 1993) and *Solenopsis* ants. A meta-analysis of the determinants of invading pine distributions around the world found that temperature and moisture exerted principal control over invasibility at climatic extremes of the invaders' distributions (Richardson and Bond 1991). At intermediate locations, biotic factors such as competition with the resident flora played a greater role in shaping the local distribution and abundance of the invader. These findings suggest that while the intensity of invasion within an exotic species' range depends to a greater extent on biotic factors, the extensiveness of invasion—the geographic range containing sites susceptible to invasion—can be reasonably predicted from climate. Studies of climate–range relationships can serve as baselines for more detailed analyses that incorporate the roles of biotic factors.

The limitations of climate-based range studies for predicting future species distributions can also be reduced in key ways. Studies of climatic controls on species distribution often involve seeking correlations between species range boundaries and a parade of climatic parameters without any basis for expecting particular climate parameters to control the distributions of particular species. This leaves little way to distinguish spurious correlations from those consistent with other evidence about abiotic controls on a species' distribution. A preferable model for range studies based on climate correlations is one that combines this search for the best-fitting climatic variables with the generation of specific hypotheses about why certain climatic variables are most likely to exert influences on individual species' distributions. These hypotheses should be based on knowledge of the biology of the species in question, field observations, and/or native habitat characteristics of an invasive species whose naturalized range is of interest. Testing this type of hypothesis proceeds by "climate-matching"—looking for a relationship between the probability of occurrence of the species in question and the climatic variable or variables expected to play a role in determining distribution. Findings consistent with the hypothesis should include not only a strong relationship between occurrence and the selected climatic variable, but also the finding that the observed relationship is described better by a unimodal bell-shaped, logistic, or similar model than by other models that cannot be as easily justified in biological terms (Saetersdal and Birks 1997).

The Current State of Knowledge

A number of studies have already deployed climate-matching and other modeling approaches to predicting future changes in impacts of individual invasive species in specific regions. The overall impact in a region of invasives is a function of several factors, only one of which is range size. The number of established invaders in a region (Lonsdale 1999), the abundance or density of a particular invader within its range, and the nature, frequency, and/or severity of a particular invader's impacts (Parker et al. 1999) may all also change in response to new anthropogenic impacts, from warming to land use change:

$$I = \sum_{i=0}^{S} a_i i_i r_i$$

where I = the overall impact of invaders in a given area; S = the number of invasive species present in the area; a_i = the abundance per unit area of the ith invasive species; i_i = the impact per individual of the ith invasive species; and r_i = the areal extent or range of the ith invasive species.

Very few studies have quantitatively addressed the potential for these factors, with the exception of range size, to respond to climate change. Two studies that we know of have estimated the potential for increases in invader abundance within current range limits through increases in the number of generations that multivoltine species can produce in single growing seasons. These studies predict that invasions of two insect species in Europe—the Colorado potato beetle, a serious crop pest; and the Far East Asian tick, a livestock parasite—will worsen in severity as well as extent (Sutherst et al. 1995a, Porter et al. 1991). A small number of others predict increases in abundance or individual impact through other means, such as estimating increases in climate suitability within an invader's current range (Lines 1995, Sutherst 1995, Baker et al. 1996, Bryan et al. 1996).

Table 7.2 summarizes the results of all known, published studies to date that model future range or intensity changes in invasive species due to climate change. Additional studies under way but not completed address the future ranges of the exotic avian malaria vec-

Table 7.2. Published Studies That Have Investigated Future Responses of Invasive Species to Warming.

Source	Species	Naturalized Range	Climate Change Considered	Approach	Now Limited By	Findings
VERTEBRATES						
Eaton & Scheller 1996	Common carp *Cyprinus carpio*, largemouth bass *Micropterus salmoides*, species (black bluegill *Lepomis macrochirus*, black crappie *Pomoxis nigromaculatus*)	Warm U.S. streams and lakes	Stream T increases modeled based on GCM with mean air T +4.4°C	Inductive climate-matching model	Minimum and maximum weekly mean temperatures correlated with species distributions	Three species increase range size 25–35%; one crappie) experiences range reduction of 26% at S (warm) end of distribution.
Eaton & Scheller 1996	Brook, rainbow, and brown trout (*Salvelinus fontinalis*, *Oncorhynchus mykiss*, *Salmo trutta*); small-mouth bass (*Micropterus dolomieu*)	Cool to cold U.S. streams and lakes	Stream T increases modeled based on GCM with mean air T +4.4°C	Inductive climate-matching model	Minimum and maximum weekly mean temperatures correlated with species distributions	All four species experience range reductions of 35–50%
Mandrak 1989	Fifty-eight common freshwater fish species	Central and eastern U.S. streams and lakes	Northward isotherm shifts based on models of Emanuel et al. 1985	Correlation of ranges with mean July Ts, terrestrial vegetation boundaries	Variable; correlated with July mean Ts, Holdridge Life Zones for vegetation distribution in E United States	Twenty-seven species become potential invaders of the Great Lakes under warming
Sutherst et al. 1995b	Cane toad *Bufo marinus*	NE Australia	+1.5 to 2°C	CLIMEX climate-matching model	Development requires sufficient degree-days, moisture	Southward spread along E coast of Australia

Beerling 1993	*Fallopia japonica*	NW Europe	+1.5 to 4.5°C	Climate matching model (degree-days and T minima)	Growing season length, minimum Ts	Potential range expands into much of Scandinavia, Iceland
Beerling 1993	*Impatiens glandulifera*	NW Europe	+1.5 to 4.5°C	Climate matching model (degree-days and T minima)	Growing season length	Potential range expands throughout Scandinavia, Iceland
Sasek & Strain 1990	Kudzu vine *Pueraria lobata*	SE United States	Average/minimum winter Ts +3°C	Climate matching with winter T isoclines	Dieback and root/ crown death caused by low winter temperatures	Northward spread of up to 400 km
Sasek & Strain 1990	Japanese honey-suckle *Lonicera japonica*	SE United States	Average/minimum winter Ts +3°C	Climate matching with winter T isoclines	Dieback and root/ crown death caused by low winter temperatures	Northward spread of up to 400 km
Sutherst 1995	Japanese honey-suckle *Lonicera japonica*	E United States	+3°C	CLIMEX climate-matching model	Unspecified (multivariate eco-climatic index)	Northward spread into Great Lakes region, increased severity in NE United States
Sutherst 1995	*Mimosa pigra*	Global tropics and subtropics	+0.1°C/ degree of latitude, +20% summer rain, −10% winter rain.	CLIMEX climate-matching model	Unspecified (multivariate eco-climatic index)	Potential range expands in SE United States, SE Asia, Madagascar, South Africa, E Australia

continues

Table 7.2. *Continued*

Source	Species	Naturalized Range	Climate Change Considered	Approach	Now Limited By	Findings
INVERTEBRATES						
Boag et al. 1991	Virus–vector nematodes *Longidorus leptocephalus*, *L. macrosoma*, *Xiphinema diversicaudatum*	British Isles	+1.5°C	Climate-matching model based on July mean T	Northern edges of ranges correlate with July T isotherm minima	Northward spread into Scotland, Scandinavia
Baker et al. 1996	Colorado potato beetle *Leptinotarsa decemlineata*	United Kingdom	+1.7 to 1.8°C	CLIMEX climate-matching model	Multivariate ecoclimatic index	Range increase of 102%, suitability increase (corresponding to higher severity) of 76% in current range
Bryan et al. 1996	Malaria vector *Anopheles farauti*	Northern Australia	+1.5°C, +10% summer rainfall	CLIMEX climate-matching model	Transmission T-sensitive; multivariate ecoclimatic index	Increased severity and spread southward in coastal Queensland
Evans & Boag 1996	New Zealand flatworm *Artioposthia triangulata*	United Kingdom and western Europe	+0.1°C/degree of latitude	CLIMEX climate-matching model	Multivariate ecoclimatic index	Range expansion into much of Scandinavia, range contraction in the United Kingdom and southern France
Francy et al. 1990	Asian tiger mosquito *Aedes albopictus*	SE United States	Natural inter-annual variability	Qualitative description/field study	Survival of overwintering eggs T-dependent	Cold winters limit survival at northern edge of

Reference	Species	Location	Scenario	Model	Factors	Predicted impact
						spreading range; warm winters allow spread
Porter et al. 1991	European corn borer *Ostrinia nubilalis*	Southern Europe	+1°C, spatially variable T increase based on GISS model (Carter et al. 1991)	Climate-matching model (degree days)	Survival, development rate, number of generations per year sensitive to low Ts	Northward spread of up to 1200 km, increases in generations/year throughout range increase severity
Sutherst et al. 1995a	Far East Asian tick, *Haemaphysalis longicornis*	South Mediterranean Europe	+0.1°C/degree of latitude, +20% summer rain, -10% winter rain.	CLIMEX climate-matching model	Sufficient summer heat to complete life-cycle and enter nymphal overwintering phase	Major range expansion across Great Britain, northwestern Europe, Scandinavia
Sutherst 1990	Far East Asian tick, *Haemaphysalis longicornis*	New Zealand—currently restricted to N island	+0.1°C/degree of latitude, +20% summer rain, -10% winter rain.	CLIMEX climate-matching model	Sufficient summer heat to complete life-cycle and enter nymphal overwintering phase	Expect range expansion to include most of the South island
Sutherst 1995	European honey bee *Apis mellifera*	United States	3°C warming	Simple climate matching	Winter minimum temperatures	Potential range would expand north from Oregon and Washington into British Columbia

continues

Table 7.2. *Continued*

Source	Species	Naturalized Range	Climate Change Considered	Approach	Now Limited By	Findings
Sutherst et al. 1995a	Colorado beetle *Leptinotarsa decemlineata*	United Kingdom	+0.1°C/degree of latitude, +20% summer rain, −10% winter rain.	CLIMEX climate-matching model	Unspecified (multivariate eco-climatic index)	Mean increase in number of generations per year of 1–1.5 at various latitudes in the United Kingdom
This study	Red imported fire ant *Solenopsis wagneri*	SE United States	+1 to 4°C	Climate-matching model (January minimum T)	Survival, development, foraging rates limited by cold Ts	Northward spread in range of 200,000–300,000 km²/degree of warming
HUMAN DISEASES						
Lines 1995	Dengue (human arboviral disease)	S & SE Asia, W. Pacific, tropical Africa, Latin America, and Caribbean	2°C	Qualitative description	Vector, and hence transmission, limited by low Ts, persistence limited by length of winter	Spread to higher latitudes, increased severity near current range limits expected
Lines 1995	American trypanosomiasis (Chagas' disease)	Mexico	Unspecified	Qualitative description	Vector is temperature-limited	Spread northward from Mexico into the U.S. possible
Lines 1995	Malaria *Plasmodium falciparum*	Global tropics	+2°C	Qualitative description	Reproductive rate very sensitive to T in low end of tolerated range	Spread and increases in severity in highlands of Madagascar, E Africa, Nepal, Papua New Guinea; former USSR

Abbreviations: T, temperature; N/S/E/W, north/south/east/west tor in Hawaii (C. Atkinson pers. comm. 1998), the Melaleuca and Brazilian pepper trees in South Florida (D. Simberloff pers. comm. 1998, Myers and Ewel 1990), and the southwestern corn borer (Baskauf 1999).

In all, published studies point to range size increases for 50 harmful invasive species, including 30 fishes, 1 amphibian, 5 plants, 11 insects and other invertebrates, and 3 human diseases; and range contractions for only 5 species. Thirty-seven of these range increases would occur in or into the United States, with the remainder affecting Europe, Australia, New Zealand, and in the cases of two diseases—malaria and Chagas' disease—the global tropics. Collectively, these studies indicate that climate does limit the current distributions of a wide range of damaging invasives across continents, biomes, and taxa; and that warming, everything else being equal,[3] will likely allow expanded invasions to occur worldwide.

Case Studies

We chose two study organisms—a plant, the exotic shrub *Tamarix ramosissima*,[4] and an animal, the red imported fire ant *Solenopsis wagneri*[5]—to illustrate some of the impacts that invasive species have on native wildlife and to explore the effects of global climate change on invasive species distributions. *Tamarix* and *S. wagneri* were selected for (1) well-studied impacts on native wildlife, (2) available, reasonably detailed information on distribution throughout the United States, (3) representativeness of different geographic areas in the United States, (4) representativeness of contrasting taxonomic groups with contrasting modes of impact on wildlife, (5) being highly prolific and noxious with observed tolerance for a broad range of biotic conditions, and (6) already noted by the scientific community as likely to have naturalized distributions in the United States limited by cold temperatures or moisture availability, two climatic variables expected to change as atmospheric concentrations of carbon dioxide increase.

We used ArcView 3.0 and ArcView Spatial Analyst 1.0 for Windows (Environmental Systems Research Institute, Inc. 1996) to construct range maps for each study species and to analyze spatial relationships among species range, climate, and, for *Tamarix*, hydrography.[6] We reconstructed range maps for the red imported fire ant (*Solenopsis wagneri*) from Federal Code of Regulations records of the infestation status of individual counties (FCR Title 7, Section 301.81) (*sensu* Callcott and Collins 1996). We initially obtained *Tamarix* range data from a 1961 range map constructed by the U.S. Bureau of Reclamation (Robinson 1965). Point samples beyond the reported 1961 range were used to update the range boundaries of the map (Zavaleta, unpublished data, and J. Gaskin, unpublished data). For *Tamarix,* we limited our analysis to riparian areas by recording presence or absence of the invader on waterways identified on a U.S. Geological Survey (USGS) map of U.S. hydrography (USGS Earth Remote Observing System Data Center).

We used climate maps developed by the Oregon Climate Service's PRISM (Parameter-Elevation Regressions on Independent Slopes Model) project to analyze relationships between species distributions and measures of temperature and moisture (Daly et al. 1994, 1997). To examine relationships between species range and climatic variables, we recorded the proportion of raster grid cells with particular values of a climatic variable that contained the species of interest. Rather than exploring them exhaustively, we chose to select climatic variables carefully, based on hypotheses about what factors were most likely to limit each species. For each species, we analyzed two to three primary and two to three secondary climatic variables. We fitted curves using least squares regression with nonlinear models selected by visual examination of data scatterplots using Systat 7.0 software. We used modeled relationships between species occurrence and climatic variables to generate predictions of future range shifts under scenarios of +1, +2, +3, and +4°C uniform changes to mean minimum and maximum temperatures throughout the United States and decreases in western regional precipitation of 50 mm/year. These temperature increases fall within the range of mean annual changes typically predicted by current climate models for a doubled CO_2 world, while the precipitation change is consistent with the magnitude and direction of change predicted by regional climate models (Giorgi et al. 1998). Changes in monthly minimum and maximum temperatures will not necessarily match changes in the annual mean but are likely to shift in the same magnitude and direction.

The Red Imported Fire Ant in the American Southeast

Red imported fire ants (*Solenopsis wagneri*) have rapidly invaded the southeastern United States in this century. Since their introduction approximately 80 years ago, they have colonized millions of hectares at densities 4 to 10 times higher than those found in their native habitat of Brazil (Porter et al. 1997). Economic losses due to agricultural damage and direct threats to human health have prompted careful monitoring and study of red imported fire ants (RIFA) since as early as 1930, providing detailed information about the history of its spread and a wealth of evidence that it seriously impacts native faunal communities when it invades. More recently, a new social form of *S. wagneri* with multiple queens per colony has been found in North America (Glancey et al. 1973). Studies show that this polygyne form is spreading within the United States (Glancey et al. 1987, Porter 1993) and suggest that it is capable of establishing in greater densities, with faster population growth and more serious impacts on native ecosystems and wildlife, than the original monogyne form (Macom and Porter 1996). Evidence from the historic spread of this species and supplementary studies strongly suggest that fire ants are climate limited. Climate change may therefore lead to an expansion in RIFA's current range in the United States, broadening its impacts on native wildlife as well as on human welfare.

The Distribution of RIFA

Data collection on the distribution and spread of imported fire ants began as early as 1930 (Creighton). Since the U.S. Department of Agriculture initiated the federal fire ant quarantine in 1958, range surveys have been conducted on a nearly annual basis. As of 1995, RIFA infested 670 counties in 11 states and Puerto Rico, covering approximately 114 million ha (Callcott and Collins 1996). Since then, an additional 55 counties have joined the list of infested counties, bringing the total invaded area to approximately 125.5 million ha.[7] RIFA have thus not only infested a vast area of the United States, they have done so at an astonishing rate. RIFA have invaded an area approximately equal in size to the state of New Hampshire every year for the past 40 years (Callcott and Collins 1996). It appears, however, that their rate of spread has declined within the last 20 years (Vinson 1991a), particularly in the northern portion of their range where they appear to be nearing an ecological limit.

Factors Contributing to the Spread of RIFA

The rapid spread of red imported fire ants throughout North America is the result of both natural and assisted dispersal. Fire ants disperse mainly when sexually reproductive male and female ants fly to mate (Markin et al. 1971). These nuptial flights take place throughout the majority of the year, although reproduction in ants peaks in spring and early fall (Markin and Dillier 1971, Lofgren et al. 1975). While most queens land within 1.6 km of their nest of origin, queens may fly or be carried by winds up to 16 miles (Banks et al. 1973, Cokendolpher and Phillips, Jr. 1989). Fire ants are also transported long distances with the aid of flowing water. During flooding, fire ants have been observed alive and floating in dense mats containing queens on the surfaces of lakes and rivers. Fire ants can survive while floating on water for up to 2 weeks (Morrill 1974). These floating colonies can be transported long distances downstream and washed ashore, allowing them to establish new nests in potentially uninfested areas. Finally, human transport may be the most important mode of long-distance fire ant dispersal. Ninety percent of all isolated colonizations between 1945 and 1955 were linked to the transport of untreated nursery stock (Markin et al. 1971, Summerlin and Green 1977). While strict measures exist to reduce anthropogenic spread, fire ants remain notorious hitchhikers. In fact, Ball et al. (1984) found that mated queens prefer to land on shiny surfaces. Fire ants have been observed clinging to freight trains (Anon. 1958) and constructing nests in automobiles (Collins et al. 1993).

RIFA's ability to disperse long distances complements a suite of characteristics that enable it to rapidly fill a wide range of habitats as it spreads. First, fire ants are clearly associated with a broad range of ecologically disturbed lands including agricultural fields and roadways (Banks et al. 1985, Tschinkel 1988). Fire ants reproduce rapidly, with as much as 30 to 40% of annual biomass production allocated to the formation of reproductive ants (Tschinkel 1986). Fire ant colonies are capable of rapid growth due to the cooperation of queens during colony foundation. Often, mutiple queens will share the excavation of the founding nest and rearing of the first brood of workers (Tschinkel and Howard 1983). While all but one of the queens will eventually be killed by the workers,[8] this strategy increases the colony's chances of surviving the foundation period and results in the colony having approximately three times as many workers, early on, as colonies founded by a single queen (Tschinkel and Howard 1983). Finally, red fire ant colonies are capable of reproducing relatively early in their development. About 50% of colonies produce reproductive ants within their first year, and all colonies produce reproductive ants after 2 years (Tschinkel 1986). All of these characteristics allow RIFA to quickly invade and thrive in available, disturbed habitats that characterize the eastern United States. They also suggest that dispersal and population growth limitations would not restrict range expansion enabled by climate change.

Factors Limiting the Spread of RIFA

Biotic factors appear to be of limited importance in determining the distribution of RIFA. Prior to colony establishment, RIFA are vulnerable to predation on newly mated queens by dragonflies, birds, spiders, earwigs, beetles, and RIFA workers as well as the workers of other ant species (Whitcomb et al. 1972, 1973, Hung 1974, Nickerson et al. 1975, Glancey 1981, Lockley 1995b).[9] However, predation has not been able to stop the spread of RIFA in most areas because of the tremendous number of queens produced per colony. Suggestions that competition with the ant *Lasius neoniger* could be a factor limiting the northward expansion of RIFA were subsequently rejected because the ranges of the two species do not coincide well (Bhatkar et al. 1972, Buren et al. 1974). No other competitors have been found to exert enough pressure to effectively control RIFA populations. Finally, Jouvenaz et al. (1977) concluded that RIFA are affected by only a few pathogens and that none of these have significantly affected RIFA populations. Predation, competition, and pathogens thus all appear to play limited roles in halting the spread of RIFA. Neither do vegetation and soil characteristics appear to strongly limit RIFA. A detailed study of RIFA distributions in east Texas found that while mound densities were positively correlated with climatic factors including minimum January temperature, growing season length, and precipitation, mound densities bore no significant relationships to vegetation type, vegetation height, soil moisture, or soil type (Porter et al. 1991). RIFA appear only to prefer to build mounds in locations where sunlight strikes directly for some portion of the day (Hung and Vinson 1978) and in clear rather than wooded or densely overgrown areas (Anon. 1958, Green 1952, 1962).

Finally, dispersal limitations cannot easily be used to explain the current northern distributional boundary of RIFA. RIFA first invaded the southern coast of Alabama in 1918, and much of its spread since has been northward away from that point of introduction. This pattern raises the possibility that RIFA's current northern range limit reflects time since introduction, rather than any abiotic or biotic barrier to expansion. However, we have already detailed how RIFA can spread extremely fast under a wide range of conditions—its dispersal capacity of more than 2 million new hectares per year far exceeds the decadal spread distances that the species has covered toward the north from its initial point of introduction. Moreover, since its introduction, RIFA have spread 1300 km west into Texas and northeast to the North Carolina coast, and 1000 km southeast into Florida. In the same period, it has spread only 400 km due north and northwest, where minimum temperature isoclines dip south as they move inland. If the distance of RIFA's northern spreading front from Alabama is a function of time since introduction for dispersal, then based on its rate of longitudinal spread it should have reached its northern limit some 30 years ago and should occur roughly 500 to 600 km farther north than it currently does.

Climatic restrictions have therefore been cited numerous times as the probable cause of RIFA's limited northward expansion. Buren et al. first suggested in 1974 that cold temperatures, leading to winter mortality, might be the key factor limiting RIFA's spread along their northern range boundary. Having evolved in a tropical setting, RIFA are not freeze tolerant[10] (Morrill et al. 1978, Francke et al. 1986), and their capacity for cold acclimation is fairly limited at best (Francke et al. 1986, Cokendolpher and Phillips 1990). RIFA therefore depend on their ability to avoid freezing for winter survival through supercooling, the removal of nucleating agents, which are necessary for ice formation, from body fluids (Landry and Phillips 1996). Studies have found that the different life stages and different castes of RIFA have different supercooling capacities (Francke et al. 1986). Adult RIFA can supercool down to temperatures between −6.0 and −11.2°C (Francke et al. 1986, Landry and Phillips 1996). Beyond supercooling, RIFA's only other freeze avoidance mechanism is vertical migration within mounds to remain within optimum temperature limits (Anon. 1958).

The growth and spread of RIFA colonies may also be affected by temperatures above their supercooling point. Development rates for immature RIFA depend on temperature (Francke and Cokendolpher 1986, Porter 1988). Temperatures not extreme enough to kill RIFA hence may impact colonies by slowing or halting brood development. Temperature also affects RIFA's foraging rate. Porter and Tschinkel (1987) found that RIFA only forage when the soil temperature at a depth of 2 cm is between 15 and 43°C, while optimal foraging occurs between 22 and 36°C. Suboptimal temperatures, if sustained over a long period of time, could thus result in the starvation of a colony.

It has also been suggested that extremely arid conditions in Texas and the southwestern states have slowed RIFA's spread into these areas. Low humidity does not significantly affect the foraging activity of RIFA (Porter and Tschinkel 1987); in fact, RIFA actually survive high temperatures better at 0% humidity than at 100% humidity.[11] RIFA nuptial flights, however, have been observed to occur only under certain conditions, one of which is high relative humidity (> 80%) (Markin et al. 1971). The ability of RIFA to reproduce and spread into or beyond arid western states marked by year-round low humidities could therefore be very limited. Despite this evidence, RIFA colonies have successfully established in arid parts of the country. In 1986, a well-established colony was discovered in Mesa, Arizona (Frank 1988). Red imported fire ants also appeared in two counties of Texas's arid Rio Grande Valley in 1991 and showed signs of spreading south into arid portions of Mexico (Allen et al. 1993). These pioneering populations may represent new or distinct ecotypes of RIFA with evolved tolerance for drier conditions: isolated populations of the ant from arid west Texas were found experimentally to be significantly more resistant to desiccation than populations from moister eastern Texas (Phillips et al. 1996). Sufficient genetic variation may exist among populations of RIFA in the United States that some variants are, in fact, not limited by the constraints of low moisture and humidity that characterize variants more commonly found in the southeastern states. Most studies predict that RIFA will eventually spread through portions of New Mexico, Nevada, and California (Anon. 1972, CAST 1976, Pimm and Bartell 1980, Francke et al. 1986, Vinson and Sorenson 1986, Stoker et al. 1994, Killion and Grant 1995). The role of moisture in limiting the spread of RIFA, therefore, may be only a partial or temporary one.

Many studies have made specific predictions about the equilibrium range of RIFA based on available knowledge of its biology and current climatic conditions in the Unites States. Since little is still known about the moisture requirements of RIFA, most of these studies have focused on determining a northern limit to the spread of RIFA. The $-12.2°C$ January thermocline has been suggested as an absolute northern limit of RIFA (CAST 1976). Studies of the ant's supercooling ability support this limit (Francke et al. 1986). More recently, some researchers have suggested that the limit is farther north at the $-17.8°C$ January thermocline (Anon. 1972, Pimm and Bartell 1980, Killion and Grant 1995). Hung and Vinson (1978) suggested that RIFA may be able to survive beyond this limit if transported by humans and able to find protection inside human-made structures as they often do (Lyle and Fortune 1948, Bruce et al. 1978, Lofgren 1986, Collins et al. 1993). A study modeling RIFA spread has even suggested that they may be able to colonize areas beyond the $-17.8°C$ January thermocline under some natural conditions (Stoker et al. 1994). Where all of these studies concur is that as minimum temperatures drop, discovery of the right conditions for survival—and correspondingly, the successful establishment of RIFA—becomes increasingly improbable. A general rise in mean minimum temperatures, everything else being equal, could therefore be expected to improve RIFA's odds of establishing farther north than its current range.

Effects of RIFA on Wildlife

Scientists have tracked the effects of RIFA on North American wildlife since their introduction approximately 60 years ago (Table 7.3). Numerous studies, especially in the last 15 years, have found that through competition, predation, and the painful stings for which they were named, RIFA have the ability to substantially harm both individual species and entire faunal communities.[12]

RIFA AS A COMPETITOR TO NATIVE WILDLIFE

Because of large overlap in resource use, RIFA's primary competitors are other ant species. RIFA are extremely successful ant competitors, mainly because of their aggressive nature and high densities. Competitor ants who are stronger on an individual basis often succumb to RIFA due to the latter's sheer numbers and persistence (Bhatkar et al. 1972). RIFA's high numbers also allow them to quickly locate and control the available food sources in an area. (Porter and Savignano 1990). According to Wojcik (1994), RIFA have an edge over many native ants because the latter have smaller

Table 7.3. The Known Impacts of the Red Imported Fire Ant on Native Wildlife

Source	Organism	Effect of RIFA
ANTS		
Bhatkar et al. 1972	*Lasius neoniger*	Displaced colonies through interference competition
Porter and Savignano 1990	Ant and arthropod community	Reduced species richness and total ant abundance (when excluding RIFA)
Morris and Steigman 1993	Ant community	Reduced species richness and total ant abundance (when excluding RIFA)
Wojcik 1994	Ant community	Reduced abundance of most other ant species
NON-ANT INVERTEBRATES		
Harris and Burns 1972	Lone Star tick (*Amblyomma americanum L.*)	Reduced survival of engorged females, engorged larvae, and eggs through predation
Fleetwood et al. 1984		Increased mortality of engorged adult females through predation
Summerlin et al. 1977	Horn fly (*Haematobia irritans L.*)	Suppressed horn fly populations
Hu and Frank 1996		Suppressed horn fly populations
Summerlin et al. 1984		Suppressed horn fly populations and some horn fly predators
Howard and Oliver 1978		Preyed upon immature forms and reduced populations
Sterling 1978	Boll weevil (*Anthonomus grandis*)	Controlled boll weevil populations by preying upon larvae
Jones and Sterling 1979		Reduced boll weevil damage to crops through predation of immature weevils
Calvert 1996	Monarch butterfly (*Danaus plexippus L.*)	Consumed larvae in central Texas
VERTEBRATES		
Drees 1994	Colonial waterbirds (great egret, great blue heron, olivaceous cormorant, snowy egret, tricolored heron, roseate spoonbill, laughing gull, gull billed tern, Forster's tern)	Reduced waterbird nestling production

continues

Table 7.3. *Continued*

Source	Organism	Effect of RIFA
Landers et al. 1980	Gopher tortoise (*Gopherus Polyphemus*)	Preyed upon hatchlings
Ridlehuber 1982	Wood duck (*Aix sponsa*)	Preyed upon pipped eggs and hatchlings
Dickinson 1995	Crested caracara (*Caracara plancus*)	Preyed upon nestlings
Allen et al. 1997	White-tailed deer (*Odocoileus virginianus*)	Reduced fawn recruitment (doe:fawn ratio)
Allen et al. 1995	Northern bobwhite (*Colinus virginianus*)	Reduce populations in areas of high RIFA density
Giuliano et al. 1996		Reduce body mass and survival of chicks with stings
Sikes and Arnold 1986	Cliff swallow (*Hirundo pyrrhonota*)	Reduced nestling success
Allen et al. 1995	Northern bobwhite (*Colinus virginianus*)	Negatively impact bobwhite populations in Texas
Lockley 1995a	Least tern (*Sterna antillarum* Lesson)	Increased chick mortality

nests and seasonally restricted mating flights and are limited in their ability to withstand habitat disturbance. Studies show that the presence and dominance of RIFA have led to regional decline of several ant species. They have completely displaced the native fire ants *Solenopsis geminata* and *Solenopsis xyloni* throughout much of the invaded range (Wilson 1951, Wilson and Brown 1958, Hung and Vinson 1978, Wojcik 1983, Porter et al. 1988) and have greatly reduced the abundance of a number of other native ant species (Whitcomb et al. 1972, Glancey et al. 1976, Wojcik 1983, Camilo and Phillips 1990, Hook and Porter 1990, Porter and Savignano 1990). Even other successful north American invaders, including the argentine ant (*Iridiomyrmex humilis*) and the black imported fire ant (*Solenopsis richteri*), are often outcompeted by RIFA (Glancey et al. 1976, Anon. 1958).

Perhaps more important than the effects that the invasive ant has on individual competitor species are its effects on entire communities. The elimination of competitor and prey species by RIFA has led to reduced arthropod species richness wherever they are present (Nichols and Sites 1989). Among invaded areas, those with higher RIFA densities have consistently lower ant species richness, indicating that as they establish, RIFA increasingly simplify the ecosystems they invade (Glancey et al. 1976). The decline of both native and nonnative ant species coupled with the rise in RIFA has led to a radical restructuring of ant communities. Areas that have been invaded by the polygyne form of *S. wagneri* show a 66 to 70% decrease in ant species richness and a 90 to 99% decrease in native ant abundance relative to similar uninvaded areas (Morris and Steigman 1993, Porter and Savignano 1990). A detailed study in eastern Texas found that native ants disappeared completely from a large proportion of sites infested with RIFA—native ants were found at only 0.8% (3 out of 376) of RIFA infested sites, but at 11% of the uninfested sites. No native ants were found in the central core of the infestation even though they were formerly common in the region. Invaded areas also showed a 4- to 30-fold increase in total ant abundance resulting from RIFA's extremely high densities (Porter et al. 1991). This shift has implications for impact to both vegetation and non-ant fauna as greatly increased ant populations forage and build networks of mounds and trails.[13] While the process of community replacement is more gradual for monogyne than for polygyne RIFA colonies, the former also negatively impact ant communities by reducing species richness (Wojcik 1994). In summary, ant communities that contain RIFA tend to contain many more individuals but to suffer serious declines in diversity.

Studies also show that RIFA's presence may impact basic ecosystem functioning. Stoker et al. (1995) found that RIFA altered the community composition and the process of succession of the carrion-decomposing community. Population levels for six families[14] of decomposers declined when RIFA were present, leading to declines in decomposition rates. On small carrion such as those of mice, decomposition halted completely because RIFA totally excluded members of the normal decomposer community. Decomposer arthropods that colonize rotting fruit in the absence of fire ants are also excluded when fire ants are present (Vinson 1991b). When fire ants encounter rotting fruit that has already been colonized by native decomposer arthropods, the ants utilize both the decomposer arthropods and the fruit as food sources. In both studies, the presence of fire ants greatly reduced the diversity and abundance of decomposer species. Fire ants are thus capable of displacing a small but functionally essential community of organisms.

RIFA AS A PREDATOR

RIFA are extremely active, aggressive, general predators. Wilson and Oliver (1969) found that they bring back an average of 48 to 56 prey items to a single tunnel entrance each hour; and that rather than seeking any particular type of prey, RIFA appear to capture whatever is available in their area. While the majority of their diet is made up of other insects, they are also attracted to plant material (Hays and Hays 1959, Wilson and Oliver 1969) and to the mucous membranes of vertebrates (Vinson and Sorenson 1986). A wide range of organisms are therefore impacted by the predation of RIFA.

The results of studies examining the impacts of predation by RIFA on invertebrate species have varied over the years. Porter and Savignano (1990) suggest that some of this variation reflects increases in RIFA density as well as changes and improvements in study methods. While earlier studies found RIFA to have no measurable impact on invertebrate species or communities (Rhoades 1962, 1963, Howard and Oliver 1978, Sterling et al. 1979), more recent studies have found that RIFA profoundly impact some invertebrate species and communities through predation. Porter and Savignano (1990) measured arthropod diversity and abundance in both infested and uninfested areas. Arthropod species richness was 40% lower in areas with RIFA, with 30% of that decline consisting of non-ant arthropod species. Meanwhile, there was a five-fold increase in the number of arthropod individuals in infested areas due primarily to the large numbers of RIFA. Abundance of non-RIFA arthropods, however, declined by 75%, with non-ant arthropods again making up most of the loss. RIFA, therefore, are radically restructuring whole arthropod communities.

More detailed studies examining the effects of RIFA predation on individual invertebrate species have focused mainly on economically important pests. RIFA do suppress many agricultural and household pest populations including ticks (Harris and Burns 1972, Fleetwood et al. 1984), sugarcane borers (Negm and Hensley 1967, 1969), cotton boll weevils (Sterling 1978, Jones and Sterling 1979), horn flies (Summerlin et al. 1977, 1984, Hu and Frank 1996), earwigs (Gross and Spink 1969), leafhoppers (Wilson and Oliver

1969), and beetles (Howard and Oliver 1978, Brown and Goyer 1982). However, RIFA have also replaced or reduced many natural predators of these pests, offsetting much of their pest control benefit (Summerlin et al. 1984, Hu and Frank 1996). RIFA are themselves supreme pests, posing their own threats to agriculture and human health (see following text). Finally, impacts on non-pest species can also be severe: A study of monarch butterflies found that not a single butterfly survived beyond the second instar larval stage in a field occupied by RIFA (Calvert 1996).

Despite their size advantage, vertebrates are not excluded from attack by RIFA.[15] Vertebrates, including humans, are aggressively attacked when they disturb mounds and foraging trails (Lofgren 1986, Giuliano et al. 1996) and are occasionally preyed upon as well. RIFA are apparently attracted to mucous membranes in vertebrates, such as the eyes and mouth (Vinson and Sorenson 1986). While most vertebrates can escape from RIFA by simply moving away, they are vulnerable when their mobility is limited. Most vertebrates are only immobile during a very short period of time when their motor skills are not fully developed, or during entrapment—vertebrates most affected by RIFA are often young, ground dwelling organisms. The majority of studies have examined the damaging effects of RIFA on bird and reptile hatchlings (Landers et al. 1980, Mount 1981, Mount et al. 1981, Ridlehuber 1982, Drees 1994, Dickinson 1995, Lockley 1995b, Guiliano 1996). One study found that the presence of RIFA led to a 26.7% increase in mortality for nestlings of the least tern, an endangered species (Lockley 1995b). RIFA have also caused the death of mammals, including trapped rodents (Masser and Grant 1986) and newborn calves and piglets (Lofgren 1986). Allen et al. (1997) found not only that RIFA were capable of killing white-tailed deer fawns, but also that RIFA infestations actually decrease white-tailed deer recruitment (doe:fawn ratio). They suggest that declines in recruitment were a result of both greater direct mortality of neonatal fawns by foraging RIFA and indirect mortality by blinding and debilitating injury. Finally, RIFA significantly depress populations of the northern bobwhite quail (*Colinus virginianus*) through a combination of predation on hatchling chicks, impacts to food sources, and stings (Allen et al. 1995). RIFA prey on 6 to 12% of hatchling chicks annually in addition to impacting the survival and weight of chicks through stings (Johnson 1961, Dewberry 1962, Guiliano et al. 1996).

RIFA AND HUMANS

RIFA impact humans both medically and economically with their sting. A study by Adams and Lofgren (1981) found that between 29 and 35% of the people living in RIFA-infested areas are stung at least once per year—a proportion that amounts to 18 to 22 million people each year. The typical reaction to a sting is burning and itching followed by the formation of a large cyst (Lockey 1974, James et al. 1976). A small percentage of people, however, have much more severe reactions including respiratory distress[16] and even death.[17] A preliminary estimate put the costs associated with stings at $3 million per year (Lofgren 1986a).

RIFA also impact humans economically through their destruction of agricultural crops, roads, and other structures. Fire ant–induced damage has been documented to numerous agricultural crops including corn (Lyle and Fortune 1948, Glancey et al. 1979), soybeans (Lofgren and Adams 1981, Adams et al. 1976, 1977, 1983, Apperson and Powell 1983), eggplant (Adams 1983), potatoes (Adams et al. 1988), and citrus trees (Adams 1986). RIFA tunnel underneath roadways, causing subsidence and collapse of asphalt. Finally, the tendency of RIFA to nest inside humanmade structures has caused damage to wiring that controls airport runway lights (Lofgren et al. 1975), air conditioning, and telephone communications (Lofgren 1986a). While some studies report the costs associated with certain kinds of damage, no study has yet provided a complete picture of economic losses associated with RIFA (but see Lofgren 1986a).

Control Efforts

Finding a safe and effective method for controlling RIFA has proven difficult. Harmful residual pesticides such as heptachlor and dieldrin were employed to control RIFA in the 1950s (Lofgren 1986b). Concern over the accumulation of pesticide residues in the environment and their effects on wildlife prompted the development of Mirex, a toxic bait. Mirex became the standard for RIFA control in the early 1960s (Banks et al. 1973). It was used until 1978, when negative impacts to nontarget organisms discovered in the 1960s (Markin et al. 1974, Spence and Markin 1974) finally led to the banning of the substance by the Environmental Protection Agency (Johnson 1976). Evaluation of numerous chemical alternatives found insect growth regulators (IGRs) a possible candidate for control (Banks et al. 1983, Williams et al. 1997, Zhakharov and Thompson 1998). However, IGRs are not fully species-specific, and few studies have addressed impacts on nontarget organisms. The potential import of competitors, predators, pathogens, or parasites to control RIFA biologically has also been examined. Several studies, however, have concluded that only strong competitors imported from RIFA's native Brazil would likely have any substantial effect on RIFA densities (Lofgren et al. 1975, Jouvenaz et al. 1977, Buren 1983). Intentional introductions for the purpose of biological control often result in unexpected, negative consequences for native systems, and the introduction of a strongly competitive ant complex seems particularly untenable. In short, efforts to control RIFA do not appear promising. The prospect of it spreading into millions of new hectares due to climate change should therefore be viewed as a serious and possibly irreversible threat.

Tamarix *Shrubs in the West's Riparian Zones*

Invasive shrubs of the genus *Tamarix*[18] do not pose the direct threats to human health or agriculture that *S. wagneri* does. As a result, far less is known about *Tamarix* and, in general, far less attention has gone to its adverse impacts as it spreads through the western United States. In spite of the serious impacts that it has on water quality and supplies (Johns 1990, Wiesenborn 1996), the frequency and severity of floods (Graf 1980, Blackburn et al. 1982) and native wildlife (DeLoach 1997) in 23 states (Zavaleta 2000a), *Tamarix* does not appear on the pest list of a single state in the United States (USOTA 1993). Its range has not been monitored since 1961 (Robinson 1965), and only in the last few years have researchers begun to investigate experimentally the abiotic and biotic controls on its spread (S. Smith, A. Sala, pers. comm. 1998).

Because of the current dearth of knowledge, less can be concluded than for *S. wagneri* about the degree to which *Tamarix* is climate limited in distribution and likely to spread as warming and drying occur in the western United States (Giorgi et al. 1998). Similarly, fewer rigorous studies of its impacts on wildlife have been carried out than for RIFA. Nevertheless, existing data and observations support the impressions of scientists and land managers in the western United States that *Tamarix* is limited by high moisture and by low growing-season temperatures. The drying predicted by regional climate models for the western states thus has the potential to assist in eastward and westward spread into the central and Pacific coastal regions of the United States. Warming may also allow northward and uphill spread. The few existing studies of *Tamarix*'s impact on indigenous wildlife all indicate significant to drastic effects at every scale—species diversity, individual species populations, and the health of individual organisms decline when the invader replaces native riparian forest and scrub habitat. For these reasons, despite knowledge gaps, *Tamarix* provides another strong case for examining the links between global climate change, species invasions, and the welfare of native ecosystems in the United States.

Distribution and Spread

Members of the genus *Tamarix* may have reached North America as early as the sixteenth century in the hands of Spanish explorers in Mexico, but the invader most likely gained its foothold when it was imported in large numbers from Asia and the Mediterranean in the mid-1800s as an ornamental, windbreak, and agent of erosion control (Robinson 1965, Baum 1978, Brock 1994, Walker and Smith 1997). Since then, *Tamarix* has spread rapidly into perennial drainages throughout the arid and semiarid regions of the western United States, benefiting in part from human alteration of the natural flow regimes of rivers and streams (Everitt 1980). To date, *Tamarix* has replaced native riparian forest and scrub communities in approximately 500,000 hectares of riparian floodplain habitat in 23 states, from sea level to 7000 feet (Christensen 1962, Robinson 1965, Everitt 1980, Zavaleta 2000a). While less is known about its naturalized range outside of the United States, it is known to occur in northwestern Mexico and along the Mexico–U.S. border from central Texas to the Sea of Cortez (GWR 1989, pers. obs.).

Range data on a national scale for widespread exotic plants in the United States generally do not exist, though some states maintain weed databases for herbaceous pest species. The limited attention that *Tamarix* has received comes from the federal agency responsible for water management in the western states, the Bureau of Reclamation, out of concern for the invader's high water consumption rates. In 1961, the Bureau sponsored a careful study of the distribution of the genus in the western states (Robinson 1965). This study remains the only national-scale source of distribution data for *Tamarix*, though more recent updates exist of the areal extent of *Tamarix* in the western United States (GWR 1989, Zavaleta 1999). No range data exist by species for the four or more distinct naturalized *Tamarix* species, in part because hybridization among them may have obscured species identities and because distinctions among the species are intrinsically difficult, but ongoing research may help to identify major species differences among subregions of the United States (J. Gaskin, pers. comm. 1998).

While the areal coverage of *Tamarix* appears to have increased significantly since 1965, the boundaries of its regional distribution have changed little—most of its expansion has consisted of increases in dominance along drainages already invaded by 1965 (GWR 1989, Salas et al. 1996, Zavaleta 2000b). Between the mid-1920s and 1960, however, rapid spread occurred simultaneously in several parts of the western United States. Local expansion rates of 24 to over 400 hectares per year were observed within individual drainages (Robinson 1965), illustrating *Tamarix*'s capacity to spread rapidly. *Tamarix* can produce as many as 17 seeds per cm^2 of ground in a single season (Warren and Turner 1975) that germinate easily and quickly in a wide range of conditions (Everitt 1980), including highly saline soils (Jackson et al. 1990). Once germinated, it grows rapidly—up to 4 cm/day (Loope 1988)—and can tolerate drought, fire, and flooding to degrees that native species cannot (Busch and Smith 1993, 1995, Cleverly et al. 1997). It can withstand submersion for up to 3 months (Warren and Turner 1975), but it can also withstand prolonged desiccation and is more able than native species to establish in areas with deep zones of permanent moisture availability (Brock 1994). *Tamarix*'s capacity for rapid dispersal, establishment, and displacement of native vegetation suggests that dispersal limitations and narrow biotic, edaphic, and hydrogeomorphic tolerances are unlikely to strongly constrain its responses to climate trends.

The invader's rapid expansion in the mid-twentieth century also suggests that *Tamarix*'s reduced spread since the 1960s reflects a collision with some range-limiting factor. *Tamarix* eradication programs have not increased, and riparian corridor disturbances in the western states such as grazing and alteration of natural flow regimes have not decreased, since the middle of this century. Alteration to flow regimes and riparian disturbance are widespread in areas of the western United States not invaded by *Tamarix* such as, respectively, the Columbia River Basin and the heavily grazed western Sierra

Nevada foothills. These factors thus cannot explain the significant decline in *Tamarix* expansion that has occurred in recent decades. If, as appears to be the case, *Tamarix* has ceased to enlarge its range in the western United States because of climatic limits, then global climate change may impact its extent in the region.

The Impacts of *Tamarix* on Native Wildlife

When *Tamarix* invades, it impacts the functioning of the riparian zone in at least two ways that harm wildlife. First, *Tamarix* stands themselves provide relatively poor habitat for many species of native terrestrial fauna, drastically reducing the abundance and diversity of riparian zone taxa (DeLoach 1997). Second, *Tamarix* consumes water more rapidly than native vegetation, drawing down water tables, drying up desert springs, and lowering the flow rates of waterways and the levels of lakes (Vitousek 1986, Loope et al. 1988, MacDonald et al. 1989, Johns 1990). This impacts both terrestrial and aquatic wildlife by removing reliable sources of water and reducing aquatic habitat quality for desert wildlife ranging from bighorn sheep and quail to endangered pupfish (Neill 1983, Loope et al. 1988, DeLoach 1997). The negative impacts on terrestrial fauna of the displacement of native riparian plant communities by dense, monospecific stands of *Tamarix* are well-documented. *Tamarix* lacks palatable fruits and seeds, fails to harbor phytophagous insects that insectivores can eat, occurs in high-density stands with little structural or microclimatic diversity, and is too small in stature and limb size to meet the habitat requirements of raptors and woodpeckers (DeLoach 1997).

Impacts of *Tamarix* on reptiles and mammals have been documented. One study found reptile densities over 10 times lower in *Tamarix* than in native stands (Engel-Wilson and Ohmart 1978). However, *Tamarix*'s effects on avian communities have been better studied than those on any other class of animals. Bird densities, diversity, and individual health have been found to be greatly reduced in *Tamarix* stands compared to native vegetation. In the most comprehensive study conducted to date on the effects of saltcedar on wildlife, Anderson and Ohmart (1982) write: "Study of the riparian vegetation types in the lower Colorado River Valley has provided ample evidence that salt cedar has little value to a large majority of the bird and rodent species in the valley" (198), and ". . . bird populations, in general, were enhanced by merely clearing salt cedar, even though the total amount of vegetation was reduced by more than 90 percent" (37).

They report that total avian abundances in *Tamarix* on the lower Colorado River range seasonally from 39 to 59% of those found in native willow, cottonwood, and mesquite habitats (Anderson and Ohmart 1977, Anderson et al. 1983). Several bird species are found only in native cottonwood habitat and never in *Tamarix* stands, including cavity nesters and timber drillers such as the gilded northern flicker (*Colaptes auratus*) and downy woodpecker (*Picoides pubescens*), and passerines like the summer tanager (*Piranga rubra*), a species of concern in California (Anderson et al.1983, Hunter et al. 1988, Ellis 1995). *Tamarix* thus significantly alters native bird communities through reductions in diversity, total abundance, and individual species abundances. In addition, interaction with climate appears to influence the severity of *Tamarix*'s impact on avian communities, which become increasingly depauperate as one moves west from the eastern limit of *Tamarix*'s range in Texas. It has been suggested that as summer temperatures rise along that gradient, low, shrubby *Tamarix* thickets fail to provide the increasingly important cooling that would occur beneath a multilayered canopy of mature trees (Hunter et al. 1988). Temperature increases alone, even if range expansion did not occur, might thereby exacerbate declines in bird diversity and abundance throughout the current range of *Tamarix*.

The effects of *Tamarix* extend to species whose current population sizes warrant protection against the risk of extinction. Three federally listed (endangered or threatened) species and one candidate for federal listing suffer clear, quantifiable negative impacts from *Tamarix* invasion: respectively, the southwestern willow flycatcher (*Empidonax trailii extimus*), bald eagle (*Haliaeetus leucocephalus*), whooping crane (*Grus americana*), and Peninsula bighorn sheep (*Ovis canadensis cremnobates*).

The Southwestern willow flycatcher relies on riparian woodland vegetation for both foraging and nesting (Brown and Trosset 1989). Damage to riparian habitat therefore directly impacts both its survival and its reproductive success. Still, much of the literature on habitat use and natural history of the flycatcher, including the U.S. Fish and Wildlife Service's listing of critical habitat, treats all riparian vegetation as equivalent in terms of habitat quality (e.g., Sogge et al. 1997a, 1997b, USFWS 1997). The flycatcher uses *Tamarix* successfully in areas where it is the only riparian vegetation available (Brown and Trosset 1989). However, wildlife scientists in the Southwest generally agree that *Empidonax trailii extimus* fares worse in *Tamarix* than in native stands (DeLoach 1997, various pers. comm. 1997–98). The existing evidence supports their observations. A large, 2-year mist-netting study in western Texas found significantly more flycatchers than expected based on chance in native willow habitat, but significantly fewer in stands containing *Tamarix* and exotic Russian olive (*Eleagnus angustifolia*) (Yong and Finch 1997). Flycatchers caught in native willow stands also had higher body mass and observable fat stores than those in the mixed habitats, suggesting that willow provides superior foraging habitat (Yong and Finch 1997). An additional study comparing bird use of *Tamarix* and native cottonwood communities found *E. trailii* only in native cottonwood stands (Ellis 1995). More study is certainly warranted to quantify the impact of *Tamarix* on this endangered songbird, especially since reproductive success rather than mere presence is the best indicator of habitat quality. Nevertheless, it appears that native riparian woodlands provide superior habitat to *Tamarix* for the willow flycatcher.

Two other federally listed birds suffer significant but less widespread impacts of *Tamarix*. The endangered whooping crane occurs mainly east of *Tamarix*'s range, but one population—at Bosque del Apache National Wildlife Refuge, New Mexico—is strongly affected by the encroachment of *Tamarix* thickets into its critical marsh habitat (GWR 1989). The bald eagle requires large-limbed trees for nesting and prefers them to smaller-statured vegetation for roosting (GWR 1989). It is impacted in areas where heavy *Tamarix* infestation has displaced all or most of the native trees of the riparian zone, such as the Cibola, Imperial, and Havasu National Wildlife Refuges of the lower Colorado River (pers. comm., USFWS Refuges 1998). However, impacts to the bald eagle in less heavily infested areas are unclear.

Finally, *Tamarix* impacts the Peninula bighorn sheep of southern California and northern Baja California, Mexico, a candidate for listing under the Endangered Species Act. In the hot deserts of the bighorn's range, *Tamarix* dries remote springs that the bighorn depends on for water (National Park Service 1981, Neill 1983, DeLoach 1997). In addition to the bighorn, desert fishes and migratory birds may decline following the invasion of remote desert water sources by *Tamarix*. In Death Valley National Monument, *Tamarix* invasion of a shallow pond at Eagle Borax Springs caused it to gradually dry and, concomitantly, to lose its historical stopovers of migratory birds. In 1971, more than 20 years after the initial appearance of the invader, the Park Service removed the *Tamarix* at the springs with fire. Within 8 weeks, the water table rose over 35 cm and the pond reappeared (Neill 1983). At a second Death Valley site that provides habitat to an endemic species of desert pupfish, *Tamarix* control in 1972 reversed gradual drying by the invader that had lowered the water level of the pond 15 cm in 15 years (National Park Service 1981).

Tamarix thus poses threats to a broad range of native taxa, including not only all major vertebrate classes but also the insect community (Stevens 1995). *Tamarix* also assists in the spread of another nonindigenous species: the exotic honeybee, *Apis mellifera*, which thrives in *Tamarix* stands and competes with native honeybees for limited resources (Lovich 1996). All of these impacts on wildlife occur because of radical change to the riparian forest communities of the western states, which are themselves becoming an endangered forest type. For example, since *Tamarix*'s introduction, Fremont cottonwood (*Populus fremontii*) and Goodding willow (*Salix gooddingii*) have declined steadily from their former roles as important constituents of gallery riparian forests throughout the region to become virtual relicts that occur mainly in isolated, protected areas (Molles et al. 1998). Biodiversity losses to *Tamarix* thus extend from wildlife to the whole ecosystems that they inhabit, with implications for such basic riparian functions as flood control, beneficial interaction with aquatic ecosystems, and the maintenance of refugia for biological diversity in harsh landscapes.

Tamarix and Humans

The invasion of western rivers, lakes, and springs by *Tamarix* has caused damage to ecosystem services other than support of native wildlife and biological diversity. Water is scarce in the arid southwestern and Great Basin states for both natural systems and human populations—in some regions, such as the lower Colorado River, water supplies are in imminent danger of falling short of the needs of growing populations (Waggoner and Schefter 1990, Morrison 1996). *Tamarix* stands, with their dense canopies and rapid growth rates, have been found by nearly 20 studies to transpire more water than the native vegetation that they replace (Johns 1990, Brock 1994). Across all of these studies, *Tamarix* causes a loss every year of an additional 3050 to 4500 m³ of water per hectare of invaded habitat. Over all of its invaded range this amounts to a total annual loss of 1.4 to 3.0 billion m³, which is accompanied by *Tamarix*'s additional impacts on the quality of remaining water through salinization (Wiesenborn 1996). In total, water losses to *Tamarix* generate a yearly economic loss for municipal water providers, farmers, hydropower, and recreationists on the order of $60 million to $180 million (Zavaleta 2000a).

When *Tamarix* invades a river corridor, it also exerts long-term effects on the hydrology and geomorphology of the river channel itself. *Tamarix*'s thick, densely rooted stands trap and stabilize enormous quantities of waterborne sediment on sandbars, riverbanks, and midstream islands (Graf 1978, 1980). Over time, the accretion of sediment narrows river channels and reduces their water-holding capacity enough to significantly reduce their ability to contain high flows (Hadley 1961, Blackburn 1982). Texas's Brazos River has narrowed nearly 90 meters since its invasion by *Tamarix* around 1940 (Blackburn 1982). Comparisons of recent photographs with photographs taken by Major John Wesley Powell's 1871 expedition show that Utah's Green River has also narrowed by nearly one-third since invasion by *Tamarix* at the turn of the century (Graf 1978). A hydrological flow model study of the reduced abilities of similarly affected rivers to contain floodwaters indicated that *Tamarix* is responsible for worsening the frequency and severity of floods as well as the economic and human costs associated with them (GWR 1989).

Evidence for Range Limitation

Tamarix has been noted for decades by its researchers as altitude-limited in its distribution on local scales. In the Southwest, Bowser (1957) observed that although *Tamarix* occurred as high as 3350 m, it spread very slowly above 1220 m. Stromberg (1998) observed that radial growth rates of *Tamarix* declined substantially with increasing elevation and attributed the pattern to cool temperatures and more frequent frosts. A survey of *Tamarix* species in the United States (J. Gaskin, unpub. data) supports the hypothesis that naturalized distributions reflect native range limits. Each of the seven species found in the United States appears limited to a different degree by cold temperatures in its native range—from *T. canariensis*, which occurs strictly in the southernmost points of Mediterranean Europe and in North Africa, to *T. ramosissima*, which occurs as far north and inland as central Mongolia (Baum 1978). In the United States, *T. canariensis* occurred only south of 30N, while *T. ramosissima* was the only one of seven species found north of 39N. Sala and colleagues at the University of Montana are currently performing the first controlled experiments on functional responses of *Tamarix* to temperature (Sala, pers. comm. 1998). They expect *Tamarix* not to tolerate freezing temperatures during the growing season, which may cause it to cavitate. Their findings may provide much greater insight into the degree to which low temperatures do limit the elevational and northern limits of *Tamarix*'s naturalized range.

In addition to temperature, patterns of water availability appear to exert strong controls on the presence and performance of *Tamarix* species. Permanence of flowing water and depth to water table strongly influenced the dominance of *Tamarix chinensis* versus native Fremont cottonwood (*Populus fremontii*) in the arid Southwest (Stromberg 1998). Deeper water tables and ephemeral surface flows favor *Tamarix*, which has deeper rooting ability, greater water use efficiency, and a higher capacity for extracting water from drier soils (Graf 1980, Busch et al. 1992, Busch and Smith 1995). Water table decline associated with anthropogenic withdrawals appears to have assisted *Tamarix*'s spread at some sites (Schwartzman 1990, Busch and Smith 1995), while shallow water tables and perennial flows have allowed native species to persist at others (Stromberg 1998). The Pacific coastal and eastern range limits of the *Tamarix* complex, where it spreads slowly or ceases to spread (Robinson 1965, N.E. West pers. comm. 1998, Zavaleta, unpub. data) are consistent with the hypothesis that it does not compete successfully with native species in areas of higher moisture availability. Moreover, *Tamarix* species widespread in the United States are limited in their native ranges to sandy and salty sites in arid parts of Asia, North Africa, and extreme southern Europe (Baum 1978). Once again, this is consistent with observations in the naturalized range that *Tamarix* is less invasive or even absent in the riparian zones of moister regions.

Tamarix's affinity for drier conditions may allow it to spread as warming drives up evapotranspiration rates from the land surface, lowering water tables and reducing flow rates in streams and rivers

(Gleick 1990). Regional climate model simulations of the impact of doubled atmospheric CO_2 concentrations predict not only higher temperatures but also decreased precipitation for the interior western United States (Giorgi et al. 1998). *Tamarix* may benefit, therefore, not only from increases in temperature per se but also from decreases in moisture availability associated with future warming and possible precipitation declines.

Efforts to Control *Tamarix*

Like the red imported fire ant, *Tamarix* has been the subject of considerable control efforts with limited success. *Tamarix* has been successfully removed from small, isolated areas of invasion such as individual desert ponds (Neill 1983) through combinations of labor-intensive measures such as clearing, root plowing, burning, and repeated hand application of herbicides (Taylor and McDaniel 2000, Taylor et al. 1999). However removal of *Tamarix* from only small portions of a larger invaded area yields a high probability of reinvasion, particularly if disturbances such as grazing and flood control continue to favor the reestablishment of *Tamarix* over native vegetation. Growing evidence suggests that permanently removing *Tamarix* on anything smaller than a continental scale will not be feasible without accompanying restoration of the natural flood regimes to which native riparian species are adapted (Molles et al. 1998). Restoring natural flood regimes in the western United States is a virtual impossibility, given the number of dams and diversion projects that now mark the region and the current trajectory of demands on impounded water for municipal use, irrigation, hydropower, and recreation (Waggoner and Schefter 1990, Morrison 1996). Consequently, as for the fire ant, further increases in the areal extent of *Tamarix* may represent further permanent losses of native wildlife habitat and riparian and riverine ecosystem function.

Prospects for Future Range Expansions

The Red Imported Fire Ant

Abundant evidence of northward temperature limitations on ant distributions led us to begin by examining relationships between January mean minimum temperatures and the occurrence of RIFA.

We selected winter temperature minima rather than other temperature-related variables such as growing season length because experimental evidence indicates that absolute temperature minima, not summer or active season minima, drive mortality and brood failure in RIFA. In addition, growing season length (frost-free period) incorporates a threshold of 0°C, while the red imported fire ant appears to encounter temperature limits only well below freezing (see earlier). We selected January mean minimum temperatures because annual lows occur more often in January than in any other month. Absolute annual temperature minima rather than minima at any particular time of the year appear most relevant to RIFA's biology.[19] We also chose to limit our analysis to the portion of the United States east of the Rocky Mountains. This helped to avoid imputing climatic factors for the absence of RIFA in areas whose climates may be suitable (e.g., central California), but which likely remain uninvaded simply because the ant has not been introduced or had the opportunity to spread to them.

As expected, RIFA occurrence probabilities[20] very closely reflected January mean minimum temperatures. Within the range of minimum January temperatures that occur in the United States, RIFA's occurrence probability peaks at the highest temperatures with a value of 1 (100% of counties infested) (Fig. 7.1). Below

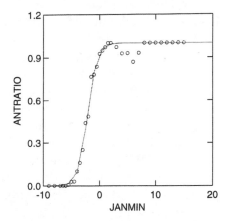

Figure 7.1. The relationship between January mean minimum temperature and the probability of red imported fire ant occurrence in the eastern United States. A logistic curve fitted through the data yields an adjusted R^2 value of 0.992. Model and parameter values are cited in the text.

+1.5°C, occurrence of RIFA declines rapidly to values near 0 (0% of counties infested) by –5.0°C.

We used data from –9 to +15°C to calculate a logistic model using ordinary least squares:

$$\text{Occurrence} = \frac{1}{1 + e^{C + \beta T}}$$

where C and ß are constants and T is January mean minimum temperature.[21] We selected a logistic model because it provides a simple fit to data falling entirely on one side of the peak of a bell-shaped curve. The range of January mean minimum temperatures encountered in the continental United States is confined to relatively cold ones. RIFA occurrence peaks at the warmest available winter temperatures in the United States. This suggests that the only portion of the theoretical bell-shaped relationship between temperature and occurrence encountered in the United States is to the left (low temperature side) of RIFA's optimum January minimum temperature (Gleason 1926, Saetersdal and Birks 1997). The scaling constant ß reflects the sensitivity of occurrence to January minimum temperature and the direction in which occurrence changes with temperature. The constant C sets the level of occurrence associated with a January minimum mean temperature of 0°C. For January mean minimum temperature, the modeled constant values were C = –2.586 and ß= –1.203. The model provided an extremely strong fit to the data, with a raw R^2 value of 0.997 and a corrected R^2 of 0.992. For comparison, a C constant-only logistic model yielded a raw R^2 of 0.691 and a corrected R^2 of zero.

We also examined the relationship between growing season length[22] and the probability of RIFA occurrence to test the hypothesis that temperature constraints on foraging and development prevent RIFA colonies from surviving in areas with long freeze periods. Once again, RIFA occurrence peaked at 100% in analysis areas with long growing seasons, declining rapidly below a growing season length of 225 days and reaching 0% at a growing season length of 135 days (Fig. 7.2). We used data from 135 to 365 days to calculate a logistic model using ordinary least squares. For growing season length, modeled constant values were C = 18.111 and β = –.088. The model provided nearly as good a fit as that of January mean minimum temperatures, with a raw R^2 value

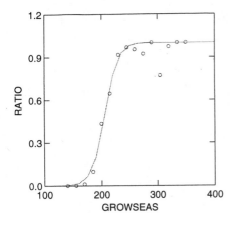

Figure 7.2. The relationship between growing season length and the probability of red imported fire ant occurrence in the eastern United States. A logistic curve fitted through the data yields an adjusted R^2 value of 0.972. Model and parameter values are cited in the text.

of 0.992 and a corrected R^2 of 0.972. Although some have argued that moisture also limits RIFA distributions, we chose not to analyze the relationship between occurrence and precipitation because available evidence counters the assumption that RIFA is already nearing the western (dry) limit of its distribution (Pimm and Bartell 1980, CAST 1976, Francke et al. 1986, Vinson and Sorenson 1986, Francke 1988, Allen et al. 1993, Stoker et al. 1994, Killion and Grant 1995).

We used occurrence values predicted by minimum January temperature model to estimate the potential effects of warming alone on RIFA distributions in the eastern United States. We chose to use minimum January temperatures rather than growing season length because (1) stronger experimental evidence supports an absolute minimum temperature limit to RIFA survival, (2) January minimum temperature allows more direct application of the findings of global climate models (GCMs), which predict mean annual temperature changes directly, and (3) model fit of January minimum temperatures was similar to that of growing season length.[23] We raised January minimum temperatures by 1, 2, 3, and 4°C and applied model occurrence probabilities to each 1°C U.S. isotherm based on its new January temperature minimum. We then added infested counties to each region by hand to reflect differences in temperature minima within each 1°C region and large differences in precipitation. Within 1°C regions, warmer counties received priority over cooler ones; in the western extreme of the range, wetter counties received priority over drier ones. We

Table 7.4. Projected Impacts of Warming on the Red Imported Fire Ant's Distribution in the Eastern United States.

	Current	At +1°C	At +2°C	At +3°C	At +4°C
Number of partially infested counties	25	86	73	96	102
Number of fully infested counties	696	813	981	1096	1268
Total number of infested counties	721	899	1054	1192	1370
Total area of infestation ($\times$ 1000 km^2)	1261	1583	1846	2038	2271

also made an effort to include counties contiguous to invaded areas before including more distant counties, though long-distance transport by vehicles or even water make noncontiguous distributions possible. The number of counties infested by RIFA increases by approximately 150 with each additional 1°C of warming, and the total area infested by RIFA increases by nearly 200,000 to over 320,000 km^2 with each additional 1°C of warming (Table 7.4). At 1°C of warming,[24] states such as Virginia and Tennessee, which currently suffer little or no RIFA infestation, become susceptible to invasion over substantial areas. With 2°C of warming, additional advances into the moister parts of western Texas, Oklahoma, much of Virginia, and coastal areas of states as far north as Maryland occur. With 3°C of warming, almost all of Tennessee and Virginia become susceptible, as well as additional areas of Kentucky, Oklahoma, Texas, and the southernmost portions of Illinois and New Jersey. By 4°C of warming, our model suggests that RIFA will be able to spread solidly into an area nearly double that of its current extent, affecting nearly twice a many as the 721 counties that it currently plagues (Table 7.4). While these figures represent only a climate-based first approximation of the potential for RIFA spread under warming, they suggest that substantial increases in the extent of RIFA's effects will occur.

Tamarix

We began our analysis of *Tamarix* distributions and climate by looking at January mean minimum temperatures and growing season length, to test hypotheses that high altitude and northern limits on *Tamarix* growth and spread reflect temperature limitations during the growing season or damage by cavitation[25] at sufficiently low temperatures. However, no relationships between *Tamarix* occurrence and either temperature parameter were sufficiently consistent to warrant use in projecting future range shifts (Figs. 7.3A, B). However, we did observe that for both January minimum temperature and growing season length, no *Tamarix* occurred below or above a certain range.

We then examined the relationship between annual precipitation and *Tamarix* occurrence to test existing hypotheses that *Tamarix* is limited in its distribution to relatively dry areas. Despite the complex suite of hydrogeomorphic factors that appear to affect *Tamarix*'s success, mean annual precipitation was a strong predictor of *Tamarix* occurrence. *Tamarix* occurrence peaked at the lowest mean annual precipitations found in the United States (47 mm/yr) and declined to zero by 1150 mm/yr (Fig. 7.4). An observed rise in occurrence between 350 and 650 mm/yr of precipitation was not explained by overrepresentation of warmer regions in that precipitation range; the rise persisted when the analysis of distribution as a function of precipitation was constrained to sites with growing season length within the range tolerated by *Tamarix*. Fitting of a logistic model to the data for *Tamarix* and annual precipitation yielded constant values of C = 1.703 and ß = 0.003. Model fit was extremely good, yielding a raw R^2 value of 0.954 and a corrected R^2 of 0.917. For comparison, a model without precipitation (setting $\beta = 0$) yielded a raw R^2 value of 0.444.

The finding that annual rainfall strongly predicts *Tamarix* occurrence is consistent with characteristics of the native habitats in which *Tamarix* occurs, observed limitations on *Tamarix* expansion at the western extreme of its range in California, and the working hypotheses of investigators of *Tamarix* in the western United States. The most recent regional climate model (RegCM) developed for the western United States (Giorgi et al. 1998) predicts a complex

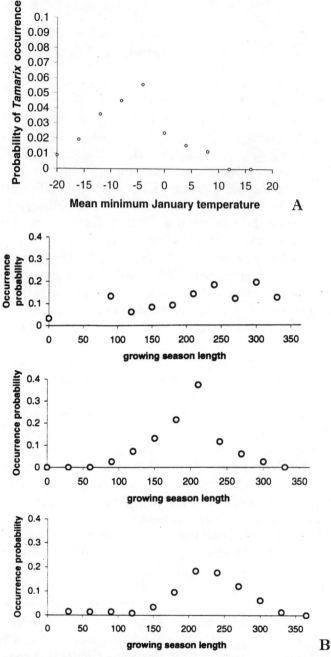

Figure 7.3. (A) The relationship between mean minimum January temperature and *Tamarix* distribution across all levels of annual precipitation. (B) The relationship between *Tamarix* distribution and growing season length in three different annual precipitation ranges. For both January temperature minima and growing season length, the relationship with *Tamarix* distribution was inconsistent across precipitation zones.

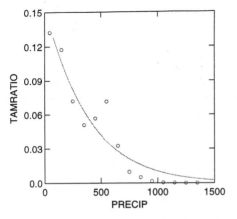

Figure 7.4. The relationship between annual precipitation totals and the probability of *Tamarix* occurrence in riparian areas in the United States. A logistic curve fitted through the data yields an adjusted R^2 value of 0.917. Model and parameter values are cited in the text.

spatial array of precipitation responses for the area under doubled atmospheric CO_2 concentrations, which, across the region infested by *Tamarix*, yields an approximate net precipitation change of –67 mm/yr (Zavaleta 1998). In the arid Southwest, a decline of that magnitude represents a 15 to 40% decrease in annual precipitation. Because of the difficulty of accounting for seasonal and spatial variation in RegCM model predictions, specific geographic predictions of *Tamarix* spread are not feasible. However, the strong negative relationship between rainfall and *Tamarix* distribution found in this study suggests that a decline in precipitation or system drying due to warming-enhanced evapotranspirational losses would allow *Tamarix* to spread to areas currently too moist for it to invade. These areas would likely include greater portions of California and Oregon and larger areas of eastern Texas, Oklahoma, Kansas, Nebraska, South Dakota, and even Arkansas, where *Tamarix* has been observed near the Oklahoma border on the Arkansas River (Holla Bend National Wildlife Refuge, pers. comm. 1998).

Conclusion

Scientists now concur that, on balance, global change bodes ill for the future of species invasions (Mooney and Hobbs 2000). In general, as atmospheric chemistry, climate, international commerce, patterns of transport and migration, land use change, and herbicide and pesticide use continue on their current trajectories, these global changes will increase rates, intensities, and extents of species invasion and will correspondingly worsen the impacts of invasions on ecosystem services, economies, and biological diversity. It is also clear, however, that more detailed research on specific global change–invasion interactions and their potential implications is badly needed.

The area of climate change–invasion interactions is no exception. Few investigations of potential range expansions of exotic species under warming or precipitation changes exist. How can ecologists choose the next case studies wisely? For the purposes of informing management and climate-related policy decisions, it is not necessary to exhaustively study the climate responses of all harmful invasives in a region. A strong portfolio of cases illustrating that damaging invaders will expand their impacts under expected future climate change is far more feasible and may be equally influential. However, specific invasives management decisions may call for in-depth studies of particular species. Invaders that threaten human health, such as exotic disease vectors, deserve high priority, as do invaders already known to have both troubling economic or ecological impacts and an apparent potential to benefit from warming. A search for patterns among those species already well studied will also help guide future research by identifying geographical and taxonomic gaps and illuminating what kinds of invaders, in what environments or regions, are most ripe for expansion.

More research is also needed on the impacts of multiple stresses on invasion trajectories. Studies focused on regional warming, such as this one, help to identify the large-scale trends that we can expect from particular invaders. However, predicted species responses will undoubtedly be modified by other large-scale changes, such as atmospheric CO_2 rise, and local changes, such as increased air pollution and land use intensification. For example, 1 to 2°C of warming enhances the ability of native shrubs to establish in annual California grasslands dominated by exotics; however, increased nitrogen deposition at levels now experienced in industrialized urban areas suppresses native shrub establishment (Zavaleta, unpub. data). While single-factor studies capture elements of the changing dynamics of an invader, their utility depends on the degree to which other anthropogenic stressors modify these dynamics. An exclusive focus on climate is useful to make rough, nonspecific predictions about the direction and magnitude of response of species that are clearly abiotically limited and likely to be less sensitive to other stresses—land-use change, for example, does not strongly affect the success of *Solenopsis* ants. To make general statements about how the invasions problem is likely to respond to multiple environmental assaults on the globe, however, studies of different factors need to be pooled (see Mooney and Hobbs 2000).

The specific findings of this chapter support the current consensus that climate change will on balance worsen the problem of exotic species and their impacts on native biota. Many more harmful exotics in the United States are likely to respond favorably to climate change than are expected to decline as warming occurs. The cases of the red imported fire ant and the invasive *Tamarix* shrub illustrate some of the mechanisms that may underlie future positive responses of harmful invasives to warming and precipitation changes. The case of the red imported fire ant, in particular, illustrates that the magnitude of change in invasive species impacts due to changing climate can be large: the expected magnitude of anthropogenic warming may extend the impacts of RIFA to hundreds of thousands of square kilometers that have so far escaped infestation. Both cases argue that absent significant and unexpected biotic changes, at least two harmful exotic species in the United States will experience range expansions under predicted climate changes. The potential for climate-driven increases in the effects of invasive species on native ecosystems and the species that they support should bolster the already vast array of arguments for confronting the invasions problem, reducing carbon emissions, and minimizing the prospects for anthropogenic climate change in the next century.

Acknowledgments

We would like to thank Heather Schoonover and John Fay for their generous assistance with the Geographic Information Systems portion of this project, Stephen Schneider and Harold Mooney for helpful advice and feedback, and the National Wildlife Federation for its support of this research.

Notes

1. For a species to be assigned to this category, it must have either a native range confined to the low elevation tropics and/or subtropics; or a known naturalized range restricted to the southern half of the United States or to Hawaii, Puerto Rico, Guam, and/or other U.S. possessions in the tropics and subtropics.

The assignment of species to categories here is based strictly on known native and naturalized ranges and does not consider the mechanisms that limit each species' range. Factors other than climate cannot be ruled out, so these data can only be considered a crude approximation of how many species will be assisted or harmed by uniform warming in the United States.

2. For a species to be assigned to this category, it must have either a native range restricted to boreal and/or arctic/antarctic regions; or a known naturalized range confined to the interior northern and/or extreme northeastern portions of the United States.

3. This analysis does not take into account potential changes in precipitation that may accompany greenhouse warming; predictions of precipitation change vary from model to model and vary spatially within single models. However, changes in water balance associated with higher temperatures are crudely captured—current, temperature-associated range limits should reflect all temperature-related processes, including net water balance.

4. Most of the *Tamarix* in the United States appears to be *T. ramosissima.* However, members of at least six other species in the genus, notably *T. chinensis,* also occur in the wild and are inconsistently differentiated from *T. ramosissima* in the existing literature. Obsolete labels, including *T. pentandra* and *T. tentandra,* also appear in older references. In practice, the invasive species complex is often simply referred to as *Tamarix.*

5. Appears as *S. invicta* in all but the most recent literature.

6. All geographical analyses were performed in a Lambert Azimuthal Equal Area projection for the Northern Hemisphere, which preserves the accuracy of areal data at the expense of directional data.

7. More detailed RIFA distribution data has been collected in certain portions of their range, including parts of Texas (Porter et al. 1991), Florida (Tschinkel 1988, Porter 1992), and North Carolina (Bass and Hays 1976).

8. This is not the case for the polygyne form of RIFA, where multiple queens persist in a single mound.

9. Predators of RIFA queens include the dragonflies *Libellula axilena, Epiaeschna heros, Pachydiplax longipennis, Tramea carolina, Anax junius,* and *Somatochlora provocans;* the birds *Cahaeura pelagica, Tyrannus tyrannus,* and *Colinus virginianus virginianus;* a wolf spider *Lycosa timuga;* an earwig *Labidura riparia;* a tiger beetle *Cicindella punctulata;* and the ants *S. wagneri* (RIFA itself), *Conomyrma insana, Conomyrma lavopecta, Lasius neoniger, Pogonomyrmex badius, Hypoponera opacior, Aphaenogaster floridana, Solenopsis geminata, Pheidole dentata, Iridiomyrmex pruinosum, Formica schaufussi, Camponotus abdominalus floridanus,* and *Paratrechina* species.

10. There are three principal components of winter survival ability in insects: freeze tolerance (the ability to survive bodily freezing), freeze avoidance (the ability to evade bodily freezing), and cold acclimation (an increase in cold-hardiness upon exposure to temperatures near a lower thermal limit) (Salt 1961).

11. This is likely due to the loss of evaporative cooling at 100% humidity (Francke et al. 1986).

12. Vinson (1997) reviews much of the material on *S. wagneri*'s impacts on wildlife.

13. The effects of mound and trail building themselves are more serious than one might imagine because of the incredibly high densities in which RIFA occurs. In eastern Texas, Porter et al (1991) found a mean RIFA density in excess of 500 mounds per hectare.

14. Calliphoridae, Muscidae, Sarcophagidae, Staphylidae, Silphidae, and Histeridae.

15. Allen et al. (1994) review the impacts of *S. wagneri* on vertebrates.

16. 7.1% of those seeking medical attention (Adams and Lofgren 1982).

17. Lockey and Fox (1979) cited five deaths in 2 years. Rhoads et al. (1977) determined that there were 3.8 deaths per 100,000 persons per year in Jacksonville, but considered the figure to be low because of incomplete survey response.

18. Commonly called saltcedar or tamarisk.

19. Subsequent analysis of December minimum temperatures and growing season length confirmed that although fairly strong relationships with ant occurrence existed (both December minima and growing season length are highly correlated with January minima), fits were not as strong or consistent as those with January minimum temperatures.

20. Calculated as $C_{I,t}/C_{I,t+}C_{N,t}$, where $C_{I,t}$ denotes number of counties invaded (*I*) within January mean minimum temperature region *t*, and $C_{I,t}$ denotes number of counties not invaded (*N*) within the same temperature region *t*. We divided up the region of analysis into temperature minima regions of 1°C each (e.g., analysis areas with January minima ≥ 7°C and < 8°C fell into one temperature region).

21. Since each data point represents a region with a 1°C range of January minimum temperatures, T is actually the lower bound of each 1°C range.

22. Mean number of consecutive frost-free days between spring thaw and fall freeze.

23. The slightly better fit of the January mean minimum temperature model does not, on its own, constitute a reason to suspect that January minima are more important. Only a substantial difference in model fit might argue for selection of the better fit.

24. For this model, degrees of warming refer to increases in mean minimum January temperature assumed to accompany general tempera-

ture increases predicted by climate models.
25. Leakage in vascular system, in this case associated with freezing.

Literature Cited

Adams, C.T. 1983. Destruction of eggplants in Marion County, Florida, by red imported fire ants (Hymenoptera: Formicidae). *Florida Entomologist* 66: 518–520.

———. 1986. Agricultural and medical impact of the imported fire ants. Pages 48–57 in C. S. Lofgren and R. K. Vander Meer, eds., *Fire Ants and Leaf-Cutting Ants: Biology and Management*. Boulder, Co.: Westview.

Adams, C. T., and C. S. Lofgren. 1981. Red imported fire ants (Hymenoptera: Formicidae): Frequency of sting attacks on residents of Sumter County, Georgia. *Journal of Medical Entomology* 18: 378–382.

———. 1982. Incidence of stings or bites of the red imported fire ants (Hymenoptera: Formicidae) and other arthropods among patients at Ft. Stewart, Georgia, USA. *Journal of Medical Entomology* 19: 366–370.

Adams, C. T., J. K. Plumley, C. S. Lofgren, and W. A. Banks. 1976. Economic importance of the red imported fire ant, *Solenopsis invicta* Buren. I. Preliminary investigations on impact on soybean harvest. *Journal of the Georgia Entomological Society* 11: 165–169.

Adams, C. T., J. K. Plumley, W. A. Banks, and C. S. Lofgren. 1977. Impact of the red imported fire ant, *Solenopsis invicta* Buren (Hymenoptera: Formicidae), on harvest of soybeans in North Carolina. *Journal of the Elisha Mitchell Scientific Society* 93: 150–152.

Adams, C. T., W. A. Banks, C. S. Lofgren, B. J. Smittle, and D. P. Harlan. 1983. Impact of the red imported fire ant, *Solenopsis invicta* (Hymenoptera: Formicidae), on the growth and yield of soybeans. *Journal of Economic Entomology* 76: 1129–1132.

Adams, C. T., W. A. Banks, and C. S. Lofgren. 1988. Red imported fire ant (Hymenoptera: Formicidae): Correlation of ant density with damage to two cultivars of potatoes (Solanum tuberosum L.). *Journal of Economic Entomology* 81: 905–909.

Allen, C. R., Phillips, S. A., Jr., and M. R. Trostle. 1993. Range expansion by the ecologically disruptive red imported fire ant into the Texas Rio Grande Valley. *Southwestern Entomologist* 18: 315–316.

Allen, C. R., S. Demarais, and R. S. Lutz. 1994. Red imported fire ant impact on wildlife: An overview. *Texas Journal of Science* 46: 51–59.

———. 1995. Red imported fire ant impacts on northern bobwhite populations. *Ecological Applications* 5: 632–638.

———. 1997. Effects of red imported fire ant on recruitment of white-tailed deer fawns. *Journal of Wildlife Management* 61: 911–916.

Anderson, B. W., and R. D. Ohmart. 1977. *Wildlife Use and Densities Report of Birds and Mammals in the Lower Colorado River Valley.* Tempe: Arizona State University and Bureau of Reclamation.

————. 1982. *Revegetation for Wildlife Enhancement along the Lower Colorado River.* Boulder City, Nev.: Bureau of Reclamation, Lower Colorado Region.

Anderson, B. W., R. D. Ohmart, and J. Rice. 1983. Avian and vegetation community structure and their seasonal relationships in the lower Colorado River Valley. *Condor* 85: 392–405.

Anon. 1958. Observations on the Biology of the Imported Fire Ant. USDA, ARS 33–49, 21 pp.

Anon. 1972. Ecological range for the imported fire ant—based on plant hardiness. *Coop. Econ. Insect Rep.* 22(7): centerfold map.

Apperson, C. S., E. E. Powell. 1983. Correlation of the red imported fire ant (Hymenoptera: Formicidae) with reduced soybean yields in North Carolina. *Journal of Economic Entomology* 76: 259–263.

Atkinson, C. T., K. L. Woods, R. J. Dusek, L. S. Sileo, and W. M. Iko. 1995. Wildlife disease and conservation in Hawaii: Pathogenicity of avian malaria (*Plasmodium relictum* in experimentally infected iiwi (*Vestiaria coccinea*). *Parasitology* 111: 559–569.

Baker, R. H. A., R. J. C. Cannon, and K. F. A. Walters. 1996. An assessment of the risks posed by selected nonindigenous pests to UK crops under climate change. *Aspects of Applied Biology* 45: 323–330.

Ball, D. E., H. J. Williams, S. B. Vinson. 1984. Chemical analysis of the male aedeagal bladder in the fire ant, *Solenopsis invicta* Buren. *Journal of the New York Entomological Society* 92: 365–370.

Banks, W. A., B. M. Glancey, C. E. Stringer, D. P. Jouvenaz, C. S. Lofgren, and D. E. Weidhaas. 1973. Imported fire ants: Eradication trials with mirex bait. *Journal of Economic Entomology* 66: 785–789.

Banks, W. A., L. R. Miles, and D. P. Harlan. 1983. The effects of insect growth regulators and their potential as control agents for imported fire ants (Hymenoptera: Formicidae). *Florida Entomologist* 66: 172–181.

Banks, W. A., D. P. Jouvenaz, D. P. Wojcik, and C. S. Lofgren. 1985. Observations on fire ants, *Solenopsis* spp., in Mato Grosso, Brazil. *Sociobiology* 11: 143–152.

Barry, J. P., C. H. Baxter, R. D. Sagarin, and S. E. Gilman. 1995. Climate-related, long-term faunal changes in a California Rocky Intertidal Community. *Science* 267, 672–675.

Baskauf, S. J. 1999. The edge of the range of the southwestern corn borer, *Diatraea grandiosella* Dyar (Lepidoptera: Pyralidae): Its characteristics and factors determine its location. In *Ecological Society of America Annual Meeting.* (Spokane, Washington, 4–10 August) Washington, D.C.: Ecological Society of America.

Bass, J. A., and S. B. Hays. 1976. Geographic location and identification of fire ant species in South Carolina. *Journal of the Georgia Entomological Society* 11: 34–36.

Baum, B. R. 1978. *The Genus* Tamarix. Jerusalem: Israel Academy of Sciences and Humanities.

Beerling, D. J. 1993. The impact of temperature on the northern distribution of the introduced species *Fallopia japonica* and *Impatiens glandulifera*. *Journal of Biogeography* 20: 45–53.

Bhatkar, A., W. H. Whitcomb, W. F. Buren, P. Callahan, and T. Carlysle. 1972. Confrontation behavior between *Lasius neoniger* (Hymenoptera: Formicidae) and the imported fire ant. *Environmental Entomology* 1: 274–279.

Blackburn, W. H., R. W. Knight, and J. L. Schuster. (1982). Saltcedar influence on sedimentation in the Brazos River. *Journal of Soil and Water Conservation* 37: 298–301.

Boag, B., J. W. Crawford, and R. Neilson. 1991. The effect of potential climatic changes on the geographical distribution of the plant-parasitic nematodes *Xiphinema* and *Longidorus* in Europe. *Nematologica* 37: 312–323.

Bowser, C. W. 1957. Introduction and spread of the undesirable tamarisks in the Pacific Southwest section of the United States and comments concerning the plants' influence upon indigenous vegetation. *American Geophysical Union Transaction* 38: 415–416.

Brock, J. H. 1994. *Tamarix* spp. (salt cedar), an invasive exotic woody plant in arid and semi-arid riparian habitats of western USA. Pages 27–43 in L. C. de Waal, L. E. Child, P. M. Wade, and J. H. Brock, eds., *Ecology and Management of Invasive Riverside Plants*. Chichester: John Wiley and Sons.

Brown, B. T., and M. W. Trosset. 1989. Nesting–habitat relationships of riparian birds along the Colorado River in Grand Canyon, Arizona. *Southwestern Naturalist* 34: 260–270.

Brown, D. W., and R. A. Goyer. 1982. Effects of a predator complex on lepidopterous defoliators of soybean. *Environmental Entomology* 11: 385–389.

Bruce, W. A., L. D. Cline, G. L. LeCato. 1978. Imported fire ant infestation of buildings. *Florida Entomologist* 61: 230.

Bryan, J. H., D. H. Foley, and R. W. Sutherst. 1996. Malaria transmission and climate change in Australia. *Medical Journal of Australia* 164: 345–347.

Buren, W. F. 1983. Artificial faunal replacement for imported fire ant control. *Florida Entomologist* 66: 93–100.

Buren, W. F., G. E. Allen, W. H. Whitcomb, F. E. Lennartz, and R. N. Williams. 1974. Zoogeography of the imported fire ants. *Journal of the New York Entomological Society* 82: 113–124.

Busch, D. E., and S. D. Smith. 1993. Effects of fire on water and salinity relations of riparian woody taxa. *Oecologia* 94: 186–194.

Busch, D. E., and S. D. Smith. 1995. Mechanisms associated with decline of woody species in riparian ecosystems of the southwestern U.S. *Ecological Monographs* 65: 347–370.

Busch, D. E., N. L. Ingraham, and S. D. Smith. 1992. Water uptake in woody riparian phreatophytes of the southwestern United States: A stable isotope study. *Ecological Applications* 2: 450–459.

Callcott, A. A., and H. L. Collins. 1996. Invasion and range expansion of imported fire ants (Hymenoptera: Formicidae) in North America from 1918–1995. *Florida Entomologist* 79: 240–251.

Calvert, W. H. 1996. Fire ant predation on monarch larvae (Nymphalidae: Danainae) in a central Texas prairie. *Journal of the Lepidopterists' Society* 50: 149–151.

Camilo, G. R., S. A. Phillips, Jr. 1990. Evolution of ant communities in response to invasion by the fire ant *Solenopsis invicta*. Pages 190–198 in R. K. Vander Meer, K. Jaffe, and A. Cedeno, eds., *Applied Myrmecology: A World Perspective*. Boulder, Co.: Westview Press.

Carter, T. R., M. L. Parry, and J. H. Porter. 1991. Climatic change and future agroclimatic potential in Europe. *International Journal of Climatology* 11: 251–269.

CAST. 1976. Fire Ant Control. Council for Agricultural Science and Technology. CAST Report No. 62, 2nd ed., Report. No. 65. Ames: Iowa State University. 24 pp.

Christensen, E. M. 1962. The rate of naturalization of *Tamarix* in Utah. *American Midland Naturalist* 68: 51–57.

Clark, J. S., C. Fastie, G. Hurtt, S. T. Jackson, C. Johnson, G. A. King, M. Lewis, J. Lynch, S. Pacala, C. Prentice, E. W. Schupp, T. Webb, and P. Wyckoff. 1998. Reid's paradox of rapid plant migration. *BioScience* 48: 13–24.

Cleverly, J. R., S. D. Smith, A. Sala, and D. Devitt. 1997. Invasive capacity of *Tamarix ramosissima* in a Mojave Desert floodplain: The role of drought. *Oecologia* 111: 12–18.

Cohen, A. N., and J. T. Carlton. 1995. *Nonindigenous Aquatic Species in a United States Estuary: A Case Study of the Biological Invasions of the San Francisco Bay and Delta*. Washington, D.C.: U.S. Fish and Wildlife Service.

Cokendolpher, J. C., and S. A. Phillips, Jr. 1989. Rate of spread of the red imported fire ant, *Solenopsis invicta* (Hymenoptera: Formicidae), in Texas. *Southwestern Naturalist* 34: 443–449.

Cokendolpher, J. C., and S. A. Phillips, Jr. 1990. Critical thermal limits and locomotor activity of the red imported fire ant (Hymenoptera: Formicidae). *Environmental Entomology* 19: 878–881.

Collins H. L., T. C. Lockley, and D. J. Adams. 1993. Red imported fire ant

(Hymenoptera: Formicidae) infestation of motorized vehicles. *Florida Entomologist* 76: 515–516.

Creighton, W. S. 1930. The New World species of the genus *Solenopsis* (Hymenopter. Formicidae). *Proceedings of the American Academy of Arts and Sciences* 66: 39–151, plates 1–8.

Daly, C., R. P. Neilson, and D. L. Phillips. 1994. A statistical-topographic model for mapping climatological precipitation over mountainous terrain. *Journal of Applied Meteorology* 33: 140–158.

Daly, C., G. H. Taylor, and W. P. Gibson. 1997. The PRISM approach to mapping precipitation and temperature. Pages 10–12 in reprints: Tenth Conference on Applied Climatology, (Reno, Nevada, 20–24 October) Boston: American Meteorological Society.

Davis, A. J., L. S. Jenkinson, J. H. Lawton, B. Shorrocks, and S. Wood. 1998. Making mistakes when predicting shifts in species range in response to global warming. *Nature* 391: 783–786.

DeLoach, J. 1997. Saltcedar: Ecological interactions and potential effects of biological control. Woody Plant Wetland Workshop. (Grand Junction, Colorado, 3–4 September).

Dewberry, O. 1962. Fire Ant Control Investigations. In Fish and Wildlife Restoration Project W-37-R-1-2. Social Circle, Ga. 44pp.

Dickinson, V. M. 1995. Red imported fire ant predation on Crested Caracara nestlings in south Texas. *Wilson Bulletin* 107: 761–762.

Drees, B. M. 1994. Red imported fire ant predation on nestlings of colonial waterbirds. *Southwest. Entomologist* 19: 355–359.

Eaton, J. G., and R. M. Scheller. 1996. Effects of climate warming on fish thermal habitat in streams of the United States. *Limnology and Oceanography* 41: 1109–1115.

Ellis, L. M. 1995. Bird use of saltcedar and cottonwood vegetation in the Middle Rio Grande Valley of New Mexico, U.S.A. *Journal of Arid Environments* 30: 339–349.

Emanuel, W. R., H. H. Shugart, and M. Stevenson. 1985. Climatic change and broad-scale distribution of terrestrial ecosystem complexes. *Climatic Change* 7: 29–43.

Engel-Wilson, R. W., and R. D. Ohmart. 1978. *Floral and Attendant Faunal Changes on the Lower Rio Grande between Fort Quitman and Presidio, Texas.* Washington, D.C.: USFS.

Evans, K. A. and B. Boag. 1996. The New Zealand flatworm: Will climate change make it a potentially European problem? *Aspects of Applied Biology* 45: 335–338.

Everitt, B. L. 1980. Ecology of saltcedar: A plea for research. *Environmental Geology* 3: 77–84.

Ewel, J. J. 1986. Invasibility: Lessons from South Florida. Pages 214–230 in H. A. Mooney and J. A. Drake, eds., *Ecology of Biological Invasions of North America and Hawaii.* New York: Springer-Verlag.

Fleetwood, S. C., P. D. Teel, and G. Thompson. 1984. Impact of imported fire ant on lone star tick mortality in open and canopied pasture habitats of east central Texas. *Southwestern Entomologist* 9: 158–163.

Francy, D. B., C. G. Moore, and D. A. Eliason. 1990. Past, present, and future of *Aedes albopictus* in the United States. *Journal of the American Mosquito Control Association* 6: 127–132.

Francke, O. F., and J. C. Cokendolpher. 1986. Temperature tolerances of the red imported fire ant. Pages 104–113 in C. S. Lofgren and R. K. Vander Meer, eds., *Fire Ants and Leafcutting Ants: Biology and Management.* Boulder: Westview.

Francke, O. F., J. C. Cokendolpher, and L. R. Potts. 1986. Supercooling studies on North American fire ants (Hymenoptera: Formicidae). *Southwestern Naturalist* 31: 87–94.

Frank, W. A. 1988. Report of limited establishment of red imported fire ant, *Solenopsis invicta* Buren, in Arizona. *Southwestern Entomologist* 13: 307–308.

Giorgi, F., L. O. Mearns, C. Shields, and L. McDaniel. 1998. Regional nested model simulations of present day and $2 \times CO_2$ climate over the Central Plains of the U.S. *Climatic Change* 40: 457–493.

Giuliano W. M., C. R. Allen, R. S. Lutz, and S. Demarais. 1996. Effects of red imported fire ants on northern bobwhite chicks. *Journal of Wildlife Management* 60: 309–313.

Glancey, B. M. 1981. Two additional dragonfly predators of queens of the red imported fire ant, *Solenopsis invicta* Buren. *Florida Entomologist* 64: 194–195.

Glancey, B. M., C. H. Craig, C. E. Stringer, and P. M. Bishop. 1973. Multiple fertile queens in colonies of the imported fire ant, *Solenopsis invicta*. *Journal of the Georgia Entomological Society* 8: 237–238.

Glancey, B. M., D. P. Wojcik, C. H. Craig, and J. A. Mitchell. 1976. Ants of Mobile County, AL, as monitored by bait transects. *Journal of the Georgia Entomological Society* 11: 191–197.

Glancey, B. M., J. D. Coley, and F. Killibrew. 1979. Damage to corn by the red imported fire ants. *Journal of the Georgia Entomological Society* 14: 198–201.

Glancey, B. M., J. C. E. Nickerson, D. Wojcik, J. Trager, W. A. Banks, and C. T. Adams. 1987. The increasing incidence of the polygynous form of the red imported fire ant, *Solenopsis invicta* (Hymenoptera: Formicidae), in Florida. *Florida Entomologist* 70: 400–402.

Gleason, H. A. 1926. The individualistic concept of the plant association. *Bulletin of the Torrey Botanical Club* 53: 7–26.

Gleick, P. H. 1990. Vulnerability of Water Systems. Pages 223–242 in P. E. Waggoner, ed., *Climate Change and U.S. Water Resources.* New York: John Wiley and Sons.

Graf, W. L. 1978. Fluvial adjustments to the spread of tamarisk in the Colorado Plateau region. *Geological Society of America Bulletin* 89: 1491–1501.

Graf, W. L. 1980. Riparian management: A flood control perspective. *Journal of Soil and Water Conservation* 35: 158–161.

Great Western Research (GWR). 1989. *Economic Analysis of Harmful and Beneficial Aspects of Saltcedar.* Mesa, Ariz.: Bureau of Reclamation.

Green, H. B. 1952. Biology and control of the imported fire ant in Mississippi. *Journal of Economic Entomology* 45: 593–597.

Green, H. B. 1962. On the biology of the imported fire ant. *Journal of Economic Entomology* 55: 1003–1004.

Gross, H. R., Jr., and W. T. Spink. 1969. Responses of striped earwigs following applications of heptachlor and mirex, and predator–prey relationships between imported fire ants and striped earwigs. *Journal of Economic Entomology* 62: 686–689.

Hadley, R. F. 1961. Influence of riparian vegetation on channel shape, northeastern Arizona. US Geological Survey Professional Paper 424-C.

Harris, W. G., E. C. Burns. 1972. Predation on the lone star tick by the imported fire ant. *Environmental Entomology* 1: 362–365.

Harte, J., M. Torn, and D. Jensen. 1992. The nature and consequences of indirect linkages between climate change and biological diversity. Pages 325–343 in R. L. Peters and T. E. Lovejoy, eds., *Global Warming and Biological Diversity.* New Haven: Yale University Press.

Hays, S. B., and K. L. Hays. 1959. Food habits of *Solenopsis saevissima richteri* Forel. *Journal of Economic Entomology* 52: 455–457.

Hook, A. W., and S. D. Porter. 1990. Destruction of harvester ant colonies by invading fire ants in South-central Texas (Hymenoptera: Formicidae). *Southwestern Naturalist* 35: 477–478.

Howard, F. W., and A. D. Oliver. 1978. Arthropod populations in permanent pastures treated and untreated with mirex for red imported fire ant control. *Environmental Entomology* 7: 901–903.

Hu, G. Y., and J. H. Frank. 1996. Effect of the red imported fire ant (Hymenoptera: Formicidae) on dung-inhabiting arthropods in Florida. *Environmental Entomology* 25: 1290–1296.

Hung, A. C. F. 1974. Ants recovered from refuse pile of the pyramid ant *Conomyrma insana* (Buckley) (Hymenoptera: Formicidae). *Annals of the Entomological Society of America* 67: 522–523.

Hung, A. C. F., and S. B. Vinson. 1978. Factors affecting the distribution of fire ants in Texas (Myrmicinae, Formicidae). *Southwestern Naturalist* 23: 205–213.

Hunter, W. C., R. D. Ohmart, and B. W. Anderson. 1988. Use of exotic saltcedar (*Tamarix chinensis*) by birds in arid riparian systems. *Condor* 90: 113–123.

Jackson, J., J. T. Ball, and M. R. Rose. 1990. Assessment of the salinity tolerance of eight Sonoran desert riparian trees and shrubs. Yuma, Ariz.: Bureau of Reclamation.

James, F. K., Jr., H. L. Pence, D. P. Driggers, R. L Jacobs, and D. E. Horton. 1976. Imported fire ant hypersensitivity: Studies of human reactions to fire ant venom. *Journal of Allergy and Clinical Immunology* 58: 110–120.

Johns, E. L. 1990. Vegetation Management Study, Lower Colorado River: Appendix 1, *Water Use of Naturally Occurring Vegetation*. Denver, Co.: Bureau of Reclamation.

Johnson, A. S. 1961. Antagonistic relationships between ants and wildlife with special reference to imported fire ants and bobwhite quail in the southeast. *Proceedings of the Annual Conference of the Southeast Association of Game Fish Comm.* 15: 88–107.

Johnson, E. L. 1976. Administrator's decision to accept plan of Mississippi Authority and order suspending hearing for the pesticide chemical mirex. *Federal Register* 41: 56,694–56,703.

Jones, D., and W. L. Sterling. 1979. Manipulation of red imported fire ants in a trap crop for boll weevil suppression. *Environmental Entomology* 8: 1073–1077.

Jouvenaz, D. P., G. E. Allen, W. A. Banks, and D. P. Wojcik. 1977. A survey for pathogens of fire ants, *Solenopsis* spp., in the Southeastern United States. *Florida Entomologist* 60: 275–279.

Killion, M. J., and W. E. Grant. 1995. A colony-growth model for the imported fire ant: Potential geographic range of an invading species. *Ecological Modelling* 77: 73–84.

Landers, J. L., J. A. Garner, and W. A. McRae. 1980. Reproduction of gopher tortoises (Gopherus polyphemus) in southwestern Georgia. *Herpetologica* 36: 353–361.

Landry, C.E., and S.A. Phillips, Jr. 1996. Potential of Ice-Nucleating Active Bacteria for Management of the Red Imported Fire Ant (Hymenoptera: Formicidae). *Environmental Entomology* 25: 859–866.

Lines, J. 1995. The effects of climatic and land-use change on the insect vectors of human disease. Pages 158–177 in R. Harrington and N. E. Stork, eds., *Insects in a Changing Environment*. London: Academic Press.

Lockey, R. F. 1974. Systemic reactions to stinging ants. *Journal of Allergy and Clinical Immunology*. 54: 132–146.

Lockey, R. F., and R. W. Fox. 1979. Allergic emergencies. *Hospital Medicine* 15(6): 64, 66–67, 71–74, 78.

Lockley, T. C. 1995a. Effect of imported fire ant predation on a population of the least tern—an endangered species. *Southwestern Entomologist* 20: 517–519.

———. 1995b. Observations of predation on alate queens of the red imported fire ant (Hymenoptera: Formicidae) by the black and yellow garden spider (Araneae: Araneidae). *Florida Entomologist* 78: 609–610.

Lofgren, C. S. 1986a. The economic importance and control of imported fire ants in the United States. Pages 227–256 in S. B. Vinson, ed., *Economic Impact and Control of Social Insects*. New York: Praeger
———. 1986b. History of imported fire ants in the United States. Pages 36–47 in C. S. Lofgren. and R. K. Vandemeer, eds., *Fire Ants and Leaf-Cutting Ants: Biology and Management*. Boulder: Westview Press.
Lofgren, C. S., and C. T. Adams. 1981. Reduced yield of soybeans in fields infested with the red imported fire ant, *Solenopsis invicta* Buren. *Florida Entomologist* 64: 199–202.
Lofgren, C. S., W. A. Banks, and B. M. Glancey. 1975. Biology and control of imported fire ants. *Annual Review of Entomology* 20: 1–30.
Lonsdale, W. M. 1999. Global patterns of plant invasions and the concept of invasibility. *Ecology* 80: 1522–1536.
Loope, L. L., P. G. Sanchez, P. W. Tarr, W. L. Loope, and R. L. Anderson. 1988. Biological invasions of arid land reserves. *Biological Conservation* 44: 95–118.
Lovich, J. E. 1996. A brief overview of the impact of tamarisk infestation on native plants and animals. Page 4 in J. DiTomaso and C. E. Bell, eds., *Saltcedar Management Workshop*. Rancho Mirage, Calif.: UC Cooperative Extension and CalEPPC.
Lyle, C., and I. Fortune. 1948. Notes on an imported fire ant. *Journal of Economic Entomology* 41: 833–834.
MacDonald, I. A. W., L. L. Loope, M. B. Usher, and O. Hamann. 1989. Wildlife conservation and the invasion of nature reserves by introduced species: A global perspective. Pages 215–255 in H. A. Mooney, J. A. Drake, F. DiCastri, R. H. Groves, F. J. Krueger, M. Rejmanek, and M. Williamson, eds., *Biological Invasions: A Global Perspective*. New York: Wiley and Sons.
Macom, T. E., and S. D. Porter. 1996. Comparison of polygyne and monogyne red imported fire ant (Hymenoptera: Formicidae) population densities. *Annals of the Entomological Society of America* 89: 535–543.
Mandrak, N. E. 1989. Potential invasion of the Great Lakes by fish species associated with climatic warming. *Journal of Great Lakes Research* 15: 306–316.
Markin, G. P., and J. H. Dillier. 1971. The seasonal life cycle of the imported fire ant, *Solenopsis saevissima richteri*, on the Gulf coast of Mississippi. *Annals of the Entomological Society of America* 64: 562–565.
Markin, G. P., J. H. Dillier, S. O. Hill, M. S. Blum, and H. R. Hermann. 1971. Nuptial flight and flight ranges of the imported fire ant, *Solenopsis saevissima richteri* (Hymenoptera: Formicidae). *Journal of the Georgia Entomological Society* 6: 145–156.
Markin, G. P., H. L. Collins, and J. Davis. 1974. Residues of the insecticide mirex in terrestrial and aquatic invertebrates following a single aerial application of mirex bait, Louisiana—1971–72. *Pesticide Monitor Journal* 8: 131–134.

Masser, M. P., and W. E. Grant. 1986. Fire ant–induced trap mortality of small mammals in East-central Texas. *Southwestern Naturalist* 31: 540–542.

Molles, M. C., C. S. Crawford, L. E. Ellis, H. M. Valett, and C. N. Dahm. 1998. Managed flooding for riparian ecosystem restoration. *Bioscience* 48: 749–756.

Mooney, H. A., and R. H. Hobbs. 2000. *Invasive Species in a Changing World.* Washington, D.C.: Island Press.

Morrill, W. L. 1974. Dispersal of red imported fire ants by water. *Florida Entomologist* 57: 39–42.

Morrill, W. L., P. B. Martin, and D. C. Sheppard. 1978. Overwinter survival of the red imported fire ant: Effects of various habitats and food supply. *Environmental Entomology* 7: 262–264.

Morris, J. R., and K. L. Steigman. 1993. Effects of polygyne fire ant invasion on native ants of a blackland prairie in Texas. *Southwestern Naturalist* 38: 136–140.

Morrison, J. I. 1996. The sustainable use of water in the Lower Colorado River Basin. Oakland, Calif.: The Pacific Institute and the Global Water Policy Project.

Mount, R. H. 1981. The red imported fire ant, *Solenopsis invicta* (Hymenoptera: Formicidae), as a possible serious predator of some native southeastern vertebrates: Direct observation. *Journal of the Alabama Academy of Science.* 52: 71–78.

Mount, R. H., S. E. Trauth, and W. H. Mason. 1981. Predation by the red imported fire ant, *Solenopsis invicta* (Hymenoptera: Formicidae), on eggs of the lizard *Cnemidophorus sexlineatus* (Squamata: Teiidae). *Journal of the Alabama Academy of Science.* 52: 66–70.

Myers, R. L., and J. J. Ewel 1990. Problems, prospects, and strategies for conservation. Pages 619–632 in R. L. Myers and J. J. Ewel, eds., *Ecosystems of Florida.* Orlando: University of Central Florida Press.

National Park Service. 1981. Proposed Natural and Cultural Resources Management Plan and Draft EIS, Death Valley NM. Washington, D.C.: U.S. Dept. of the Interior.

Negm, A. A., and S. D. Hensley. 1967. The relationship of arthropod predators to crop damage inflicted by the sugarcane borer. *Journal of Economic Entomology.* 60: 1503–1506.

Negm, A. A., and S. D. Hensley. 1969. Evaluation of certain biological control agents of the sugarcane borer in Louisiana. *Journal of Economic Entomology.* 62: 1008–1013.

Neill, W. M. 1983. *The Tamarisk Invasion of Desert Riparian Areas.* Spring Valley, Calif.: Desert Protective Council, Inc.

Nichols, B. J., and R. W. Sites. 1989. A comparison of arthropod species within and outside the range of *Solenopsis invicta* Buren in central Texas. *Southwestern Entomologist* 14: 345–350.

Nickerson, J. C., W. H. Whitcomb, A. P. Bhatkar, and M. A. Naves. 1975. Predation on founding queens of *Solenopsis invicta* by workers of *Conomyrma insana*. *Florida Entomologist* 58: 75–82.

Niemela, P., and W. J. Mattson. 1996. Invasion of North American forests by European phytophagous insects. *BioScience* 46: 741–753.

Ohmann, J. L., and T. A. Spies. 1998. Regional gradient analysis and spatial patterns of woody plant communities of Oregon forests. *Ecological Monographs* 68: 151–182.

Parker, I. M., D. Simberloff, W. M. Lonsdale, K. Goodell, M. Wonham, P. M. Kareiva, M. H. Williamson, B. Von Holle, P. B. Moyle, J. E. Byers, and L. Goldwasser. 1999. Impact: Toward a framework for understanding the ecological effects of invaders. *Biological Invasions* 1: 3–19.

Phillips, S. A., Jr., R. Jusino-Atresino, and H. G. Thorvilson. 1996. Desiccation resistance in populations of the red imported fire ant (Hymenoptera: Formicidae). *Environmental Entomology* 25: 460–464.

Pimm, S. L., and D. P. Bartell. 1980. Statistical model for predicting range expansion of the red imported fire ant, *Solenopsis invicta*, in Texas. *Environmental Entomology* 9: 653–658.

Porter, J. H., M. L. Parry, and T. R. Carter. 1991. The potential effects of climatic change on agricultural insect pests. *Agricultural and Forest Meteorology* 57: 221–240.

Porter, S. D. 1988. Impact of temperature on colony growth and developmental rates of the ant, *Solenopsis invicta*. *Journal of Insect Physiology* 34: 1127–1133.

Porter, S. D. 1992. Frequency and distribution of polygene fire ants (Hymenoptera: Formicidae) in Florida. *Florida Entomologist.* 75: 248–257.

Porter, S. D. 1993. Stability of polygyne and monogyne fire ant populations (Hymenoptera: Formicidae: *Solenopsis invicta*) in the United States. *Journal of Economic Entomology* 86: 1344–1347.

Porter, S. D., and D. A. Savignano. 1990. Invasion of polygyne fire ants decimates native ants and disrupts arthropod community. *Ecology* 71: 2095–2106.

Porter, S. D., and W. R. Tschinkel. 1987. Foraging in *Solenopsis invicta* (Hymenoptera: Formicidae): Effects of weather and season. *Environmental Entomology* 16: 802–808.

Porter, S. D., B. Van Eimeren, and L. E. Gilbert. 1988. Invasion of red imported fire ants (Hymenoptera: Formicidae): Microgeography of competitive replacement. *Annals of the Entomological Society of America* 81: 913–918.

Porter, S. D., A. P. Bhatkar, R. Mulder, S. B. Vinson, and D. J. Clair. 1991. Distribution and density of polygyne fire ants (Hymenoptera: Formicidae) in Texas. *Journal of Economic Entomology* 84(3): 866–874.

Porter, S. D., D. F. Williams, R. S. Patterson, and H. G. Fowler. 1997. Intercontinental differences in the abundance of *Solenopsis* fire ants (Hymenoptera: Formicidae): Escape from natural enemies? *Environmental Entomology* 26: 373–384.

Prentice, I. C., P. J. Bartlein, and T. Webb. 1991. Vegetation and climate change in eastern North America since the last glacial maximum. *Ecology* 72: 2038–2056.

Rejmanek, M., and D. M. Richardson. 1996. What attributes makes some plant species more invasive? *Ecology* 77: 1655–1661.

Rhoades, W. C. 1962. A synecological study of the effects of the imported fire ant eradication program. I. Alcohol pitfall method of collecting. *Florida Entomologist* 45: 161–173.

Rhoades, W. C. 1963. A synecological study of the effects of the imported fire ant eradication program. II. Light trap, soil sample, litter sample and sweep net methods of collection. *Florida Entomologist* 46: 301–310.

Rhoades, R. B., W. L. Schafer, M. Newman, R. Lockey, R. M. Dozier, P. F. Wubbena, A. W. Townes, W. H. Schmid, G. Neder, T. Brill, and H. J. Wittig. 1977. Hypersensitivity to the imported fire ant in Florida: Report of 104 cases. *Journal of the Florida Medical Association* 64: 247–254.

Richardson, D. M., and W. J. Bond. 1991. Determinants of plant distribution: Evidence from pine distributions. *American Naturalist* 37: 639–668.

Ridlehuber, K. T. 1982. Fire ant predation on wood duck ducklings and piped eggs. *Southwestern Naturalist* 27: 222.

Robinson, T. W. 1965. *Introduction, Spread, and Areal Extent of Saltcedar (Tamarix) in the Western States.* Washington, D.C.: U.S. Geological Survey.

Root, T. 1988a. Energy constraints on avian distributions and abundances. *Ecology* 69: 330.

———. 1988b. Environmental factors associated with avian distributional boundaries. *Journal of Biogeography* 14: 489–505.

Saetersdal, M., and H. J. B. Birks. 1997. A comparative ecological study of Norwegian mountain plants in relation to possible future climatic change. *Journal of Biogeography* 24: 127–152.

Salas, D. E., J. R. Carlson, B. E. Ralston, D. A. Martin, and K. R. Blaney. 1996. *Riparian Vegetation Mapping of the Lower Colorado River from the Davis Dam to the International Border.* Denver, Colo.: Bureau of Reclamation.

Salt, R. W. 1961. Principles of insect cold-hardiness. *Annual Review of Entomology* 6: 55–74.

Sasek, T. W., and B. R. Strain. 1990. Implications of atmospheric CO_2 enrichment and climatic change for the geographical distribution of two introduced vines in the USA. *Climatic Change* 16: 31–51.

Schwartzman, P. N. 1990. A hydrogeologic resource assessment of the

Lower Babocomari watershed, Arizona. M.S. Thesis, University of Arizona, Tucson.

Sikes, P. J. and K. A. Arnold. 1986. Red imported fire ant (*Solenopsis invicta*) predation on cliff swallow (*Hirundo pyrrhonota*) nestings in east-central Texas. *Southwestern Naturalist* 31: 105–106.

Simberloff, D., D. C. Schmitz, and T. C. Brown. 1997. *Strangers in Paradise: Impact and Management of Nonindigenous Species in Florida.* Washington, D.C.: Island Press.

Sogge, M. K., R. M. Marshall, S. J. Sferra, and T. J. Tibbitts. 1997a. *A Southwestern Willow Flycatcher Natural History Summary and Survey Protocol.* Flagstaff, Ariz.: National Park Service.

Sogge, M. K., T. J. Tibbitts, and J. R. Petterson. 1997b. Status and breeding ecology of the southwestern willow flycatcher in the Grand Canyon. *Western Birds* 28: 142–157.

Spence, J. H., and G. P. Markin. 1974. Mirex residue in the physical environment following a single bait application—1971–72. *Pesticide Monitor Journal* 8: 135–139.

Sterling, W. L. 1978. Fortuitous biological suppression of the boll weevil by the red imported fire ant. *Environmental Entomology* 7: 564–568.

Sterling, W. L., D. Jones, and D. A. Dean. 1979. Failure of the red imported fire ant to reduce entomophagous insect and spider abundance in a cotton agroecosystem. *Environmental Entomology* 8: 976–981.

Stevens, L. E. 1995. Invertebrate herbivore community dynamics on *Tamarix chinensis* Loueiro and *Salix exigua* Nuttall in Grand Canyon, Arizona. M.S. Thesis, Northern Arizona University, Flagstaff, Arizona.

Stoker, R. L., D. K. Ferris, W. E. Grant, and L. J. Folse. 1994. Simulating colonization by exotic species: A model of the red imported fire ant (*Solenopsis invicta*) in North America. *Ecological Modelling* 73: 281–292.

Stoker, R. L., W. E. Grant, and S. B. Vinson. 1995. *Solenopsis invicta* (Hymenoptera: Formicidae) effect on invertebrate decomposers of carrion in central Texas. *Environmental Entomology* 24: 817–822.

Stromberg, J. 1998. Dynamics of Fremont Cottonwood (*Populus fremontii*) and Saltcedar (*Tamarix chinensis*) populations along the San Pedro River, Arizona. Draft manuscript.

Summerlin, J. W., and L. R. Green. 1977. Red imported fire ant: A review on invasion, distribution, and control in Texas. *Southwestern Entomologist* 2: 94–101.

Summerlin, J. W., J. K. Olson, R. R. Blume, A. Aga, and D. E. Bay. 1977. Red imported fire ant: Effects on Onthophagus gazella and the horn fly. *Environmental Entomologist* 6: 440–442.

Summerlin, J. W., R. L. Harris, R.L., H. D. Petersen. 1984. Red imported fire ant (Hymenoptera: Formicidae): Frequency and intensity of invasion of fresh cattle droppings. *Environmental Entomologist* 13: 1161–1163.

Sutherst, R. W. 1990. Impact of climate change on pests and disease in Australasia. *Search* 21: 230–232.

———. 1995. The potential advance of pests in natural ecosystems under climate change: Implications for planning and management. Pages 88–98 in J. Pernetta, R. Leemans, D. Elder and S. Humphrey, eds., *The Impact of Climate Change on Ecosystems and Species: Terrestrial Ecosystems.* Gland, Switzerland: IUCN.

Sutherst, R. W., G. F. Maywald, and D. B. Skarrat. 1995a. Predicting Insect Distributions in a Changing Climate. Pages 60–93 in R. Harrington and N. E. Stork, eds., *Insects in a Changing Environment.* London: Academic Press.

Sutherst, R. W., R. B. Floyd, and G. F. Maywald. 1995b. The potential geographical distribution of the Cane Toad, *Bufo marinus* L. in Australia. *Conservation Biology* 9: 294–299.

Taylor, J. P., and K. C. McDaniel. 2000. Restoration of saltcedar infested flood plains on the Bosque del Apache National Wildlife Refuge. *Weed Technology* 12: 345–352.

Taylor, J. P., D. B. Wester, D. B., and L. M. Smith. 1999. Disturbance and riparian woody plant establishment in the Rio Grande floodplain. *Wetlands* 19: 372–382.

Tschinkel, W. R. 1986. The ecological nature of the fire ant: Some aspects of colony function and some unanswered questions. Pages 72–87 in C. S. Lofgren and R. K. Vander Meer, eds., *Fire Ants and Leaf Cutting Ants: Biology and Management.* Boulder: Westview.

———. 1988. Distribution of the fire ants *Solenopsis invicta* and *S. geminata* (Hymenoptera: Formicidae) in northern Florida in relation to habitat and disturbance. *Annals of the Entomological Society of America* 81: 76–81.

Tschinkel, W. R., and D. F. Howard. 1983. Colony founding by pleometrosis in the fire ant, *Solenopsis invicta. Behavioral Ecology and Sociobiology* 12: 103–113.

U.S. Congress, Office of Technology Assessment (USOTA). 1993. Harmful Nonindigenous Species in the United States, OTA-F-565. Washington, D.C.: U.S. Government Printing Office.

U.S. Fish and Wildlife Service (USFWS). 1997. Endangered and threatened wildlife and plants, final determination of critical habitat for the southwestern willow flycatcher. *Federal Register* 62: 39129–39147.

Vinson, S. B. 1991a. Population growth of the imported fire ant. Pages 70–73 in *Texas Environmental Guide.* Austin: Holt, Rinehart and Winston and Harcourt, Brace, Jovanovich.

———. 1991b. Effect of the red imported fire ant (Hymenoptera: Formicidae) on a small plant-decomposing arthropod community. *Environmental Entomology* 20: 98–103.

———. 1997. Invasion of the red imported fire ant (Hymenoptera: Formicidae): Spread, biology, and impact. *American Entomologist* 43: 23–39.

Vinson, S. B., and A. A. Sorensen. 1986. Imported Fire Ants: Life History and Impact. Austin: Texas Dept. Agric. 28 pp.

Vitousek, P. 1986. Biological invasions and ecosystem properties: Can species make a difference? Pages 163–176 in H. A. Mooney and J. A. Drake, eds., *Ecology of Biological Invasions of North America and Hawaii.* New York: Springer-Verlag.

Vitousek, P. M., C. M. D'Antonio, L. L. Loope, M. Rejmanek, and R. Westbrooks. 1997. Introduced species: A significant component of human-caused global change. *New Zealand Journal of Ecology* 21: 1–16.

Waggoner, P. E., and J. Schefter. 1990. Future water use in the present climate. Pages 19–40 in P. E. Waggoner, ed., *Climate Change and U.S. Water Resources.* New York: Wiley and Sons.

Walker, L. R., and S. D. Smith. 1997. Impacts of invasive plants on community and ecosystem properties. Pages 69–86 in J. O. Luken and J. W. Thieret, eds., *Assessment and Management of Plant Invasions.* New York: Springer-Verlag.

Warren, D. K., and R. M. Turner. 1975. Saltcedar seed production, seedling establishment, and response to inundation. *Arizona Academy of Science* 10: 131–144.

Whitcomb, W. H., H. A. Denmark, A. P. Bhatkar, and G. L. Greene. 1972. Preliminary studies on the ants of Florida soybean fields. *Florida Entomologist* 55: 129–142.

Whitcomb, W. H., A. Bhatkar, and J. C. Nickerson. 1973. Predators of *Solenopsis invicta* queens prior to successful colony establishment. *Environmental Entomology* 2: 1101–1103.

Wiesenborn, W. D. 1996. Saltcedar impacts on salinity, water, fire frequency, and flooding. Pages 9–12 in J. DiTomaso and C. E. Bell, eds., *Saltcedar Management Workshop.* Rancho Mirage, Calif.: UC Cooperative Extension and CalEPPC.

Williams, D. F., W. A. Banks, and C. S. Lofgren. 1997. Control of *Solenopsis invicta* (Hymenoptera: Formicidae) with Teflubenzuron. *Florida Entomologist* 80: 84–91.

Wilson, E. O. 1951. Variation and adaptation in the imported fire ant. *Evolution* 5: 68–79.

Wilson, E. O., and W. L. Brown, Jr. 1958. Recent changes in the introduced population of the fire ant *Solenopsis saevissima* (Fr. Smith). *Evolution* 12: 211–218.

Wilson, N. L., and A. D. Oliver. 1969. Food habits of the imported fire ant in pasture and pine forest areas in southeastern Louisiana. *Journal of Economic Entomology* 62: 1268–1271.

Wojcik, D. P. 1983. Comparison of the ecology of red imported fire ants in North and South America. *Florida Entomologist* 66: 101–111.

Wojcik, D. P. 1994. Impact of the red imported fire ant on native ant species in Florida. Pages 269–281 in D. F. Williams, ed., *Exotic Ants: Biology, Impact, and Control of Introduced Species.* Boulder: Westview Press.

Woodward, F. I. 1987. *Climate and Plant Distribution.* Cambridge: Cambridge University Press.

Yong, W., and D. M. Finch. 1997. Migration of the willow flycatcher along the Middle Rio Grande. *Wilson Bulletin* 109: 253–268.

Zavaleta, E. S. 2000a. Valuing ecosystem services lost to *Tamarix* invasion in the United States. Pages 261–300 in H. A. Mooney and R. Hobbs, eds., *Invasive Species in a Changing World.* Washington, D.C.: Island Press.

———. 2000b. The economic value of controlling an invasive shrub. *Ambio* 29: 462–467.

Zhakharov, A. A., L. C. Thompson. 1998. Effects of repeated use of fenoxycarb and hydramethylnon baits on nontarget ants. *Journal of Entomological Science* 33: 212–220.

CHAPTER 8

Climate Change, Whitebark Pine, and Grizzly Bears in the Greater Yellowstone Ecosystem

LAURA KOTEEN

The subalpine tree species whitebark pine (*Pinus albicaulis*) faces multiple threats to its existence. Predominant among them is the risk posed by the exotic fungus white pine blister rust, *Cronartium ribicola* . Throughout much of its range (Fig. 8.1), whitebark pine has suffered considerable losses; approaching 90% in some locations. Blister rust, mountain pine beetle (*Dendroctonus ponderosae*) and successional replacement by subalpine cohorts due to fire exclusion account for the bulk of its decline. The seeds of the white-bark pine tree are of primary importance as a food source for wildlife, most notably the threatened grizzly bear (*Ursus arctos horribilis*). In the Greater Yellowstone Ecosystem (GYE), where the grizzly's range and that of whitebark pine coincide, many biologists have concluded that the fate of the bear is linked to the survival of the tree.

According to the most recent survey conducted in 1995, blister rust infection levels in the GYE are still relatively low. Estimates suggest that 5% or less of the trees are infected, with another 7% standing dead trees due to all causes (Kendall 1996b). Biologists previously assumed that the northwestern Wyoming climate was

343

generally unfavorable for the spread of the fungus and would thus prevent whitebark population collapse (Carlson 1978, Arno 1986). The first aim of this study was to investigate this assumption and to describe the current profile of conditions for blister rust spread that exist in the GYE, with specific reference to the region's climate. Toward this end, I analyzed available weather station data, distributed across the landscape and spanning the last 50 years, to assess the frequency of occurrence of the climatic conditions likely to produce pine infection. Secondly, this chapter discusses the prospects for future blister rust spread through a sensitivity analysis of altered climatic conditions representing predicted consequences of global climate change. Also considered here are alternative means by which whitebark pine population contraction could result from climate change, as well as the impacts of the tree's loss on grizzly bears and other wildlife.

Climate change is generally expected to hasten the recession of whitebark pine communities in the region with three separate mechanisms contributing to its decline. The first mechanism concerns the shifting of pathogen ranges into new regions or into locations that are now only marginally suitable climatically for infestation. This topic will be further elaborated and explored as an important mechanism of species loss associated with climate change. In the case of whitebark pine, climate changes that tend to increase summer moisture availability would serve to accelerate the rate of blister rust spread. Secondly, warming, associated with global climate change, may also limit the range of whitebark pine. Because whitebark pine is found among high-elevation subalpine communities, warming may lead to its competitive replacement by lodgepole pine at lower, warmer elevations (Mattson and Reinhart 1989). Through initiating upward and latitudinal migration of subalpine communities, and replacement by species better adapted to the new set of climatic conditions, the tree's range may, over time, shift up and off the region's mountain peaks (Romme and Turner 1991). Thirdly, drier conditions are predicted to accompany a warmer climate and to increase the frequency of stand-replacing wild fires. Although *P. albicaulis* is adapted to the effects of small fires, vast conflagrations, such as those that burned through Yellowstone National Park in the summer of 1988, reducing whitebark pine populations by 26% (Tomback 1995, P. Farnes, pers. comm. 1999) increase the tree's overall vulnerability.

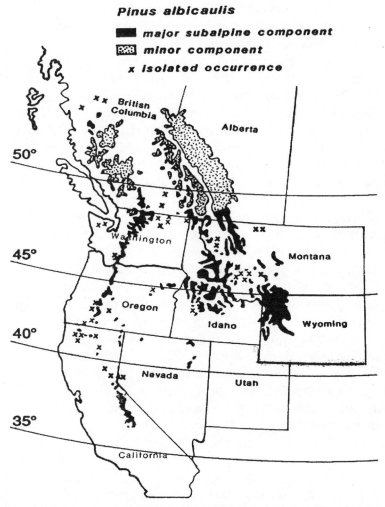

Figure 8.1. The range of whitebark pine in North America. The Greater Yellowstone Ecosystem is found at the southeasternmost portion of whitebark pine's range in Wyoming, southern Montana, and southeastern Idaho. (Diagram from Arno and Hoff 1989.)

Whitebark Pine: A Waning Constituent of Western Subalpine Communities

Whitebark pine is the only North American representative of the five Stone pine species, subsection Cembrae, which occupy the subalpine forests of the Northern hemisphere (Bingham 1972, Schmidt

1992). These pines claim a unique history of coevolution with three corvid (Corvidae) species; a relationship that allows them to exploit and pioneer recent stand openings and fresh burns (Tomback 1982, Mattes 1992). In the case of whitebark pine, Clark's nutcracker (*Nucifraga columbiana*) is the tree's avian mutualist. Nutcrackers harvest the tree's seeds from indehiscent cones and cache them by the thousands at the bases of trees and rocks, in open and recently disturbed areas and in dense moss mats at the optimal 1 to 3 cm depth needed for their germination (Hutchins and Lanner 1982, Tomback 1982, Arno and Hoff 1989). In a given year, one nutcracker may plant from 50,000 to 120,000 seeds in 8000 caches across the landscape (Tomback 1982).

Whitebark pine's range in Wyoming extends from a low of 5000 feet, where it grows in stunted clusters dominated by Lodgepole pine (*Pinus contorta*) to a high of above 10,000 feet (McCaughey and Schmidt 1989). Although a pioneer species at elevations below 7500 feet, whitebark pine is generally outcompeted by more shade tolerant species and species that are faster growing at midsuccessional stages (Mattson and Reinhart 1989). At higher elevations, *P. albicaulis* occurs as a seral component of mixed-species stands of Engelmann spruce (*Picea engelmannii*) and subalpine fir (*Abies lasiocarpa*). It is in these mixed species stands, constituting 12% of Yellowstone's area, where most of the wildlife, including grizzly bears, collect and consume the tree's seeds (Mattson et al. 1991). Above 9000 feet, the tree is capable of colonizing the harsh, windy mountain crags just below timberline. At these highest elevations whitebark pine occurs in pure stands, or tree islands amidst shrub and grasslands in phenotypes ranging from rugged individual trees to wind-contorted, multistemmed clusters and prone clonal krumholtz thickets (Arno and Weaver 1989).

Whitebark pine is important for maintaining ecosystem function and is a producer of mast for wildlife. It is also considered a highly pleasing tree aesthetically. Its ability to occupy the highest elevations and the most inhospitable sites increases the extent of forested area at timberlines. As an early colonizer of recently disturbed high elevation sites, whitebark pine provides shelter for the co-establishment of subalpine fir. Because Clark's nutcracker preferentially plants whitebark pine seeds alongside rocks and other loose material, *P. albicaulis* can also improve the stability of steep slopes (Arno and Hoff 1989, Kendall 1996a). Additionally its expanse of branches blocks wind and prolongs the snowmelt period through shading of surrounding snow, influencing hydrological processes, reducing peak and storm flows and the threat of flooding (Troendle and Kaufmann 1987). Because the date of last snowmelt often extends to mid-July in Wyoming, shading by whitebark pine contributes to the supply of mid–growing season water for irrigation to farms situated along mountain valleys (Farnes 1989).

In addition to grizzly bears, whitebark pine is an important food source to numerous wildlife species, as well as a source of shelter. Avian seed consumers include Stellar's jays (*Cyanocitta stelleri*), ravens (*Corvus corax*), pine grosbeaks (*Pinnacle enucleator*), Williamson's sapsuckers (*Sphyrapicus thyroideus*), white-headed woodpeckers (*Picoides albolarvatus*), hairy woodpeckers (*Picoides vilosus*), red- and white-breasted nuthatches (*Sitta canadensis* and *Sitta carolinensis*), red crossbills (*Loxia curvirostra*), and mountain chickadees (*Parus gabeli*). Chipmunks (*Eutamias* spp.), golden-mantled ground squirrels (*Spermophilus lateralis*), mice, voles, and black bears (*Ursus americanus*) also use whitebark pine as a food source (Hutchins and Lanner 1982).

Red squirrels (*Tamiasciuris hudsonicus*) are avid consumers of whitebark pine cones and play a critical role in rendering cones easily accessible to grizzly bears. Red squirrels harvest whitebark pine cones directly from the tree and store them in vast middens, which grizzly bears and other seed predators raid. As a buffer against annual fluctuations in seed crop availability, red squirrels predominantly occupy mixed-species stands that include lodgepole pine, Engelmann spruce, and subalpine fir, harvesting seeds from all four conifer species, but preferring those of whitebark pine (Kendall 1983, Mattson et al. 2001). Consequently, it is from these stands that grizzly bears also harvest most of the whitebark pine seeds that they consume.

Whitebark Pine and Grizzly Bears

The last remaining grizzly bears in the contiguous United States reside in two major populations—one in northwestern Montana and the other in the GYE. All in all, fewer than 1000 grizzly bears now exist where 100,000 once roamed the entire western portion of the United States and Mexico (Craighead et al. 1995). Because of their small size, the existing remnant populations are vulnerable to extirpation and are currently the focus of recovery efforts under the Endangered Species Act (Shaffer 1980). Because the GYE is bounded on all sides by human development and the recreational and economic activities that surround and isolate the region, continued grizzly bear occupancy is reliant on the quality of the habitat within the immediate region, and the biological and behavioral requirements of grizzly bears (Pease and Mattson 1999).

Several factors combine to make *P. albicaulis* seeds of critical importance to grizzly bears. First, whitebark pine seeds are much larger and more durable than the seeds of other available conifers, requiring less energy expenditure to obtain them. Second, they are highly nutritious (Mattson and Jonkel 1990, Lanner and Gilbert 1992, Mattson et al. 2001); the high fat content of *P. albicaulis* seeds allows bears to acquire significant adipose reserves (Mattson et al. 2001). Third, whitebark pine seeds are usually ripe by mid-August, at which time nutcrackers, red squirrels, and other creatures are harvesting the cones en masse (Mattson and Reinhart 1992). This temporal availability of seed coincides with hyperphagia, a physiological state in which bears become almost wholly consumed with the search for food, ensuring they will survive the long winter months of hibernation. Fourth, when whitebark pine seeds are abundant, they are the highly preferred food source. At these times, bears eat whitebark pine seeds almost exclusively, especially adult females whose ability to store fat is linked to reproductive success and better overall fitness (Mattson et al. 2001).

In reference to their location, the remote subalpine distribution of whitebark pine trees makes whitebark pine a refuge for bears and reduces the risk of higher mortality at lower elevations where human densities are higher (Mattson and Reinhart 1997). Overwhelmingly, mortality of grizzly bears can be attributed to conflicts with humans. A higher frequency of encounters with humans occurs when remote food sources become unavailable, forcing grizzly bears to seek out foods distributed close to human settlements (Knight et al. 1988, Mattson 1990, Mattson et al. 1995, Mattson 1997). The relationship between grizzly bears, human encounters, and whitebark pine crop size variability has been established empirically, and is highly correlated (Fig. 8.2). Studies have revealed that the frequency of grizzly bear captures in the GYE increases sixfold during years of low whitebark pine seed production, and bear mortality rates are two to three times higher (Mattson and Reinhart 1997).

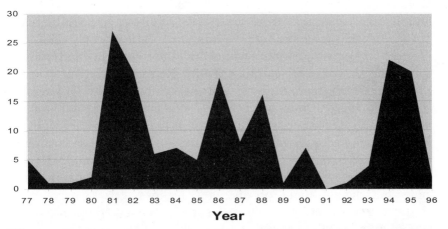

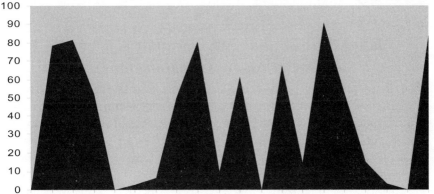

Figure 8.2. (A) Management trappings of grizzly bears. (B) Whitebark pine seed crop size (represented by percent of pine seeds in grizzly bear scat in the months of September and October). Here, the close association between grizzly bear distribution in years of good and poor seed availability is revealed. In years when mast is low, bears roam more widely, requiring greater management efforts to avoid conflicts with humans. In high mast years, bears are more often aggregated in the high elevations, remote from centers of human concentration, requiring less management effort and incurring less mortality. (Data provided by David J. Mattson, U.S. Geological Survey, Biological Resources Division, pers. comm. Used here with permission.)

Moreover, greater nonlethal contact with humans encourages bears to become habituated to humans. Once bears become habituated, they are 3.1 times as likely to die as a result of conflict with humans than bears that are more wary of humans (Mattson et al. 1992).

Seed crop production by whitebark pine trees is highly variable, however (Morgan and Bunting 1992). Reflecting seed crop variability, years of whitebark pine seed use by bears occur with almost equal frequency to nonuse years. Below a certain threshold value, estimated at 20 cones per tree, bears shift their foraging activities to alternate food sources and different locations. This behavior implies that pine seeds may actually be unavailable to bears long before *P. albicaulis* disappears from the Yellowstone landscape. As whitebark pine and overall stand productivity declines, grizzly bears may desert the whitebark pine zone, deeming the energetic profitability of remaining in those stands too low.

Effects of Global Environmental Change on Pathogen Transmission

The distribution of pathogenic organisms and the hosts that support them is largely controlled by the climatic tolerances of the individual species involved, particularly at large spatial scales (Cammell and Knight 1992, Lonsdale and Gibbs 1994). In the case of white pine blister rust, the extent of the fungus in western North America is confined to the locations where the climate and meteorological conditions are favorable for rust transmission. Currently, few regions exist within the range of *P. albicaulis* that are entirely blister rust free. However, some areas, such as the GYE, still maintain relatively low levels of infection. A closer examination of blister rust dynamics reveals that a combination of climate, topography, wind patterns, and the spatial distribution of the host species have served to impede, but not prevent, fungal transmission.

Periodic pathogen outbreaks can function as a natural regulator of successional processes, promoting ecosystem diversity and health (Castello et al. 1995). Rapid climate change on the scale forecasted by general circulation models for the next century, however, can bring pathogens into contact with host species in ways that could potentially degrade ecosystems and diminish ecosystem resilience to disturbance. Climate change is predicted to initiate expansion or contraction of pathogen distributions as climate becomes more or less favorable to their occupancy. Shifts in pathogen ranges can also result when climate change (1) alters the dispersal, reproductive, or developmental processes of the pathogen directly; (2) increases pathogen virulence or growth to host populations; or (3) increases pathogen predation of host species by mediating pathogen competition with symbiotic organisms, such as mycorrhizae, that protect plants against pathogens (Cammell and Knight 1992, Manning and Tiedemann 1995). Additionally, changes in climate can modify the dynamics of host–pathogen interactions in areas where the pathogen already exists. For example, a changing climate can increase the environmental stresses that the host organism is exposed to (i.e., drought or frost damage), heighten host disease susceptibility, or result in higher mortality rates from the pathogenic association (through ameliorating the climatic conditions that cause disease progression) (Cammell and Knight 1992, Lonsdale and Gibbs 1994, Manning and Tiedemann 1995, Brasier 1996). In those regions characterized by long-term stability in environmental conditions, and where pathogens have been excluded due to climatic constraints, host populations could exhibit limited resistance to pathogen invasions. Similar to circumstances in which pathogens have been excluded from entering a region because of physical barriers (e.g., oceans or mountain chains), impacts from pathogen introductions can lead to significant degradation (Lonsdale and Gibbs 1994).

The direct effects of climate warming on pathogen spread may already have been realized in marine systems along the eastern seaboard of the United States. The recent expansion of the range of the endoparasite *Perkinsus marinus,* causing Dermo disease in the eastern oyster (*Crassostrea virginica*) have been shown to track both the short- and long-term trends of increasing winter and annual sea surface and air temperatures. This parasite had previously been confined to regions extending south from Chesapeake Bay with occasional incursions northward into Delaware Bay during short-term warming episodes. However, recent surveys have detected significant new infection centers along Long Island Sound in Connecticut and off the coast of Cape Cod, Massachusetts, with lower infection levels extending as far north as Maine (Ford 1996, Brousseau et al. 1998, Cook et al. 1998, Musante 1998).

Similarly, research supports the relationships between the health of plants and increased incidence of pathogens due to changes in climatic patterns. Climate change may impart greater physiological stress to some species, increasing their susceptibility to pathogen attack. For example, recent declines in oak in southern Spain and Portugal have been attributed to the invasive root pathogen *Phytophthora cinnamomi* and are associated with periods of unusual winter droughts and summer moisture, but are thought to be exacerbated by drought stress (Brasier et al. 1993).

In all, the impact of a shifting climate on pathogen–host relationships is likely to be highly individualized and will differ according to spatial criteria, the temporal scale of concern, the degree and direction of the climatic change, additional pressures from environmental change that may be present, and the physiological requirements of the species involved.

In the case of white pine blister rust, climate change impacts will be realized most through alterations in summer and autumn rainfall patterns, but tree mortality will also be enhanced by additive environmental stressors. Any changes that serve to increase the frequency or persistence of rainfall events, generally increase seasonal rainfall amounts, or increase the frequency of summers in which multiple rainfall events occur, may lead to the heightened presence of white pine blister rust among the region's whitebark pine populations.

Past and Current Blister Rust Spread in the GYE

White pine blister rust pathology is initially manifest in whitebark pine through the appearance of a yellow-orange swelling on the tree's bark (Van Arsdel et al. 1956a, Hunt 1983, Tomback et al. 1995). Cankers, which burst through the bark 2 to 4 years after initial infection, grow incrementally both longitudinally and radially over the course of several years (Lachmund 1926, Mielke 1943). Tree mortality arises through girdling of the bole, or after the loss of branches from multiple cankers (Hoff and Hagle 1989). Depending on the size of the tree and the distance of the blister rust canker from the bole, tree mortality can take many years to occur. In seedlings and saplings, however, the interval between initial infection and death is generally much more rapid. Most seedlings die within 3 years of infection (Tomback et al. 1993). Mortality of all trees due to blister rust is often hastened by cumulative factors. The rate at which trees die is accelerated by the impacts of mountain pine beetle, root diseases, and other pathogens, to which whitebark pine exhibits a heightened susceptibility once infected by *C. ribicola* (Krebill and Hoff 1994).

Natural resistance of *P. albicaulis* to blister rust infection is very low, and may be conditioned by environmental as well as genotypic factors of both *P. albicaulis* and the blister rust fungus (McDonald 1992, Krebill and Hoff 1994). Cankers that originate in stems or branches more than 24 inches from the bole are rarely lethal. Girdling of the bole is prevented through natural canker inactivation, natural branch death from inadequate light, or branch death as stems are girdled by the blister rust canker (Hagle et al. 1989). Consequently, trees younger than 20 or 30 years of age are more vulnerable to death by blister rust infection because cankers rarely occur far from the bole, given their smaller size (Mielke 1943, Hagle et al. 1989).

Whitebark pine cone production and thus seed availability for wildlife consumption often ceases long before tree mortality when blister rust infects the top of the tree first, where the tree's cones are concentrated. In the GYE where most inoculum probably blows up to whitebark pine from lower elevations, this pattern of fungal infection is the primary mode by which mortality occurs (J. Smith, pers. comm. 1998).

White pine blister rust development and the epidemiology of pine infestation encompasses five spore stages (Table 8.1). Three stages occur on pine where *C. ribicola* is perennial, and two on the pathogen's primary or telial host, *Ribes* spp.(shrubs of currants and gooseberries), where it is an annual (Lachmund 1926, Lloyd et al. 1959). For pine infection to occur, it is necessary that susceptible *Ribes* species are located relatively near to pine trees. *R. petiolare/ hudsonianum* (western black currant), *R. lacustre* (prickly currant), *R. viscosissimum* (sticky currant), *R. cereum* (squaw currant), and *R. inerme* (white-stemmed currant) all inhabit the GYE, mainly lining riparian zones and mesic valley bottoms. *R. lacustre* and *R. viscosissimum*, however, can also be found growing on hillsides (Hagle et al. 1989). Mountain gooseberry (*R. montigenum*), a sixth local species, is commonly found growing in areas of whitebark pine. Although extensive studies have not been conducted on all GYE *Ribes* species, most investigations rank their susceptibility as ranging from low to high (Lachmund 1926, 1934, Mielke et al. 1937, Kimmey 1935, 1938, Kimmey and Mielke 1944, Maloy 1997, J. Smith, pers. comm. 1998).

Blister Rust Life Cycle

The dynamics of the blister rust life cycle reveal the major climatic limitations for the spread of blister rust to whitebark pine. In pines, the blister rust cycle (Fig. 8.3) commences in the summer a year or more after pine infection when spermatia, the initial spore stage, form in nectarlike droplets that ooze from the bark. These spores are transmitted by rain or insects that are attracted to the nectarlike exudate and travel from canker to canker. If successfully spermatized, the aeciospores push through the bark of the tree the following spring to form cankers with the onset of warmer weather (Van Arsdel et al. 1956a, Hunt 1983). Saturated atmospheric conditions persisting for 5 or more hours are required for these spores to mature and disseminate, conditions that commonly occur in the GYE during the spring and summer following the initiation of snowmelt when moisture from evaporating snow is freely available (P. Farnes, pers. comm. 1998). These spores, which are capable of long-distance travel, infect *Ribes* spp. through the leaf stomates, leading to the production of uredinia on the undersides of leaves. Because of the frequency of occurrence of these climatic conditions, and the durability of aeciospores, it is unlikely that aeciospore production is limiting in the GYE.

Rust damage to *Ribes* leaves varies considerably among species and even individuals, with some *Ribes* leaves exhibiting early necrosis and others revealing little or no effect from the infection (Mielke 1943, Kinloch and Dulitz 1990). Significantly, leaves that die early in the season are unavailable to support the later spore stages in the blister rust cycle that infect pines. In the event of favorable climatic conditions, several generations of urediniospores can be produced on the same leaves over the course of a single summer, and it is the intensification of this stage that in part controls the degree of pine infection if the rust cycle is completed. Abundant uredinial and telial multiplication is more a function of the number of repeated moist periods distributed over the summer (thereby allowing several generations of urediniospores to be formed) than net seasonal moisture (Mielke et al. 1937).

Urediniospore formation requires from 7 to 24 days and the rapidity of formation is temperature regulated. For development to occur, daytime temperature needs to fall between 12 and 28 °C, with maximum development of fertile spores occurring at 20 °C,

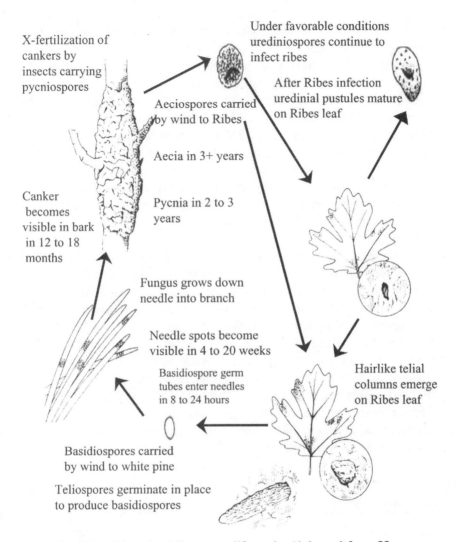

X-fertilization of
cankers by
insects carrying
pycniospores

Under favorable conditions
urediniospores continue to
infect ribes

Aeciospores carried
by wind to Ribes

After Ribes infection
uredinial pustules mature
on Ribes leaf

Aecia in 3+ years

Canker
becomes
visible in bark
in 12 to 18
months

Pycnia in 2 to 3
years

Fungus grows down
needle into branch

Needle spots become
visible in 4 to 20 weeks

Basidiospore germ
tubes enter needles
in 8 to 24 hours

Hairlike telial
columns emerge
on Ribes leaf

Basidiospores carried
by wind to white pine

Teliospores germinate in place
to produce basidiospores

Figure 8.3. The white pine blister rust life cycle. (Adapted from Hunt 1983.)

and higher temperatures generally leading to faster development (Lachmund 1934, Van Arsdel 1965). These spores multiply and some spread to nearby *Ribes* leaves and bushes during periods of near atmospheric saturation or in the presence of free water on the leaves, which occurs many nights in valley troughs. Urediniospore conversion to the telial structure requires from 4 to 12 hours of saturated conditions (Van Arsdel 1965).

The formation of teliospores is likewise governed by temperature and moisture. Cooler temperatures result in maximum spore
production and fertility, such as those commonly found in the GYE
in late summer and early autumn. At least 2 weeks are required
between the time of urediniospore conversion to telia and teliospore
formation (Van Arsdel et al. 1956a). Three consecutive days where
temperatures are in excess of 28 °C during the period of development are sufficient to prevent telial formation. This factor explains
why blister rust is more commonly found at sites with low daily
maximum temperatures, or with negative radiation budgets, common to small forest gaps (Van Arsdel 1961). Sites at high latitudes
and elevations, including many within the whitebark pine zone in
the GYE, fit these criteria.

The final spore stage involves the production of basidia from
mature teliospore columns, which dangle from beneath *Ribes* leaves.
In the same period, basidia must also germinate and release
basidiospores, which infect pines and complete the blister rust cycle
(Kimmey and Wagener 1961). Following germination of the sporid-
ium on the surface of the needle, a germ tube is produced and
enters the pine through the needle stomates. Inside the needle, a
mass of hyphae is produced that grows down the length of the needle and enters the bark through the needle fascicle (Mielke 1943).
From teliospore germination to pine infection, free water or relative
humidity in excess of 97% must be present for a duration of 24 to
48 hours, and temperatures must fall below 20°C. However, longer

Table 8.1. Spore Stage Development: Time, Temperature, and Humidity
Requirements.

Spore Stage	Time Required for Formation	Temperature Range for Spore Formation	Required Precipitation Conditions for Spore Formation
Aeciospores–Urediniospores	5 hours	8–24°C	Humidity≥97% or free water on *Ribes* leaves
Urediniospores–Teliospores	4–12 hours	12–28°C	Humidity≥97% or free water on *Ribes* leaves
Teliospores–Basidiospores–Pine infection	24–48 hours	0–20°C	Humidity≥97% or free water on *Ribes* leaves and pine needles

periods of available moisture lead to greater pine infection. Because the temporal requirements for formation of the final spore stage and pine infection are greatest, this phase of the blister rust cycle is probably most limiting in the GYE.

Where prolonged periods of rain are most likely to lead to spore germination and pine infection (Spaulding and Rathbun-Gravatt 1926, Hirt 1942, Mielke 1943), heavy rains may prevent sporidia transport or may wash away sporidia before the mycelium has had a chance to attach to the pine needle. The timing of the critical moist period is also important in relation to the diurnal cycle. Van Arsdel et al. (1956) found that the majority of extremely fragile blister rust sporidia are released at night and thus are reliant on gentle nighttime air currents for their spread. Sporidia viability has also been shown to be damaged by direct sunlight (Van Arsdel 1967).

Spatial and Temporal Scale of Blister Rust Spread

Large-Scale Spatial Dynamics of Blister Rust Spread

A deeper exploration into the history of blister rust spread in the United States and Canada since its initial introduction may shed light on the prospects for future whitebark pine infection and mortality within the GYE, and is therefore examined here. Specifically, consideration of the spatial and temporal dimensions of past *C. ribicola* invasion into regions of different vegetation, topographic, and climatic regimes yields insights into the factors that control and limit blister rust spread. Based on these insights, I predict that the rate of blister rust spread in the GYE will continue to rise, as it has over recent decades, even in the absence of climate change. If correct, these insights may also account for the different patterns of historically observed blister rust spread. They also indicate that factors that have slowed widespread establishment in the past in the GYE will be less limiting in the future. Furthermore, I conclude that the slow but steady buildup of blister rust inoculum regionally over past decades has set the stage for larger transmission events in the future.

From its point of introduction into western North America in 1910 the spread of blister rust has been highly variable when examined from regional to continental scales. Local and regional climatic conditions most certainly play a large role in the spatial distribution and intensity of infection. Very moist areas generally correspond with locations of abundant and highly susceptible *Ribes* species and experience a higher frequency of the weather events necessary for completion of the blister rust cycle. Drier areas are characterized by having more sparsely distributed *Ribes* and receive fewer of the requisite weather events that allow blister rust to complete its cycle. Based on this climatic evidence, some regions had been considered relatively safe from the ravages of significant blister rust spread (Carlson 1978, Arno 1986). However, findings of increasing infection levels in intercontinental regions where blister rust incidence was historically low indicate that mechanisms in addition to climate may also be at work.

Blister rust progression into a region can occur as a slow cumulative process over many years or in rapid, extensive "wave-like" events. Wave-like events occur when the climatic conditions are ideal for blister rust spread. Historically, these events have occurred in locations with cooler, moister climates, such as the coastal regions of British Columbia, Washington, and Oregon, where infestation levels have attained epidemic proportions in just a few years. The maritime weather systems characteristic of these regions provide a surfeit of years in which conditions for wave-like spread can occur. Yet, dissemination dynamics at a finer spatial resolution suggest that factors in addition to climate may have facilitated the spread of *C. ribicola* throughout these regions.

In the Sierra Nevada and Cascade ranges of the coastal United States, blister rust progression has followed a different history. Sugar pine (*Pinus lambertiana*), which grows as a mid-elevation species not far from riparian *Ribes* populations, was decimated a relatively short time after blister rust's initial introduction. Historically, these mid-elevation stands were infected first (Mielke 1943). High-elevation coastal populations of subalpine western white pine (*Pinus monticola*) and *P. albicaulis* were infected subsequently, and are increasingly the subject of new infection centers today. It is noteworthy that blister rust in the subalpine zone was largely delayed until some time after significant levels of infection had established among the sugar pine communities (Bedwell and Childs 1943). The infected sugar pines appear to have served as a source of inoculum for the upland *Ribes* and pines (J. Smith, pers. comm. 1998).

Based on empirical findings such as these, it appears that the degree of spatial integration of *Ribes* and pines can strongly influence the temporal scale over which *C. ribicola* becomes established in pine communities. In regions where *Ribes* and pines are contiguously distributed, blister rust may establish first where the conditions for blister rust spread are most optimal, and then disperse into adjacent regions. Conversely, at locations where the distribution of *Ribes* and pines are not continuous, the spread of blister rust from infection centers may be much slower, relying on wind to carry delicate spores long distances. However, once these distances are bridged and new infection centers are sufficiently entrenched, further spread of blister rust could be relatively rapid.

The relative hardiness of aeciospores and basidiospores provides a plausible explanation for the importance of *Ribes* and pine proximity. Aeciospores infect *Ribes,* and the sporidia produced by basidiospores infect pines. Aeciospores only require a few hours of moisture availability for production. They are also hardy and capable of long-distance travel—up to several hundred kilometers in some cases (Kinloch et al. 1998). However, because *Ribes* leaves senesce each year, if the blister rust cycle is not completed and pines are not infected within the year of initial *Ribes* infection by aeciospores, no further spores will be present in that location the following year. Conversely, the release of sporidia by mature basidiospores can be very limiting to the establishment of new infection centers when requisite climatic conditions occur infrequently. Basidiospores are very fragile, easily damaged by light, and infrequently travel more than a few hundred meters (Van Arsdel 1967, Kinloch et al. 1998). However, because sporidia infect pines, if the infection is successfully established, new aeciospores capable of infecting local *Ribes* will be available in just a few years, once blister rust cankers develop. Therefore, new infections will be established over long distances only if the conditions for completion of the blister rust cycle occur the same year as aeciospore infection of *Ribes.* Otherwise, the establishment of new infection centers will be limited by the distance to which sporidia can travel.

Within the interior regions of the Intermountain West and the GYE where whitebark pine is found, the absence of mid-elevation susceptible pine species may offer an additional explanation for the much slower spread of *C. ribicola* into these areas. This theory is somewhat confirmed by the status of limber pine (*Pinus flexilis*) in these regions. Limber pine is a five-needle pine very similar to whitebark pine. It can be found growing at lower elevations at some locations, adjacent to riparian *Ribes* species. Where this occurs, lower elevation stands support more severe infection levels than higher elevation whitebark pine stands (J. Smith, pers. comm. 1998), despite a lower frequency of occurrence of the climatic conditions necessary for completion of the blister rust cycle.

Micro- to Macro-Scale Dynamics of Blister Rust Spread

Additional factors also influence the spread of white pine blister rust on a micro to macro scale. Three phenomena primarily control the degree of infection that will develop among white pine populations within a given year. The first is the initial *Ribes* infection levels, which are mainly dependent on supply of blister rust inoculum on a local to regional scale. The second mechanism is the intensification, or multiplication, of the blister rust fungus on *Ribes* leaves, manifest through the spread of *C. ribicola* spores from *Ribes* to *Ribes* during the uredinial stage. The third concerns completion of the entire blister rust cycle and is a function of the amount and spacing of the precipitation events, the intensity and duration of these events, wind direction at the times of spore release, and the distance spores must travel.

Initial *Ribes* infection, which can be a micro-, meso-, or macro-scale event, is a function of several factors: within stand aeciospore release in locations where *Ribes* and infected pines occur on the same site; local wind patterns, where pines and *Ribes* occur within the same or adjacent drainages; or long-distance travel of aeciospores, which is governed by regional scale circulation patterns. The potential for initial *Ribes* infection will increase with the buildup of inoculum associated with increasing levels of pine infections. In the GYE, previously low levels of whitebark pine infection by blister rust help explain the very low rate of blister rust spread that occurred prior to the 1970s. These low levels were a result of limited local and regional inoculum sources. Surveys of the GYE performed by Brown and Graham (1967), although not extensive, discovered blister rust levels of 1.1% on average (Brown and Graham 1967). At that time, the highest incidence of blister rust on whitebark pine in any location was 2.3% (Brown and Graham 1967). According to surveys conducted in 1995, mean estimated infection levels have increased several-fold to 5% in Yellowstone National Park, and 15% in Grand Teton National Park (Kendall

1995). In the GYE as a whole, mean mortality of whitebark pine is approximately 7% due to blister rust and other causes. Correspondingly, increases in the rates of blister rust spread have also increased (Kendall 1995). Stands with infection levels as high as 40 to 44% have been found at several sites in Grand Teton National Park (Kendall 1996b, Kendall 1999). These increases have coincided with a buildup of blister rust infection levels both within the GYE and in the surrounding areas (J. Smith, pers. comm. 1998).

Blister rust intensification is a local development occurring within an individual *Ribes* patch or among adjacent patches, and which is primarily determined by the frequency and duration of nightly dews. In the presence of free water on *Ribes* leaves, uredinia multiply and infect nearby *Ribes*. Cold air drainage, or the tendency for cooler, heavier air masses to pool in valley bottoms, where most *Ribes* are located, contributes to the relatively more frequent occurrence of nightly dews in these locations. Because more valley bottoms and more *Ribes* are found in the lower elevations of mountainous regions, infection intensification is likewise found to occur more often at lower elevations.

Completion of the blister rust life cycle can be a local- to meso-scale phenomenon and consists of the transfer of spores from pine to *Ribes* to pine within a given year. In order for the cycle to progress, each successive stage with the associated weather conditions must occur, and must occur according to the proper sequence and timing. Because both the frequency of precipitation occurrence and precipitation amounts increase with elevation, the climatic conditions necessary for the completion of the blister rust cycle generally occur more frequently at higher elevations. However, most stream-type *Ribes* likely to experience infection intensification can be found in the low elevations of the GYE, while most pines are found at high elevations. Under this scenario, the conditions for completion of the blister rust cycle must occur simultaneously at both low and high elevations for pine infection to arise. Here, the transfer of spores from low to high elevations is a meso-scale process. However, once *C. ribicola* becomes significantly established at higher elevations, cycle completion becomes a local phenomenon where blister rust spores need only transfer among high elevation *Ribes* and pines.

Additionally, for meso-scale spore transfer to occur requires that the prevailing winds successfully transfer spores between the two hosts. Because the dispersal of aeciospores occurs frequently throughout the spring and summer, if supply is ample, suitable winds generally exist to transport these spores to *Ribes* (Smith 2000). Basidiospore transport from *Ribes* to pines, however, may be hampered even when basidiospore production is plentiful if winds of sufficient magnitude and direction are not available to transfer these spores from *Ribes* to pines at the time of spore release. In such a case, the proximity of *Ribes* and pine hosts is probably of greatest importance in producing pine infection. Because of the fragility of sporidia, where infected *Ribes* are located very near to pines and the climatic conditions are suitable, pine infection is almost a certainty. However, when infected *Ribes* and pines are spatially disjunct, basidiospore transport is more uncertain.

Once blister rust is well established at upper elevations however, additional obstacles may complicate the rate at which new infections occur as a local phenomenon. The persistence of snowpack into the summer at high elevations may delay emergence of *Ribes* leaves. Similarly, early frost may lead to early *Ribes* leaf senescence in the autumn, reducing the time available for completion of the blister rust cycle. In addition, the less frequent occurrence of nightly dews may produce less disease intensification on *Ribes* leaves, slowing the rate of new infections.

Perhaps there are additional mechanisms that serve to control the character of blister rust spread. More recent analyses of the history of blister rust transmission in individual locations have revealed the complexity of dissemination dynamics. In the Great Basin Ranges, nighttime katabatic breezes have been implicated in transporting spores to pines growing at midslope and ridgetops (Van Arsdel and Krebill 1995). Recent surveys in the Intermountain West report expansion of blister rust into areas never before impacted by the disease and considerable increases in infection levels over those documented in the 1960s (Smith and Hoffman 2000). Epidemic infestation levels have now been recorded for the sugar pine regions of the southern Sierra Nevadas of California that were once thought to be too warm and dry to support the fungus (McDonald 1992, R. Hoff, pers. comm. 1998). Data from Glacier National Park and Bob Marshall Wilderness Complex of northern Montana offer yet other salient examples of the unpredictability of the dissemination patterns of this fungus. Although blister rust had long been present in the northern Rocky Mountain latitudes of this region, extensive blister rust spread did not occur until very recently. Surveys completed in the region in 1971 revealed low infection levels at that time. The same sites revisited in 1991, however, contained "prominent" blister rust levels on all revisited sites, and had sustained an overall decrease of whitebark pine basal area of 42%. Today, more than 90% of the live whitebark trees in the area are estimated to be dying as a result of white pine blister rust (Kendall and Arno 1990, Keane and Arno 1993).

The infection of whitebark pine by blister rust is an intricate process, with many subtle factors playing into the spread of infection from pine to *Ribes* to pine. The character of the weather events that contribute to the progression of the disease must fall within specific parameters of duration, spatial and temporal distribution, and intensity. Excessively wet summers can cause *Ribes* to drop its leaves early. Inordinately dry summers may produce a similar effect. Very heavy rains in the early spring are known to prevent aecial dissemination, while unusually dry summers can curtail the period of aecial production (Mielke 1943). The prevailing winds can be unpredictable and highly variable in mountainous regions such as the GYE. Topoedaphic factors likewise influence *Ribes* distribution and abundance. Regional adaptations of the fungus to its environment have also been noted (McDonald 1992, Van Arsdel and Krebill 1995). For example, in some cases, the urediniospore stage has been skipped entirely, and prolific teliospores formation occurred at the end of the season. These phenomena have been noted in central Idaho as well as the southern Sierra Nevadas (McDonald 1992, Van Arsdel and Krebill 1995).

An additional observation includes consideration of the spatial and temporal scale at which invasive species impact their environment. Various species that lay dormant for many years have subsequently become widely distributed through several different mechanisms, many of them human-mediated. Expansion of blister rust into previously uncolonized regions could follow a similar course; especially in areas where the fungus is present but has not yet become epidemic, such as the GYE. One species, smooth cordgrass (*Spartina alterniflora*), native to the estuaries of the Gulf and Atlantic Coasts of North America but an exotic along the Pacific coast, is an example of a species that spread slowly in Willapa Bay of Washington State for the first 50 years after its introduction, yet later colonized large areas (Lantz 1995). Prior to 1945, environmental conditions had restricted *S. alterniflora* to reproduction through vegetative sprouting. Thereafter, warmer bay temperatures and increased sedimentation are thought to be responsible for the flowering and seed production that enabled much faster dispersal and colonization (Lantz 1995). In another example, a weedy invader, the cut-leaved teasel (*Dispsacus laciniatus*) enacted a similar history of broad colonization that followed a period of earlier dormancy. Introduced into Albany, New York, sometime before the turn of the century, this weed's extensive invasion of the Midwest over the last 30 years was facilitated by construction of the interstate highway system, along which it dispersed (Crooks and Soule 1999). Many additional invasive species such as garlic mustard (*Alliaria officinalis*) in the Midwest, the paper-bark tree (*Melaleuuca quinquenervia*) in the Florida Everglades, mimosa (*Mimosa pigra*) and prickly acacia (*Acacia nilotica*) in Australia, to name but a few, exhibit similar histories of large-scale invasion after substantial lag times.

Blister rust spread through western North America has a slightly different history. Introduction of blister rust into northwestern North America has been traced to a single location in Vancouver, British Columbia, in 1910 (Maloy 1997). From there, blister rust spread rapidly and over long distances to infect populations of white pines growing in areas highly favorable to fungal dissemination. From an early vantage point and preliminary understanding of blister rust transmission dynamics, the conclusion that some regions would remain free of the disease is understandable. Indeed, some areas may still permanently elude large-scale infestations by blister rust. However, recent evidence of increasing infection levels in areas such as the GYE, and new advances in the understanding of the mechanisms that contribute to blister rust and other invasive species dissemination events, and the spatial and temporal scales at which they operate, leave little room for optimism.

Climatic Regimes of the Greater Yellowstone Ecosystem

Generally, the climate of the GYE is characterized by long, cold winters and warm, dry summers. Spring and autumn in Yellowstone tend to be short, transitional periods rather than full-fledged seasons (Dirks and Martner 1982, Despain 1987, 1991). Weather systems travel into the area from several locations. Moist air approaches from the Gulf of Mexico or the southern Pacific. Drier weather systems reach the area from the west, as maritime air masses shed their moisture over the steep terrain of the northwestern ranges, or when the cold polar air mass from the Canadian plains migrates to the south. The interplay of these weather systems accounts for the complex and variable nature of the climate found in this region. On both an annual and a seasonal basis, climate can be highly variable in the GYE, with summer exhibiting the most climatic consistency (Douglas and Stockton 1975). Years characterized by the frequent passage of fronts across the region can result in relatively cool, wet summers. Warm summers generally occur under anticyclonic conditions when warm upper atmospheric air masses move over the continental United States. Wet periods can occur with the advection of moisture from the southeast or southwest (Douglas and Stockton 1975).

Of particular relevance to the spread of *C. ribicola* is the short autumn period of September through mid-October, as this is predominantly when the final spore stages and tree infection occur in the blister rust cycle. Autumn weather is highly variable from year to year, with some years being more characteristic of summer and others more typical of winter. Dendrochronological reconstructions reveal frequent anomalous precipitation and temperature conditions in the autumn. Key departures from average conditions that favor blister rust spread would incorporate a greater frequency of cool, wet summers, with temperatures above freezing extending into October. Alternatively, extremely hot, dry summers would increase the likelihood of large stand-replacing fires. These fires would favor conversion of the forested area toward younger age classes of trees, and facilitate the successional replacement of whitebark pine by lodgepole pine and subalpine fir (Mattson and Reinhart 1989).

Methods

Examination of Baseline Data

To determine the regional trends in climate for the GYE, weather station data that have been recorded for several years on a region-wide basis were used. Of the weather stations that exist, many have been in place in excess of 50 years. These stations are generally found at the lower elevations, and are termed "valley sites" for the purpose of this study. Beginning in 1981, snow survey telemetry (SNOTEL) sites were established in the higher elevations where *P. albicaulis* is generally found. These stations recorded data on daily minimum and maximum temperature values and daily precipitation amounts. This weather data is available from the Natural Resources Conservation Service in Portland, Oregon.

Rather than attempting to determine the exact locations of whitebark pine forests in the GYE, I assumed that the numerous existing weather stations located above 7500 feet accurately portray the weather patterns that exist within the elevational belt where *P. albicaulis* grow (Despain 1991). In choosing not to make this study spatially explicit, I was able to use the SNOTEL and valley site data available. Similarly, the valley stations were assumed to be situated at locations where the highest concentration of *Ribes* bushes grow, although some *Ribes* spp. also exist at higher elevations. These assumptions were generally verified through ground truthing efforts in this region (D. Reinhart, pers. comm. 1998). For this study, valley sites were differentiated from SNOTEL sites on the basis of elevation. Data on the distribution of *Ribes* do not exist for the GYE; however, because most *Ribes* are found growing along streams and in valley bottoms, more commonly situated at lower elevations, this assumption was made.

Because of the need to extend the length of climatic data for high elevation sites in the GYE and because of the need to generate daily relative humidity values for the purpose of analysis, the program MTCLIM was used (Running and Nemani 1987, Hungerford et al. 1989, Kimball et al. 1997, Thornton et al. 1997). MTCLIM, developed to examine synoptic patterns over large mountainous regions, estimates daily weather values for high elevation locations when actual measurements do not exist. The high elevation estimated values are computed using recorded minimum and maximum daily temperatures, and daily precipitation amounts from nearby low elevation weather stations and from computed or estimated lapse rates and global coordinates. Because the high elevation point for estimating daily weather values was chosen to correspond with an actual SNOTEL site, it was possible to compute lapse rates using the existing short-term SNOTEL weather record and to verify the accuracy of this method. On average, MTCLIM provides an accurate estimate of high elevation weather in the GYE. Daily maximum and minimum temperature, daily precipitation amount, and vapor pressure values were then generated using MTCLIM to compute a record of equal length to the low elevation station. Using vapor pressure and temperature values, maximum and minimum relative humidity values were computed, as well as average relative humidity values for daytime hours, and for the 12 coldest hours of the day.

The following is an example of the methodology employed: The Madison Plateau SNOTEL station, (el. 7750 ft.) has an existing data record for weather from 1989 through 1997. A record of 49 years in length was available for the nearby low elevation West Yellowstone station (el. 6670 ft.), beginning in 1948. The average lapse rates for each month from June through October were determined using the available monthly data beginning in 1989 from each of the two stations. Next, this information was entered into MTCLIM to generate weather data for the Madison Plateau station for the 1948 to 1989 period. This procedure was repeated for each of the SNOTEL stations, and the generated data were added onto the existing SNOTEL data set. Relatively complete data exists for most of the valley stations starting in 1948. A few stations, (i.e., Cooke City, and Old Faithful) have shorter records. SNOTEL station values computed from these stations with shorter periods of recorded data likewise have abridged records. Missing data for all the valley Grand Teton stations and many of the Yellowstone stations were supplied by the National Park Service. These values were computed based on empirical relationships to neighboring weather stations (Farnes et al. 1998).

Once existing data sets for the SNOTEL sites in addition to the valley sites were calculated and synthesized, a blister rust event was defined. A blister rust event is defined by the occurrence of weather conditions that could cause movement from the uredinial to the telial spore stage, or infection of whitebark pine by blister rust sporidia (i.e., formation of basidiospores and transfer of sporidia to pine needles and infection of pine needles). An event was defined conservatively, aiming to differentiate a passing afternoon thunderstorm from a wet front moving through the region. It is assumed that a passing thunderstorm would be of too short duration to cause pine infection. However, a wet front of a few days' duration could satisfy time requirements for progression of the blister rust cycle to occur. As hourly data were unavailable, a coarse filter approach was applied in which criteria chosen to identify assumed weather events corresponded to the available data of precipitation amounts and relative humidity values. For instance, one set of values that would satisfy the criteria for a blister rust event would be precipitation in excess of at least 1 mm for at least 3 days in a row, in which at least 2 of the days have rain greater than 1 cm. Similarly, a blister rust event could be defined by at least 2 consecutive days in which nighttime relative humidity was in excess of 90%. The value of 90% was chosen based on the assumption that MTCLIM underestimates vapor pressure. (See Appendix for complete list of criteria chosen to designate a blister rust event.) These values are very conservative from the viewpoint of completing a blister rust cycle, given that 4 to 12 hours are generally considered adequate for production of telia, and 24 hours are needed for basidial germination and pine infection. Ideally, survey data would have been available that had been recorded during years of known blister rust spread and which linked those years with actual weather events. However, because no such surveys existed, decisions concerning the specific criteria that constituted a blister rust event were based on the greenhouse and field experiments conducted from the 1920s through the 1960s (Spaulding 1922, Spaulding and Rathburn-Gravatt 1922, Pennington 1925, Hirt 1935, 1942, Mielke 1938, 1943, Van Arsdel et al. 1956a, 1956b, Van Arsdel 1961, 1965).

Next, the years were selected for each station in which at least two blister rust events occurred for the period from mid-June to mid-October. The following criteria had to be satisfied: at least 2 weeks had to have transpired from the initiation of the first blister rust event and the completion of the second blister rust event, and there could be no more than 28 days between the end of the first event and the initiation of the second blister rust event. In addition, I determined for each station the number of blister rust events that occurred in each year in which the minimum (two) was exceeded. In this way it was possible to determine a profile of the climatic conditions that occurred at each station with specific reference to blister rust dissemination. In addition, the simplifying assumption was made that large-scale freezing does not occur before 15 October, which would cause *Ribes* leaves to senesce and derail the blister rust transmission process. Freezing events do occur during some years, especially at the higher elevation stations where snow can begin to accumulate as early as late August when the season is particularly cold. An additional assumption made during this study was that high temperatures, which can cause spores to lose their viability, do not interfere with the blister rust cycle at these locations. This assumption should hold for most years because late summer and early autumn periods are generally cool in the GYE and coincide with periods of precipitation. However, during particularly warm years, this assumption would be invalid.

Weather Generation and Sensitivity Analysis
To examine the climatic conditions for blister rust transmission in the region under climate change conditions, a sensitivity analysis was performed. In this analysis, the observed weather data sets were varied by statistical percentages in accordance with climate change predictions and then reanalyzed. For the sensitivity analysis the stochastic weather generator Met&Roll, developed by Dubrovsky (Dubrovsky 1997), and based on the Richardson weather generation model, WGEN (Richardson and Wright 1984) was employed. The weather generator produces daily weather series for solar radiation, minimum and maximum daily temperature, and precipitation. For the purpose of this study, precipitation values were of primary interest, and therefore will mainly be discussed here. (For a more detailed description of weather generation models see Richardson and Wright 1984, Wilks 1992, Katz 1996, Mearns et al. 1996.) Weather generation in Met&Roll follows a first-order, two-state Markov chain process with different parameterizations conditional on the occurrence of a wet or dry day. The model is defined by the transitional probabilities between the two states, in which P_{01} represents the probability of a wet day following a dry day and P_{11} the probability that a wet day will succeed a wet day. The parameters for stochastic precipitation generation may be defined alternatively by the unconditional probability of wet day occurrence:

$$\pi = P_{01} / (1 + P_{01} - P_{11}),$$

and the persistence parameter,

$$d = P_{11} - P_{01}$$

Precipitation amount on wet days is modeled by the gamma distribution, with shape parameter alpha (α) and scale parameter beta (β) (Wilks 1992, Katz 1996, Mearns et al. 1996, Dubrovsky 1997).

Met&Roll reads in the observed data set and calculates the four precipitation parameters on a monthly basis for the series. The set of unmodified data may then be used to produce a synthetic series that is statistically similar to the observed data. Alternatively, the parameters may be modified to produce a synthetic series that varies according to user specifications (Dubrovsky 1997). For the purpose of this study, statistical properties of the data were altered to reflect predicted precipitation changes (i.e., increased interannual variability of monthly precipitation amounts, and increased mean monthly precipitation). In order to examine some of the range of precipitation changes that could occur, precipitation statistics were varied by a constant 25%. Toward that end, three scenarios were examined for possible climate change for two valley stations, Hebgen Dam and Snake River, and two SNOTEL stations, Lake Yellowstone and Gros Ventre Summit:

1. Twenty-five percent increase in interannual variability of monthly precipitation amounts; accomplished through varying monthly precipitation amounts for each year of the synthetic series.

2. Twenty-five percent increase in interannual variability of monthly precipitation amounts and 25% increase in mean monthly precipitation. In this scenario changes in mean monthly precipitation were obtained by increasing the product ($\alpha \times \beta$ by 25%, thereby increasing the mean precipitation amount by 25% on the days that precipitation occurs, but keeping the number of days that precipitation occurs within a given month constant.

3. Twenty-five percent increase in mean monthly precipitation through increasing the frequency of days in which precipitation occurs. In this scenario a 25% increase is applied to both the value π, the probability of wet day occurrence, and P_{01}, the transition probability of a wet day following a dry day (M. Dubrovsky, pers. comm. 1999).

In all cases, 50 years of data were generated in order to produce data sets of roughly equal length to the observed record. Upon completion of the weather generation process, the synthetic series were examined in an identical manner to the observed data. For both data sets, the number of years in which the climatic conditions for completion of the blister rust cycle occurred and the number of blister rust events which occurred in each year that the blister rust cycle was completed were examined. The only difference in how both data sets were analyzed was that relative humidity values were not considered for the generated weather. This step was omitted because selection of blister rust events was better explained by specified precipitation criteria than relative humidity values computed by MTCLIM in examination of the historical climate.

Results of Weather Station Analysis of Baseline Data

The high correlation of the results across elevational gradients obtained (Figs. 8.4 and 8.5) indicate that the described methodology provides an accurate assessment of the historical climatic favorability for blister rust transmission in the GYE. The two indices computed for the analysis of 14 SNOTEL and 11 valley stations produced equivalent results and were effective in predicting blister rust events. One predictive index was the number of years in which all the conditions for blister rust cycle completion were satisfied (Fig. 8.4). The second predictive index was the number of blister rust events per year (Fig. 8.5). (Only years in which cycle completion occurred were considered when developing the second index.) These results were then pooled based on location and elevation, and averaged over 5-year periods. The analyses revealed that the climatic conditions for completion of the blister rust cycle did occur and occurred often in the GYE for the observed time period (Figs. 8.4 and 8.5).

An additional conclusion from these analyses was that the climatic conditions for blister rust spread occurred more frequently as elevation increases. This finding matches intuitive expectations, as precipitation is known to increase with elevation in mountainous regions (Ives and Hansen-Bristow 1983). The confirmation of this hypothesis is significant in regard to blister rust transmission because it helps explain its slow rate of initial spread within the region. Where blister rust spread occurs as a meso-scale phenomenon and is dependent on the movement of blister rust spores from low to high elevations, climatic conditions required for the completion of the blister rust cycle must occur at the elevations where both

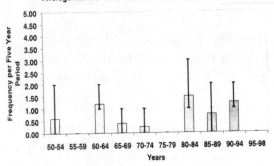

Average Number of Blister Rust Years, 6,000-6,500

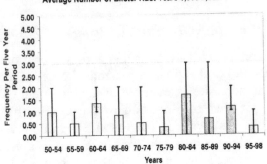

Average Number of Blister Rust Years 6,500-7,000

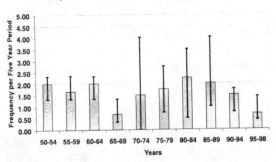

Average Number of Blister Rust Years 7,000-7,500

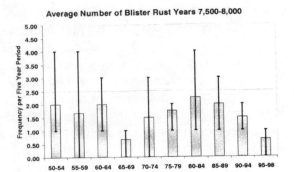

Average Number of Blister Rust Years 7,500-8,000

Average Number of Blister Rust Years 8,000-8,500'

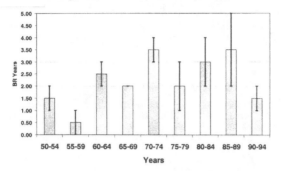

Average Number of Blister Rust Years 8,500-9,000

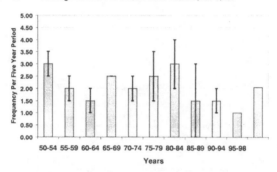

Average Number of Blister Rust Years, >9,000

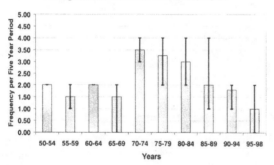

Figure 8.4. Graphs representing the number of years in which the blister rust cycle was completed, on average, for each 5-year period, computed for elevational bands of 500 feet. Error bars represent the range of values obtained for the pooled stations making up the individual bands.

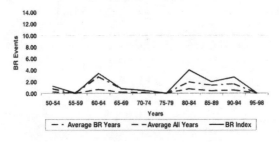

Average Number of Blister Rust Events per Five Year Period 6,000' - 6,500'

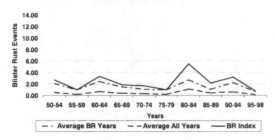

Average Number of Blister Rust Events Per Five Year Period 6,500' - 7,000'

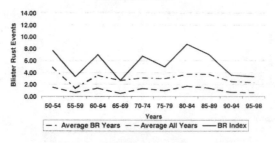

Average Number of Blister Rust Years per Five Year Period: 7,500'-8,000'

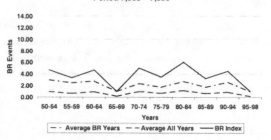

Average Number of Blister Rust Events per Five Year Period 7,000' - 7,500'

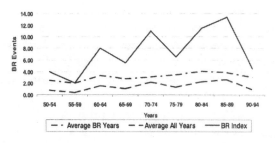

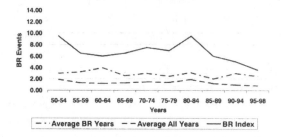

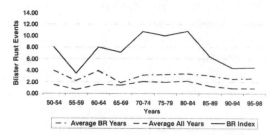

Figure 8.5. Illustration of: (1) The average number of blister rust events that occurred in each of the years of the 5-year period that the blister rust cycle was completed, computed for elevational bands of 500 feet from pooled valley and SNOTEL stations. (2) The average number of blister rust events that occurred per year, per 5-year period, computed for elevational bands of 500 feet from pooled valley and SNOTEL stations. (3) The blister rust index, which is the product of the number of years in which the blister rust cycle is completed and the average number of blister rust events that occurred in each of these years. This value is computed for each of the pooled stations and averaged to obtain the value that appears in the graph. (See Appendix.) For example: Four stations fall in the elevational band from 7000 to 7500 feet: Base Camp, Coulter Creek, Grassy Lake, and Old Faithful. Average Blister Rust Years, Average All Years, and the Blister Rust Index values are computed for each of these four stations individually. Then, these values are averaged to obtain the values that appear in the figures.

Ribes and whitebark pine are abundant. It is likely that this is the primary way in which blister rust becomes established in regions where *Ribes* and pines are not continuously distributed. It is probable that this continues to be the dominant mode of infection establishment in the GYE today.

Moreover, these findings imply that until inoculum levels build up in the high elevations to the point where further blister rust spread in the GYE will be a primarily local, high elevation phenomenon, blister rust transmission will continue to be limited by the relative infrequency of the climatic conditions for blister rust completion at the low elevations. However, as the infection levels continue to rise at high elevations, and cycle completion can occur locally between high elevation *Ribes* and pines, the rate of blister rust spread will likely increase significantly as well. Once infection levels have increased at high elevations, new infections will occur primarily as a result of spore transfer between high elevation ribes and pines, and the climatic conditions existing at low elevations will no longer regulate or significantly limit blister rust dynamics at high elevations.

No significant trends were apparent among the different regions of the GYE. This result is somewhat contrary to expectations, as blister rust infection levels are known to be higher in the southern part of the ecosystem (Kendall 1996a). It may be that the criteria chosen to classify the blister rust cycle are too coarse to portray small-scale regional differences. As noted in the preceding section, the criteria chosen to qualify a blister rust event were defined conservatively in order to discern the presence of an actual weather system passing through a particular location. Given this approach, it is unlikely that small regional differences would be detected. Additionally, the lack of regional trends may be related to data disparities between the number of stations averaged for each region, or a function of the precise elevations of the stations pooled for each location. For instance, the results for six stations are pooled for the central SNOTEL location, while only a single station (Phillip's Bench) was available to represent the southwestern quadrant of the region. Lastly, higher blister rust infection levels in Grand Teton National Park may be caused by factors unrelated to climate, such as local wind patterns, or proximity of susceptible *Ribes* to pines.

To obtain additional insights into the dynamics controlling blister rust transmission, a blister rust index was developed. Because

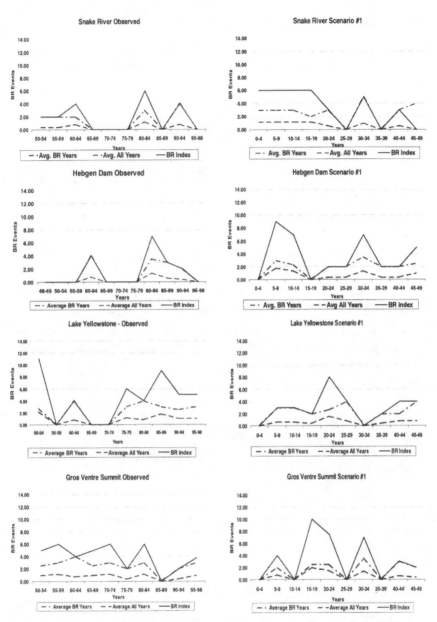

Figure 8.6. The same values calculated in Figure 8.5—average number of blister rust events occurring in years the blister rust cycle is completed, average number of blister rust events in all years, and the blister rust index. However, here these values are calculated and represented graphically for the observed weather data, and for three climate change scenarios for each of four stations (valley stations: Snake River and Hebgen Dam; SNOTEL stations: Lake Yellowstone and Gros Ventre Summit). The details of each of the three scenarios considered are provided in the text.

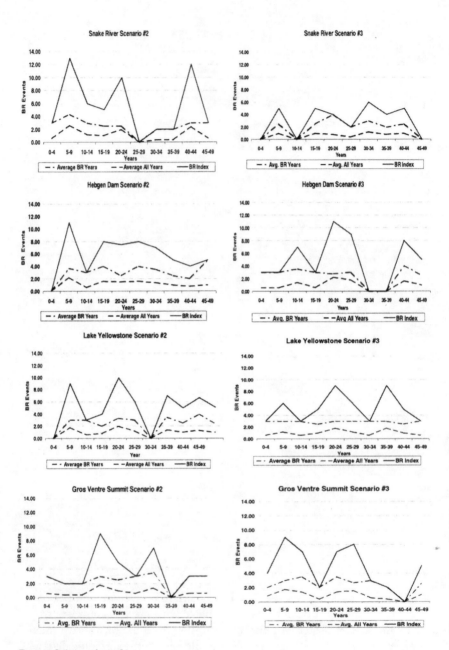

Figure 8.6 *continued*

the magnitude of transmission occurrence within a given season is a function of the number of blister rust events that contribute to *Ribes* spore intensification, and because a sufficient number of events must occur to complete the blister rust cycle, the blister rust index is simply the product of these two factors within a 5-year period. The blister rust index is first computed individually for each station, and then averaged for all stations that make up a regional or elevational division. In addition, the average number of blister rust events is computed for blister rust years alone, and for all years for the 5-year grouping and presented in Figure 8.6. These calculations are displayed graphically with the blister rust index. Most graphs revealed high numbers of blister rust events clustered around the years 1960–64, 1970–74, and 1980–84. Additional peak periods of blister rust outbreaks were obtained for individual quadrants and elevational groupings. Again, these periods are influenced by the number of stations pooled for an individual division. The period between 1995 and 1998 is compromised by differing endpoints available for weather station records and should be used with caution. For example, some years had data that extended through the summer of 1998, while others were truncated at 1995 or 1996.

Taken together, the previous observations made concerning the complex dynamics of blister rust transmission and the results of the data analysis lead to the prediction that blister rust may build up to levels where widespread dissemination will occur in the GYE. Furthermore, climate change may serve to accelerate the rate at which this process will occur.

Met & Roll Climate Change Sensitivity Analysis and Discussion of Results

To examine the potential effects of climate change on the spread of white pine blister rust to whitebark pine in the GYE, a sensitivity analysis was performed by varying monthly mean and interannual variability in precipitation. These variables were chosen because an increase in these statistics is anticipated to hasten the widespread dissemination of blister rust and because these changes match the direction of those predicted by most general circulation models (GCMs) for the region. Moreover, the changes examined reflect the need to consider the correct temporal and spatial scale of climate change and the specific variables that may be affected (Ojima et al.

1991). For example, annual or decadal scale changes in temperature and precipitation may not be meaningful in this study. However, on a seasonal or daily time scale, precipitation changes induced by climatic changes could be critical to blister rust dispersion.

Any increases in summer and/or early autumn precipitation would be expected to increase both the occurrence of years in which all the climatic conditions for blister rust cycle completion are met, and the frequency of blister rust events within an individual season, contributing to disease intensification. Regional variability in precipitation change is projected by GCMs, with some locations (i.e., coastal regions), potentially experiencing dramatic increases in precipitation, while others will experience precipitation declines. At finer spatial resolutions than most GCMs are capable of reproducing, however, predictability of precipitation changes are poor. Notwithstanding, several studies conclude that heightened precipitation conditions on an annual or seasonal basis have occurred in recent years, or are likely to occur globally, for the northwestern United States region as a whole, and at mid to high latitudes as a result of climate change (Bradley et al. 1987, Rind et al. 1989, Rosenzweig 1989, Idso and Balling 1991, Balling et al. 1992a, IPCC 1996, Zwiers and Viatcheslav 1998).

Focusing on the Yellowstone area in particular, three of four general circulation models (developed by the Goddard Institute for Space Studies, the United Kingdom Meteorological Office, and Oregon State University) predicted summer precipitation levels would increase by 5 to 24% under doubled CO_2 scenarios (Balling et al. 1992a). Only the model developed by the Geophysical Fluid Dynamics Laboratory predicted a 15% decrease in summer precipitation. Similarly, an investigation of historical trends by Idso and Balling (1991) for the conterminous United States revealed statistically significant increases in annual and autumn precipitation for the 1955–87 period compared to 1901–54. Climatologists have attributed precipitation increases globally to the rising atmospheric concentration of greenhouse gases from anthropogenic sources and associated warming (Schneider 1992). Warmer conditions are expected to influence hydrological activity by increasing the moisture holding capacity of the atmosphere, increasing evaporation from the oceans and raising overall precipitation (Rind et al. 1989, Schneider 1992).

Heightened variability in precipitation and atmospheric moisture content are also predicted as a result of climate change and will likely lead to more frequent weather extremes (Rind et al. 1989, Liang et al. 1995). Statistical analysis of extreme events, specifically applied to climate variables, concluded that even small changes in the mean can herald a highly nonlinear increase in the frequency of extreme events (Mearns et al. 1984, Mearns et al. 1990, Hennessy and Pittock 1995). Moreover, the frequency of extreme events is even more sensitive to changes in variance than to changes in the mean (Katz and Brown 1992). More frequent extreme precipitation events would provide more opportunities for widespread dispersion of blister rust. On the ground, when the climatic conditions for blister rust spread are met continually over the course of a summer, the uredial, telial, and basidial stages of blister rust development may cycle through numerous generations, increasing in intensity with each cycle and producing large-scale episodes of blister rust spread.

Because our understanding of the natural oscillations in Earth's climate that occur over longer time scales and how these oscillations may interact cumulatively is still rudimentary, detection of fluctuations in climate variables may be more complex than the interpretation of perceived general trends. However, climatologists are continuing to refine indices to measure climatic trends, including the reoccurrence of extreme events. The Climate Extremes Index (CEI) as developed by the National Climatic Data Center found extreme events for the United States as a whole to be on average 1.5% higher from the period after 1976 than for the average of the previous 65 years (Karl et al. 1996). Similarly, GCM simulations under CO_2 doubling predicted increases in the magnitude and variance in precipitation throughout the globe, while increases in extreme rainfall events have been discerned globally (Rind et al. 1989, IPCC 1996, Zwiers and Viatcheslav 1998). This variability is sometimes attributed to the strong El Niño–Southern Oscillation (ENSO) events that have been occurring with greater frequency since the 1976/77 ENSO (IPCC 2001a, Rind et al. 1989, Trenberth and Hoar 1996, 1997).

For the purpose of comparing the baseline data with the daily precipitation values generated for the climate change scenarios, the two indices developed earlier were used and pooled for 5-year intervals. The differences in the two indices for selected stations under the range of conditions examined are summarized in Figure 8.6. Overall, the climate change scenarios considered produce higher values for the blister rust index, implying a greater frequency of completion of the blister rust cycle, as well as greater disease intensification. Additionally, a higher frequency of extreme precipitation events are realized, and these values are more extreme than those of the observed data.

A great potential for the proliferation of blister rust within the GYE is demonstrated by the magnitude of change in precipitation patterns that were obtained for the two valley stations (Snake River and Hebgen Dam; Fig. 8.6). Under scenarios 1, 2, and 3, the average number of blister rust events per year during the years in which the blister rust cycle is completed, are increased by 100%, 95%, and 42%, respectively, for the Snake River weather station. Also at the Snake River station, the blister rust index increased by 94%, 211%, and 72% under the same scenarios. For the Hebgen Dam weather station, the average number of blister rust events per year of blister rust completion increased by 69%, 165%, and 116% for scenarios 1, 2, and 3, respectively, while the blister rust index for Hebgen Dam increased by 150%, 303%, and 238%.

For both stations, the increases of the largest magnitude occurred under the second scenario, of 25% increases in both interannual variability and 25% increase in the magnitude of precipitation on the days that precipitation occurred. For the Hebgen Dam station, more extreme events also resulted for each of the three scenarios than for the observed data set, and for both stations extreme events occurred more frequently. It is also important to note that in all cases, 5-year periods of no blister rust events were also generated. Increases in potential evapotranspiration that would accompany projected increases in temperature, discussed further in a later section, may heighten the severity of these drought periods (Romme and Turner 1991). These results, although computed based on the daily statistics for individual stations, should not be considered indicative of the actual changes specific to the local climate of the individual station. Instead, they should be understood as reflecting the range of changes that would likely befall the GYE as a whole under the given scenarios analyzed.

For the two SNOTEL stations, changes are more unique to the individual scenarios examined (Fig. 8.6). In fact, for both the Lake Yellowstone and the Gros Ventre Summit stations, the average blister rust index decreased for the first scenario even though interannual variability was increased by 25%. The average number of blister rust events during years of blister rust occurrence also decreased at the Gros Ventre Summit station and remained the same for the Lake Yellowstone station. However, the blister rust events were more severe at the Gros Ventre Summit station, and the oscillations among 5-year intervals were more extreme. Because severity of the blister rust season may have more bearing on blister rust spread than the number of years in which cycle completion occurs, this finding is also significant. Here, severity, as measured by the number of events that occur within a given year, can contribute to disease intensification and lead to larger dissemination events.

For the second scenario, in which both precipitation amount and interannual variability are increased by 25%, the number of blister rust years that occurred and the number of blister rust events that occurred during these years were generally very similar to the observed data for the GVS station; however, once again, more extreme events were generated, and variability was visibly increased. For the Lake Yellowstone station, the second scenario resulted in an increase in the blister rust index by 15% and higher variability among the 5-year periods.

In the third scenario, a 25% increase in the number of days in which precipitation occurs resulted in an 18% increase in the blister rust index for both the Lake Yellowstone and the GVS stations. Additionally, the average number of blister rust events was higher. As with the valley stations, the SNOTEL stations are meant to convey the magnitude and direction of the changes that may be realized under each scenario regionwide and should not be considered location specific.

Most consequentially, as presented in Table 8.2, relatively small changes in the attributes of seasonal precipitation patterns resulted in large increases in the blister rust index for the valley sites. Because observed historical precipitation at the lower elevation sites was found to limit blister rust dissemination, climate change that increases either the magnitude or the variability of summer precipitation could produce large increases in blister rust spread. In contrast, at the SNOTEL stations, where blister rust events already occur at a higher relative frequency, the changes produced under the three scenarios are small. However, because periods of larger blister rust transmission are associated with more extreme and more frequent rainfall events, changes in the way rainfall is aggregated within as well as between seasons, could also significantly impact

Table 8.2. Changes in the Blister Rust Index Value for Each Scenario vs. Observed Values by Station.

Station	Scenario 1 (%)	Scenario 2 (%)	Scenario 3 (%)
Snake River	94	211	72
Hebgen Dam	150	303	238
Lake Yellowstone	-35	15	18
Gros Ventre Summit	-16	-7	18

the rate at which whitebark pine is infected and its mortality over time.

An additional point concerns the effect of temperature increases on snowfall at higher elevations in mountainous regions, which is not captured by the previous analysis where increases in precipitation are applied uniformly to high and low elevation stations. Higher temperatures tend to produce greater snowfall at high elevations in proportion to low elevation increases due to orographic uplifting (P. Farnes, pers. comm. 1999). Such an effect may alter hydrological processes in ways that favor the spread of blister rust, through influencing soil moisture down slope and boosting *Ribes* populations, for example, or by providing an additional source of free moisture to facilitate the spread of blister rust at higher elevations. In addition, in the short term, higher temperatures and higher precipitation at high elevations may produce more-frequent years of good whitebark pine cone crops, making more seeds available to grizzly bears and other wildlife (P. Farnes, pers. comm. 1999).

Climate Change Impacts on Species Ranges

Because climate is a primary determinant of vegetation range at large spatial scales, as climate begins to shift, species will migrate in pursuit of optimal conditions. Generally, species are expected to expand their ranges toward the poles and upward in altitude, tracking rising temperatures and altered precipitation patterns, and to contract along their southern and lower elevational margins. Migration patterns will be more complex than simple range shifts, however, as climate is but one of many environmental variables shaping forest composition and structure (Bond and Richardson 1990,

Franklin et al. 1992). As temperature and precipitation adjust, physiological stress and altered competitive interactions will reduce species reestablishment along current range boundaries, while colonization of new areas is initiated by dispersal (Peters and Darling 1985, Tausch et al. 1993, Sanford, in this volume). The direct effect of higher ambient CO_2, an important plant macronutrient, will similarly influence biotic interactions.

Paleoclimatic studies of pollen records from lake and peat sediments and wood rat (*Neotoma* spp.) middens reveal that communities become disassociated in times of climatic change, and that species adjust individually according to their unique physiological tolerances and competitive abilities (Peters and Darling 1985, Brubaker 1986, Webb 1986, Franklin et al. 1992, Prentice 1992). Such studies additionally indicate that migration generally lags behind climate change by a few years to several centuries. Because of the long life span of most trees and the relatively greater phenological plasticity to climatic shifts of adult individuals compared to those required by seedlings for tree establishment and development, mature forests often remain extant long beyond the time when reproduction is possible (Brubaker 1986, Davis 1989b, Hanson et al. 1989, Huntley and Webb 1989, Bond and Richardson 1990, Davis and Zabinski 1991, Huntley 1991). When adjustments take place, they can occur over long time periods as a result of natural succession, or with relative rapidity when facilitated by disturbance.

As with range contraction, range expansion is similarly bounded by lags inherent to individual species, such as dispersal, reproduction, and population growth rates, or by additional indirect factors such as loss of mycorrhizal symbionts or increased herbivory associated with a new location (Davis and Botkin 1985, Hanson et al. 1989, Huntley and Webb 1989, Davis and Zabinski 1991). As the rate of climate change predicted to occur over the next century is of a magnitude several times greater than the average since the last glacial maximum, it is unlikely that species will be capable of successfully tracking such changes (Davis and Botkin 1985, Hanson et al. 1989). Moreover, the fragmented nature of our current landscapes will slow and in some cases prevent species migration to regions suitable for their establishment, as will natural barriers such as mountain ranges or large bodies of water (Peters and Darling 1985, Hanson et al. 1989). Narrowly endemic species, species confined to a small portion of their former range, poor dispersers, and highly fragmented species will be in particular danger of extinction when subjected to stochastic events and declining genetic variation (Peters and Darling 1985, Webb 1986, Davis and Zabinski 1991).

Populations growing at treelines on mountain peaks and at the margins of their range are faced with unique challenges to migration. As vegetation shifts northward, populations at the southern extreme will be among the first to face local extinction. Treeline populations will initially shrink due to the smaller area available with increasing elevation in mountainous regions, increasing their susceptibility to genetic and environmental ills (Peters and Darling 1985). Further challenges may be posed by the lag time required for soil development to reach a higher stage of abiotic resource availability in areas where vegetation was previously absent and nutrient availability is limited (Davis 1989a, 1989b, Huntley and Webb 1989, Ryan 1991). The harsh treeline topography will also reduce seedling establishment and success (Brubaker 1986).

Whitebark pine, as a treeline species found at its southern Rocky Mountain extreme in the GYE, will be subject to range shrinkage characteristic of treeline species and populations growing at southerly margins. In general however, climatic responses of Yellowstone taxa are likely to be spatially heterogeneous and unpredictable due to the topography of the region, which creates variable climatic regimes (Bartlein et al. 1997). In interpolating one set of GCM predictions onto the regional scale of the GYE, Bartlein and Whitlock hypothetically simulated the adjusted vegetation domains liable to result with CO_2 doubling and concluded that whitebark pine will become regionally scarce (Bartlein et al. 1997). Romme and Turner took a different approach and considered the range of possible responses to climate change and the resulting compositional reorganizations. In the three scenarios considered, substantial losses of whitebark pine in the GYE were predicted, and even regional extirpation was considered possible, independent of the blister rust threat (Romme and Turner 1991).

An examination of the autecology of *P. albicaulis* yields similar conclusions. Whitebark pine's upper altitudinal limit, ostensibly set by temperature, would be expected to rise under warmer conditions. At lower elevations, whitebark pine appears to be excluded from site occupancy by competition. On sites with high wind exposure and steep inclines, *P. albicaulis* and lodgepole pine occupy a higher percentage of stand area. At lower, warmer elevations however, whitebark pine is outcompeted by lodgepole pine (Mattson and Reinhart 1989, Romme and Turner 1991). On less exposed sites, successional processes lead to the replacement of whitebark pine over time by the relatively shade tolerant subalpine fir and Engelmann spruce (Mattson and Reinhart 1989). According to these observations, lodgepole pine would be expected to replace whitebark pine after disturbance as the climate warms, unless lodgepole pine is unable to disperse into areas of large burns.

Observations of Species Migrations from Paleoclimatic Studies

Stratigraphic history from within and outside the Yellowstone region, dating from the late Pleistocene, reveal the contribution of large-scale climatic forcings in shaping vegetation provinces over long time scales (Waddington and Wright 1974, Gennett and Baker 1986, 1989, Wigand and Nowak 1992, Whitlock et al. 1994, Whitlock and Bartlein 1991, 1997). Glacial retreat, changes in the cycle of solar irradiance, and atmospheric circulation patterns have set the altitudinal boundaries of major biomes from grassland to forest to tundra in the GYE, through governing changes in relative warmth and aridity (Waddington and Wright 1974, Baker 1989, Whitlock 1993). These impacts may be further influenced by the steep relief of the Yellowstone region, which shapes regional precipitation and wind patterns across the landscape (Whitlock 1993, Bartlein et al. 1997).

Such global controls have been found to govern the location of whitebark pine and associated species over the millennia within the GYE. Initial deglaciation in the GYE was soon followed by the establishment of forests of first *Picea engelmannii* by approximately 11,500 years before present, and then *P. albicaulis* and *A. lasiocarpa* as the region warmed and precipitation increased (Whitlock 1993). Subsequently, as conditions became warmer and drier by mid-Holocene, lodgepole pine entered the region and rose in elevation so that whitebark pine was confined to timberline sites (Baker 1976, Whitlock 1993). These patterns of vegetation distribution were overturned approximately 7000 years before present in the north where drier conditions prevailed, and 5000 years before present in the south where a cooler, moister period was initiated, and vegetation came to resemble its current configuration (Whitlock and Bartlein 1991, Whitlock 1993). Such findings point to the conclusion that if warmer and drier conditions than present again prevail, whitebark pine will be confined to isolated peaks across the region.

Impacts on Genetic Response of Fragmented Populations

As *P. albicaulis* becomes increasingly fragmented regionally and throughout its range due to ravages from blister rust and fire exclusion, environmental stress and competitive pressures due to climate change will further imperil whitebark pine's survival on the landscape. Over the long term, if whitebark pine continues to decline, its gene pool will shrink, subjecting the species to additional losses associated with low genetic variation (Williams and Kendall 1998). In fact, in anticipation of this consequence, measures are already under way to prevent whitebark pine genetic erosion through the collection of seeds (Williams and Kendall 1998).

In light of expected climate changes, range adjustment and genetic adaptation are *P. albicaulis*'s only options for survival. However, evidence from quaternary pollen records indicate that migration and not adaptation has been the standard response of tree species to climate change in the past (Huntley 1991). Because climate changes of the past occurred over time scales of much longer duration than those predicted for future changes, it is unlikely that future responses will differ. In fact the much faster rates of climate change anticipated for the coming centuries may lead to unprecedented extinction among a variety of species (Bradshaw and McNeilly 1991).

The consequences of increasing fragmentation include inbreeding depression due to loss of heterozygosity, outbreeding depression, and genetic drift (Tomback and Schuster 1992, McCauley 1993, Tausch et al. 1993). Whitebark pine may be increasingly subject to outcomes such as these given its existing fragmented population structure, as an occupant of remote, discontinuous mountain peaks. Whitebark pine's genetic character within and among populations is a function of the tree's dispersal by Clark's nutcracker (Brussard 1989). Because the nutcracker occasionally caches seeds at a considerable distance from source trees, some populations may be quite isolated. Among these more remote populations, gene flow within a population may be much more likely than among isolated populations, conditions in which genetic drift is likely (Brussard 1989). As a result, when migrant genes from distant sources are introduced into isolated populations, location-specific adaptive processes may be interrupted leading to outbreeding depression and rendering the population more prone to extinction (Ellstrand 1992).

It is also important to note that the ecological attributes of a species, such as dispersal mechanism and generation length, further constrain its possible evolutionary response (Bradshaw and McNeilly 1991). Although the field testing to confirm this hypothesis for whitebark pine is lacking, because of the way Clark's nutcracker harvests and plants whitebark pine seeds, biologists have speculated the tree may already be prone to low genetic diversity. Nutcrackers will harvest seeds simultaneously that originate from the same parent tree and then plant these seeds in multistemmed clusters. Under these circumstances, their proximity may increase the likelihood of cross-pollination within clusters and lowered genetic diversity of the population, leading to inbreeding (Tomback and Schuster 1992). As range shrinkage is initiated through climate change, this process would be enhanced. Inbreeding on even a moderate scale can have considerable deleterious effects on a population's prospects for success, through increasing its susceptibility to environmental stress and disease (Brussard 1989).

Additional implications of treeline migration for the GYE involve the loss of alpine species and possible heightened blister rust hazard. As treeline rises, gradual erosion of the alpine zone will occur. Obligate alpine species facing local extirpation over time include the arctic gentian (*Gentiana algida*), alpine chaenactis (*Chaenactis alpinus*), rosy finches (*Leucosticte* spp.), and the water pipit (*Anthus spinoletta*) (Romme and Turner 1991). Moreover, army cutworm moths aggregate among talus fields in large numbers each summer where they feed on the nectar of alpine flowers. These moths serve as an additional high protein food source for grizzly bears located in areas remote from human settlements (French et al. 1992, O'Brien and Lindzey 1994). If the alpine zone is subsumed, the moths will disappear and an additional food source of great value to bears may be lost. Lastly, because *Ribes* spp. are shrubs and can easily establish and quickly reproduce, their response to climate change will not be muted by the mechanisms that prevent rapid tree dispersal. Therefore, with warming, *Ribes* may increase in abundance at the same elevations as whitebark pine, facilitating *P. albicaulis* infection by blister rust.

Disturbances can also speed the response times of ecosystems to climate change by removing long-lived species from the landscape. In fact, the instability of ecosystems as well as forest susceptibility to disturbance may be heightened by climate change. Species already under physiological stress as a result of climate change will exhibit increased susceptibility to disease and insect outbreaks (Davis and Zabinski 1991, Franklin et al. 1992, Tausch et al. 1993). Additionally, a changing climate may increase vegetation flammability, produce a greater frequency of storms (Brubaker 1986, Prentice 1992), and facilitate pathogen dispersal and the spread of invasive species.

Climate Change and Fire Regime Changes

Climate change is expected to affect forested ecosystems through changes in disturbance regimes, including the frequency, intensity, and extent of fire, windstorms, flooding, avalanches, pathogen outbreaks, and more (Attiwill 1994). Specifically, the establishment of a more intense fire regime may diminish whitebark pine populations in the GYE. The frequency and intensity of fire within an ecosystem are governed by climate, supply of ignition sources, and accumulated forest fuels, which vary in accordance with successional status (Renkin and Despain 1991, Romme and Despain 1989a, Sapsis and Martin 1993, Attiwill 1994). The stand-replacing fires that burned through Yellowstone National Park in the summer of 1988 and affected approximately 400,000 hectares (1 million acres) are an example of what results when these criteria are all optimally suited for fire occurrence. A tendency toward drier conditions, as projected by general circulation models, may lead to altered fire regimes in which larger fires occur with greater frequency.

Fire Ecology of Whitebark Pine Forests

The northern Rocky Mountain whitebark pine forests currently support fire regimes with typical fire return intervals of 50 to 300 years (Arno 1986). Low intensity fires tend to favor *P. albicaulis* in stands where it is seral, as it exhibits relative fire resistance compared to stand cohorts (Bradley et al. 1992, Morgan et al. 1992). Thus fire suppression practices throughout the Northwest, in operation for much of this century, are blamed in part for whitebark pine declines (Kendall and Arno 1990, Keane and Arno 1993, Keane et al. 1994). Fires of higher intensity can kill the relatively thin-barked whitebark pine through excessive heating of the bole or roots, or when flames migrate to foliage or spread through tree crowns (Bradley et al. 1992, Morgan et al. 1992). Yet, even surface fires are generally sufficient to kill whitebark pine trees younger than 80 years of age if they have not grown beyond the lethal scorch height (Keane et al. 1990). In stands where subalpine fir is also present, the tree's low-trailing branches or developing fir understory can serve as a fuel ladder that facilitates the spread of flames to the forest canopy (Bradley et al. 1992).

Stand replacing fires are also thought to advance *P. albicaulis* over associated species, which has an advantage in pioneering recently burned sites, through its association with Clark's nutcracker (Morgan et al. 1992). In forests of pure whitebark pine, however, large fires are currently rare. The combination of snow retention, which can last well into summer at the high elevations where these stands occur, their open parklike structure, and the slow rate of fuel accumulation contribute to this effect. Still, when the climatic conditions for large fires are met, these forests will burn freely, as is evidenced by the almost 27,500 hectares of pure whitebark pine forests that burned in the 1988 fire season in Yellowstone National Park (Renkin and Despain 1991, Bradley et al. 1992, P. Farnes, pers. comm. 1999).

A higher frequency of drought occurrence, coupled with high winds, could lead to fire regimes characterized by more frequent fire and endanger whitebark pine persistence on the landscape. Because significant cone production of *P. albicaulis* generally does not occur until trees attain 100 years of age, a conversion to younger age classes as a result of more frequent and intense fires may cause seed sources to dwindle. A compromised seed source could lead to declines in the heterozygosity of individual whitebark pine populations, increasing the tree's vulnerability to blister rust and other pathogens, and reducing seed production for bears and other wildlife. Moreover, whitebark pine succession in the wake of a fire can be slow despite early seedling establishment, as the microclimatic conditions that further regeneration may occur infrequently (Bradley et al. 1992, Mattson and Reinhart 1989). However, because whitebark pine is bird-dispersed, trajectories of decline or recovery are not entirely clear. Depending on the character of resulting fire regimes under climate change, and the future behavior of blister rust, in the short or long term, whitebark pine may enjoy some advantage over other tree species in regenerating in burned areas.

Mechanisms of Increasing Fire Frequency

Warmer mean global temperatures, higher precipitation, greater frequency of extreme weather events, increased incidence of convective thunderstorms, and decreases in temperature variability are all features of predicted climate change. If such predictions hold true, fire regimes with shorter fire return intervals will certainly occur in some ecoystems. First, fire weather will be enhanced through temperature increases and concomitant increases in potential evapotranspiration. Second, trends toward greater storminess will raise the number of lightning ignitions that may occur over a given season (Davis 1989a, Clark 1990, Franklin et al. 1992, Price and Rind 1994a, 1994b). Additional indirect effects on fire regimes may come into play through unanticipated mechanisms such as a lengthening of the fire season, elevated CO_2 concentration, or higher autocorrelation of daily temperature leading to prolonged droughts (Flannigan and Wagner 1990, Balling 1996).

General circulation models forecast that global temperature increases of 1.4 °C to 5.8°C will accompany a doubling of atmospheric CO_2 (IPCC 2001b). With higher temperatures, the severity and duration of drought will be enhanced, impacting fire regimes (Balling 1996). The sequence of atmospheric and surface interactions promoting aridity will proceed as follows. Rising temperatures will lead to increases in the water holding capacity of the atmosphere, which in turn will exert higher moisture demands on the hydrosphere. These demands are likely to be of a magnitude to more than compensate for expected increases in precipitation, as atmospheric water-holding capacity increases nonlinearly with temperature (Overpeck et al. 1990, Rind et al. 1990, Price and Rind 1994a). Consequently, evapotranspiration will increase, and soil water content will decrease at all but the wettest sites (Franklin et al. 1992). This effect may be at least partially mediated by the tendency for snowfall to increase at higher elevations with increasing temperature (P. Farnes, pers. comm. 1999). Yet once the snowpack has melted, the generally low water-holding capacity of the soils that support whitebark pine communities will contribute to the rapidity in which drought conditions develop in these forests (Farnes 1989). Desiccation of forest fuels will likewise occur.

Similar mechanisms are expected to contribute to more frequent lightning ignitions and higher percentages of area burned. Elevated temperatures coupled with high concentrations of atmospheric water vapor can generate large convective thunderstorms. Higher wind speeds and lightning activity are expected to accompany such storms, increasing forest ignition and area burned (Franklin et al. 1992, Torn and Fried 1992). Focusing on the northern Rocky Mountains specifically, a 30% increase in wild fires was predicted for the region under doubled CO_2 conditions (Price and Rind 1994a, 1994b).

Lastly, climate change can be expected to influence the flammability and accumulation of forest fuels. Physiological stress associated with climate change may speed vegetation mortality, accelerating the rate of fuel buildup. In addition, rising CO_2 slows decomposition rates of the litter of some species by increasing their carbon to nitrogen ratios, therefore prolonging the forest floor residence time of decaying matter. Furthermore, these processes alter forest structure and chemistry, affecting their flammability characteristics. Taken together, these factors are likely to increase overall wildfire risk.

Region-Specific Findings Concerning Climate Variability, Historical Climatic Trends, and Fire Regimes

Two studies conducted by Balling et al. (1992a, b) and focusing directly on the GYE, concluded that trends toward both warming and increased drought have already occurred in this region. Through examination of weather station data, a mean annual increase in summer temperatures of 0.87°C over the last century was detected (Balling et al. 1992a, 1992b). In addition, employing the Palmer Drought Severity Index (PDSI), Balling et al. (1992a,b) were able to correlate years of large fires with periods of severe to extreme drought, confirming the relationship between assumed fire-related climate variables and on the ground burns. When the index was computed seasonally for the GYE, statistically significant trends toward increasing drought were discovered for both the summer fire season and the antecedent season spanning the period from January to June of each year. This trend occurred despite concurrent increases in summer precipitation (Balling et al. 1992a, 1992b). Such a scenario may also arise when precipitation events are concentrated into fewer, more intense bouts, with long stretches of dry weather in between. An additional century-long study employing the PDSI revealed the five-state area including Montana, North Dakota, Colorado, Nebraska, and Wyoming had exhibited the largest nationwide trend toward increased aridity. Of these states, Wyoming displayed the most significant trend (Balling 1996). This finding is consistent with climate change predictions for warming and drying of mid-latitude continental interiors (Overpeck et al. 1990, Rind et al. 1990).

The PDSI is a measure developed by the National Weather Service for estimating drought (Rind et al. 1990). The index is computed monthly for a particular location, and incorporates temperature, potential evapotranspiration, and precipitation values to determine relative moisture levels in relation to historical means. Values predominantly fall within the range from 6.0 to –6.0, with values of –4.0 corresponding to an extreme drought, and values of 4.0 marking an unusually wet year. An examination of drought trends conducted as part of this study for the Yellowstone National Park area also revealed a tendency toward increased drought in recent years. This trend is illustrated from 1895 through 2000 for the months of January and July (Figs. 8.7a and 8.7b). By establishing thresholds at 2 and –2, especially in January, it becomes evident that at least from the mid-1970s, and possibly earlier, there have been fewer very wet years, and more years of moderate to extreme drought. In July, with the exception of 1997, a wet year, a similar trend is evident. A longer period of observation and PDSI values computed for additional weather stations after the 1970s might confirm or reject this trend.

Paleoclimatic investigations lend additional evidence to forecasts of increased wildfire hazard with rising temperatures. The link

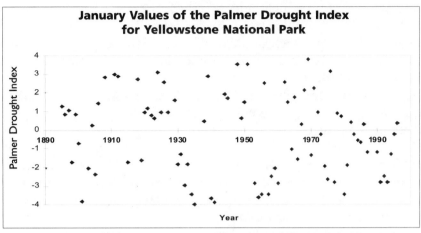

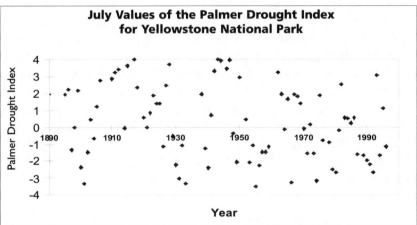

Figure 8.7. (A) January and (B) July values of the Palmer Drought Severity Index for the years 1895 to 1996 for Yellowstone National Park.

between climate and fire cycles is supported by stratigraphic methods examining the abundance of charcoal particles in varved lake sediment layers over a single or multiple drainages, and dendrochronological studies of fire-scarred trees that allow for approximate regional fire history reconstruction (Patterson and Backman 1988, Clark 1990). In their investigation of the lake sediment record from several small lakes in Yellowstone National Park, and the West Thumb portion of Lake Yellowstone, Millspaugh and Whitlock (1995) discovered intervals of large fires mimicked recorded climate fluctuations (Millspaugh and Whitlock 1995). The findings of Millspaugh and Whitlock (1995) correspond well to those determined from dendrochronological investigations conducted by Romme and Despain for the same time period (Romme and Despain 1989b).

Additional and unpredictable hazards can arise in the GYE through a variety of fire-mediated mechanisms that may be unpredictable. One such potential consequence considers an increase in blister rust susceptibility for the region through a proliferation of *Ribes* bushes in recently burned areas. Post-fire succession can favor the establishment of herbaceous cover and species that reproduce vegetatively from root systems unaffected by fire (Turner et al. 1997). *Ribes* seeds can remain viable for up to 200 years in cool moist duff until liberated by disturbance. *Ribes lacustre,* capable of establishing on hillsides in addition to valley bottoms, is able to reproduce vegetatively (Hagle et al. 1989). In the wake of disturbance, *Ribes* population presence generally peaks within 3 to 4 years. In the event that this limited period of increased *Ribes* population presence corresponds to a period of climatic conditions especially conducive to blister rust spread, large increases in infection levels of whitebark pine could result. *Ribes* presence in the GYE may currently be near historical highs as a result of the 1988 fires. More research is needed to examine the population-level changes in *Ribes* in relationship to whitebark pine and climate change.

Conclusion

This study highlights three potential mechanisms of species loss as a result of global climate change. Acting singly or cumulatively, they may accelerate the decline of an important tree species, whitebark pine, in the Greater Yellowstone Ecosystem. These mechanisms are: modified patterns of spread of harmful pathogens, adjustment of species ranges to track locations of preferred climate, and changes in species composition as a result of altered fire regimes. In highlighting the ecosystem reverberations of such changes, it becomes clear that both dependent species and ecosystem processes are affected in addition to the direct effects to whitebark pine. Moreover, this discussion points to the conclusion that climate change may be the agent of widespread biodiversity decline.

Specifically, this study reveals that small changes in climate can produce relatively large consequences for whitebark pine, for grizzly bears, and for the ecosystem they inhabit. In particular, small increments of precipitation and a tendency toward more extreme precipitation events could hasten the widespread infection of *P. albicaulis* by the blister rust fungus; an outcome that may already be on the horizon without implicating climate change. In addition, the warming and enhanced evapotranspiration predicted for the region would render the GYE an increasingly inhospitable landscape for whitebark pine, confining the tree to ever-higher elevations and facilitating its loss through fire.

Because the boundaries of the GYE are drawn and enforced according to political priorities, a fall in grizzly bear numbers may track a decline in whitebark pine. Such a consequence however is not inevitable. Grizzly bear populations could be increased through changes in a number of important policies governing human and grizzly bear interactions. These include changes in human recreational and other land uses, greater educational efforts regarding the seasonal distribution of bears across the landscape, and the establishment of corridors linking the GYE with other ecosystems capable of supporting grizzly bears (e.g., the Selkirk Mountains in Idaho), among others. Alternatively, if grizzly bear populations diminish, they will be rendered vulnerable to genetic drift, the impacts of stochastic events and possible extirpation over long time scales.

It is my hope that this study clearly reveals the dangers facing whitebark pine, the ecosystem-level functional processes it upholds, and especially the species that rely on its seed (e.g., grizzly bears). If rigorous policies are duly implemented, the population contractions that have already been recorded may not be a herald of approaching and accelerating species loss, but an impetus to further conservation efforts. What I believe this study highlights best is the complexity of future interactions that could result from global environmental change and the elusivity of discerning their outcomes. Moreover, it underscores that the consequences of the range of changes taking place will have direct as well as indirect components and are likely to be cumulative in nature. As demonstrated here, a changing climate may impart direct physiological stress, alter disturbance frequencies, initiate migrations, modify nutrient cycling, displace native species, usher in invaders, launch competitions, and, through additional untold mechanisms, impact our species and ecosystems.

The ostensible links of the system at hand, from climate to fungus to tree to squirrel to bear, reveal the complex interrelationships of our natural systems. What they hide is the role of humans in this ecological equation. In actuality, the human presence is on either end of the spectrum, from consumer of fossil fuels to bear predator, and everywhere in between. As the agent of pathogen introduction, fire suppression, landscape fragmentation, and national park establishment, and as the developer of its boundaries, we have impaired the self-sustaining functions of this natural system so that it is now reliant on our management. It may still be possible to conserve the whitebark pine tree and the complex of fauna that depend on its persistence. Let us hope we are equal to the challenge.

Acknowledgments

I am deeply grateful to Kristiina Vogt, Daniel Reinhart, and Xuhui Lee for their generous support and guidance over the course of this study. In addition, I am indebted to Jonathan Smith, Phil Farnes, and Martin Dubrovsky for their invaluable creative input and advice. Warm thanks are also in order for Patty Glick, Raymond Hoff, Vanessa Johnson, Kate Kendall, Merle Koteen, Robert Koteen, David Mattson, Larry McDaniel, Linda Mearns, Roy Renkin, Stephen Schneider, Tom Siccama, Peter Thornton, Louisa Willcox, two anonymous reviewers, and the National Wildlife Federation for their financial support of this research.

Literature Cited

Arno, S. F. 1986. Whitebark pine cone crops: A diminishing source of wildlife food? *Western Journal of Applied Forestry* 1: 92–94.

Arno, S. F., and R. J. Hoff. 1989. Silvics of whitebark pine (*Pinus albicaulis*). Ogden, Utah: U.S. Dept. of Agriculture, Forest Service.

Arno, S. F., and T. Weaver. 1989. Whitebark pine community types and their patterns on the landscape. Pages 97–105 in W. C. Schmidt and K. J. McDonald, eds., *Symposium on Whitebark Pine Ecosystems: Ecology and Management of a High Mountain Resource*. Ogden, Utah: USDA, Forest Service, Intermountain Research Station.

Attiwill, P. M. 1994. The disturbance of forest ecosystems: The ecological basis for conservative management. *Forest Ecology and Management* 63: 247–300.

Auclair, A. N. D., J. T. Lill, and C. Revenga. 1996. The role of climate variability and global warming in the dieback of Northern Hardwoods. *Water, Air, and Soil Pollution* 91: 163–186.

Ayres, P. G., T. S. Gunasekera, M. S. Rasanayagam, and N. D. Paul. 1994. Effects of UV-B radiation (280–320 nm) on foliar saprotrophs and pathogens. Pages 90–101 in J. C. Frankland, N. Magan, and G. M. Gadd, eds., *Fungi and Environmental Change: Symposium of the British Mycological Society*. Cambridge: Cambridge University Press.

Baker, R. G. 1976. Late Quaternary Vegetation History of the Yellowstone Lake Basin, Wyoming. U.S. Geological Survey Professional Paper 729-E 168: 1449–1450.

Baker, R. G. 1989. Late Quaternary history of whitebark pine in the Rocky Mountains. Pages 40–47 in W. C. Schmidt and K. J. McDonald, eds., *Symposium on Whitebark Pine Ecosystems: Ecology and Management of a High Mountain Resource*. Ogden, Utah: USDA, Forest Service, Intermountain Research Station.

Balling, R. C. 1996. Century-long variations in United States drought severity. *Agricultural and Forest Meteorology* 82: 293–299.

Balling, R. C., G. A. Meyer, and S. G. Wells. 1992a. Climate change in Yellowstone National Park: Is the drought-related risk of wildfires increasing? *Climatic Change* 22: 35–45.

Balling, R. C., G. A. Meyer, and S. G. Wells. 1992b. Relation of surface climate and burned area in Yellowstone National Park. *Agricultural and Forest Meteorology* 60: 285–293.

Bartlein, P. J., C. Whitlock, and S. L. Shafer. 1997. Future climate in the Yellowstone National Park region and its potential impact on vegetation. *Conservation Biology* 11: 782–792.

Baskin, Y. 1998. Home on the range. *Bioscience* 48: 245–251.

Bedwell, J. L., and T. W. Childs. 1943. Susceptibility of Whitebark Pine to Blister Rust in the Pacific Northwest. *Journal of Forestry* 41: 904–912.

Bingham, R. T. 1972. Taxonomy, crossability and relative blister rust resistance of 5-needle white pines. *Biology of Rust Resistance in Forest Trees*, Miscellaneous Publications, 1221: 271–280. USDA, Forest Service.

Bond, W. J., and D. M. Richardson. 1990. What can we learn from extinctions and invasions about the effects of climate change? *South African Journal of Science* 86: 429–433.

Bradley, R. S., H. F. Diaz, J. K. Eischeid, P. D. Jones, P. M. Kelly, and C. M. Goodess. 1987. Precipitation fluctuations over Northern Hemisphere land areas since the mid-19th century. *Science* 237: 171–175.

Bradley, A. F., N. V. Noste, and W. C. Fischer. 1992. Fire ecology of forest habitat types of eastern Idaho, western Wyoming. Ogden, Utah: USDA, Forest Service, Intermountain Research Station. 128 pp.

Bradshaw, A. D., and T. McNeilly. 1991. Evolutionary response to global climate change. *Annals of Botany* 67: 5–14.

Brasier, C. M. 1996. Phytophthora cinnamomi and oak decline in southern Europe. Environmental constraints including climate change. *Annales des Sciences Forestieres* 53: 347–358.

Brasier, C. M., F. Robredo, and J. F. P. Ferraz. 1993. Evidence for *Phytopthora cinnamoni* involvement in Iberian oak decline. *Plant Pathology* 42: 140–145.

Brousseau, D., J. C. Guedes, C. A. Lakatos, G. R. Lecleir, and R. I. Pinsonneault. 1998. A comprehensive survey of Long Island Sound oysters for the presence of the parasite, *Perkinsus marinus. Journal of Shellfish Research* 17: 255–258.

Brown, D., and D. Graham. 1967. White pine blister rust survey in Wyoming, Idaho, and Utah. Missoula, Mont.: USDA, Forest Service, Northern Region, State and Private Forestry.

Brubaker, L. B. 1986. Responses of Tree Populations to Climatic Change. *Vegetatio* 67: 119–130.

Brussard, P. F. 1989. The role of genetic diversity in whitebark pine conservation. Pages 315–318 in W. C. Schmidt and K. J. McDonald, eds., *Symposium on Whitebark Pine Ecosystems: Ecology and Management of a High Mountain Resource.* Ogden, Utah: USDA, Forest Service, Intermountain Research Station.

Cammell, M. E., and J. D. Knight. 1992. Effects of climatic change on the population dynamics of crop pests. *Advances in Ecological Research* 22: 117–161.

Carlson, C. 1978. Noneffectiveness of *Ribes* as a control of white pine blister rust in Yellowstone National Park. Missoula, Mont.: USDA , Forest Service, Northern Region, State and Private Forestry.

Castello, J. D., D. J. Leopold, and P. J. Smallidge. 1995. Pathogens, patterns, and processes in forested ecosystems. *BioScience* 45: 16–24.

Clark, J. S. 1990. Fire and climate change during the last 750 years in northwestern Minnesota. *Ecological Monographs* 60: 135–159.

Cook, T., M. Folli, J. Klinck, S. Ford, and J. Miller. 1998. The relationship between increasing sea-surface temperature and the northward spread of *Perkinsus marinus* (dermo) disease epizootics in oysters. *Estuarine, Coastal and Shelf Science* 46: 587–597.

Craighead, J. J., J. S. Sumner, and J. A. Mitchell. 1995. *The Grizzly Bears of Yellowstone, Their Ecology in the Yellowstone Ecosystem, 1959–1992.* Washington, D.C.: Island Press.

Crooks, J. A., and M. E. Soule. 1999. Lag times in population explosions of exotic species: Causes and implications in O. T. Sandlund, P. J. Schei, and A. Viken, eds., *Invasive Species and Biodiversity Management.* Dordrecht: Kluwer.

Davis, M. B. 1989a. Insights from paleoecology on global change. *Bulletin of the Ecological Society of America* 70: 222–228.

————. 1989b. Lags in vegetation response to greenhouse warming. *Climatic Change* 15: 75–82.

Davis, M. B., and D. B. Botkin. 1985. Sensitivity of cool temperature forests and their fossil pollen record to rapid temperature change. *Quaternary Research* 23: 327–340.

Davis, M. B., and C. Zabinski. 1991. Changes in geographical range resulting from greenhouse warming: Effects on biodiversity in forests in Peters and Lovejoy, eds., *Consequences of Greenhouse Warming to Biodiversity.* New Haven: Yale University Press.

Despain, D. G. 1987. The two climates of Yellowstone National Park. *Proceedings of Montana Academy of Sciences (Biological Sciences)* 47: 11–19.

Despain, D. G. 1991. *Yellowstone Vegetation: Consequences of Environment and History.* Boulder: Roberts Rinehart.

Dirks, R. A., and B. E. Martner. 1982. *The Climate of Yellowstone and Grand Teton National Parks.* Washington, D.C.: U.S. Department of the Interior, NPS.

Dobson, A., and M. Meagher. 1996. The population dynamics of brucellosis in the Yellowstone National Park. *Ecology* 77: 1026–1036.

Douglas, A. V., and C. W. Stockton. 1975. *Long-Term Reconstruction of Seasonal Temperature and Precipitation in the Yellowstone National Park Region Using Dendroclimatic Techniques.* Tucson: University of Arizona, Laboratory of Tree-Ring Research.

Dubrovsky, M. 1997. Creating daily weather series with use of the weather generator. *Environmetrics* 8: 409–424.

Ellstrand, N. C. 1992. Gene flow by pollen: Implications for plant conservation genetics. *Oikos* 63: 77–86.

Farnes, P. E. 1989. SNOTEL and snow course data: Describing the hydrology of whitebark pine ecosystems. Pages 302–304 in W. C. Schmidt and K. J. McDonald, eds., *Symposium on Whitebark Pine Ecosystems: Ecology and Management of a High Mountain Resource.* Ogden, Utah: USDA. Forest Service, Intermountain Research Station.

Farnes, P. E., C. Heydon, and K. Hansen. 1998. Snowpack distribution in Grand Teton National Park, Wyoming. Complete data files available from Department of Earth Sciences, Montana State University, Bozeman, MT 59717.

Flannigan, M. D., and C. E. V. Wagner. 1990. Climate change and wildfire in Canada. *Canadian Journal of Forest Research* 21: 66–72.

Ford, S. E. 1996. Range extension by the oyster parasite *Perkinsus marinus* into the northeastern United States: Response to climate change? *Journal of Shellfish Research* 15: 45–56.

Forrest, D. M., and B. Gushulak. 1997. Emerging pathogens: Threat and opportunity. *Perspectives in Biology and Medicine* 40: 119–124.

Franklin, J. F., F. J. Swanson, M. E. Harmon, and D. A. Perry. 1992. Effects of global climatic change on forests in northwestern North America. Pages 244–257 in R. L. Peters and T. E. Lovejoy, eds., *Global Warming and Biological Diversity*. New Haven: Yale University Press.

French, S. P., M. G. French, and R. R. Knight. 1994. Grizzly bear use of army cutworm moths in the Yellowstone Ecosystem. Pages 389–400 in Ninth International Conference for Bear Research and Management.

Gennett, J. A., and R. G. Baker. 1986. A late Quaternary pollen sequence from Blacktail Pond Yellowstone National Park Wyoming USA. *Palynology* 10: 61–72.

Hagle, S. K., G. I. McDonald, and E. A. Norby. 1989. Western white pine and blister rust in northern Idaho and western Montana: Alternatives for integrated management. Ogden, Utah: USDA, Forest Service, Northern Region, State and Private Forestry.

Hanson, J. S., G. P. Malanson, and M. P. Armstrong. 1989. Spatial constraints on the response of forest communities to climate change. Pages 1–23 in G. P. Malanson, ed., *Natural Areas Facing Climate Change*. The Hague: SPB Academic.

Hennessy, K., and A. B. Pittock. 1995. Greenhouse warming and threshold temperature events in Victoria, Australia. *International Journal of Climatology* 15: 591–612.

Hirt, R. R. 1935. Observations on the Production and Germination of Sporidia of *Cronartium ribicola*. Syracuse: NY State College of Forestry, Syracuse University.

———. 1942. The Relation of Certain Meterological Factors to the Infection of Eastern White Pine by the Blister-Rust Fungus. Syracuse: NY State College of Forestry, Syracuse University.

Hoff, R., and S. Hagle. 1989. Diseases of whitebark pine with special emphasis on white pine blister rust. Pages 179–190 in W. C. Schmidt and K. J. McDonald, eds. *Symposium on Whitebark Pine Ecosystems: Ecology and Management of a High Mountain Resource*. Ogden, Utah: USDA Forest Service, Intermountain Research Station.

Hungerford, R. D., R. R. Nemani, S. W. Running, and J. C. Coughlan. 1989. MTCLIM: A mountain microclimate simulation model. Ogden, Utah: USDA, Forest Service, Intermountain Research Station. 53 pp.

Hunt, R. S. 1983. White pine blister rust in British Columbia. Victoria, B.C.: Pacific Forest Research Center.

Huntley, B. 1991. How plants respond to climate change: Migration rates, individualism and the consequences for plant communities. *Annals of Botany* 67: 15–22.

Huntley, B., and T. Webb. 1989. Migration: Species' response to climatic variations caused by changes in the earth's orbit. *Journal of Biogeography* 16: 5–19.

Hutchins, H. E., and R. M. Lanner. 1982. The central role of Clark's nutcracker in the dispersal and establishment of whitebark pine. *Oecologia* 55: 192–201.

Idso, S. B., and R. C. Balling. 1991. Recent trends in United States precipitation. *Environmental Conservation* 18: 71–73.

Intergovernmental Panel on Climate Change (IPCC). 1996. *Climate Change 1995: The Science of Climate Change*, eds., J. T. Houghton, L. G. Meira Filho, B. A. Callander, N. Harris, A. Kattenberg, and K. Maskell. Contribution of working group I to the second assessment report of the IPCC. Cambridge: Cambridge University Press.

———. 2001a. *Climate Change 2001: The Scientific Basis*, ed. J. T. Houghton, Y. Ding, D. J. Griggs, M. Noguer, P. J. van der Linden, and D. Xiaosu. Contribution of working group I to the third assessment report of the Intergovernmental Panel on Climate Change. Cambridge: Cambridge University Press.

———.2001b. *Climate Change 2001: Impacts, Adaptation, and Vulnerability*, ed. J. J. McCarthy, O. F. Canziani, N. A. Leary, D. J. Dokken, and K. S. White. Contribution of working group II to the third assessment report of the Intergovernmental Panel on Climate Change. Cambridge: Cambridge University Press.

Ives, J. D., and K. J. Hansen-Bristow. 1983. Stability and instability of natural and modified upper timberline landscapes in the Rocky Mountains, USA. *Mountain Research and Development* 3: 149–155.

Karl, T. R., R. W. Knight, D. R. Easterling, and R. G. Quayle. 1996. Indices of climate change for the United States. *Bulletin of the American Meteorological Society* 77: 279–292.

Katz, R. W. 1996. Use of conditional models to generate climate change scenarios. *Climatic Change* 32: 237–255.

Katz, R. W., and B. G. Brown. 1992. Extreme events in a changing climate: Variability is more important than averages. *Climatic Change* 21: 289–302.

Keane, R. E., and S. F. Arno. 1993. Rapid decline of whitebark pine in western Montana: Evidence from 20-year remeasurements. *Western Journal of Applied Forestry* 8: 44–70.

Keane, R. E., S. F. Arno, J. K. Brown, and D. F. Tomback. 1990. Modelling stand dynamics in whitebark pine (*Pinus albicaulis*) forests. *Ecological Modelling* 51: 73–96.

Keane, R. E., P. Morgan, and J. P. Menakis. 1994. Landscape assessment of the decline of whitebark pine (*Pinus albicaulis*) in the Bob Marshall Wilderness Complex, Montana, USA. *Northwest Science* 68: 213–229.

Kendall, K. C. 1983. Use of pine nuts by black and grizzly bears in the Yellowstone area. International Conference on Bear Research and Management 5: 166–173.

Kendall, K. C. 1995. Status and Distribution of Whitebark Pine in Glacier, Grand Teton, and Yellowstone National Parks. US Geological Survey, Glacier National Park.

Kendall, K. C. 1996a. Whitebark pine ecosystems: Status and trends. Pages 88–97 in R. Walker and R. Kubian, eds., Fourth Annual Alberta–BC Intermountain Forest Health Workshop. Radium Hot Springs, British Columbia: Parks Canada.

———. 1996b. Whitebark pine health in northern Rockies National Park ecosystems. Nutcracker Notes 7. Available online at http://www.nrmsc. usgs.gov/nutnotes/number7.htm.

———. 1999. Whitebark Pine Communities Home Page. US Geological Survey. Available online at http://www.nrmsc.usgs.gov/research/whitebar.htm.

Kendall, K. C., and S. F. Arno. 1990. Whitebark pine: An important but endangered wildlife resource. Pages 264–273 in W. C. Schmidt and K. J. McDonald, eds., *Symposium on Whitebark Pine Ecosystems: Ecology and Management of a High Mountain Resource.* Ogden, Utah: USDA Forest Service, Intermountain Research Station.

Kimball, J. S., S. W. Running, and R. Nemani. 1997. An improved method for estimating surface humidity from daily minimum temperature. *Agricultural and Forest Meteorology* 85: 87–98.

Kimmey, J. W. 1935. Susceptibility of principal *Ribes* of southern Oregon to white pine blister rust. *Journal of Forestry* 33: 52–56.

———. 1938. Susceptibility of *Ribes* to *Cronartium ribicola* in the West. *Journal of Forestry* 36: 312–320.

Kimmey, J. W., and J. L. Mielke. 1944. Susceptibility of white pine blister rust of *Ribes cereum* and some other *Ribes* associated with sugar pine in California. *Journal of Forestry* 42: 752–756.

Kimmey, J. W., and W. W. Wagener. 1961. Spread of white pine blister rust from *Ribes* to sugar pine in California and Oregon. Washington, D.C.: USDA Forest Service.

Kinloch, B. B., and D. Dulitz. 1990. *White Pine Blister Rust at Mountain Home Demonstration State Forest: A Case Study for the Epidemic and Prospects for Genetic Control.* Berkeley, Calif.: USDA Forest Service, Pacific Southwest Research Station.

Kinloch, B. B., R. D. Westfall, E. E. White, M. A. Gitzendanner, G. E. Dupper, B. M. Foord, and P. D. Hodgkiss. 1998. Genetics of *Cronartium ribicola.* IV. Population structure in western North America. *Canadian Journal of Botany* 76: 91–98.

Knight, R. R., B. M. Blanchard, and L. L. Eberhardt. 1988. Mortality patterns and population sinks for Yellowstone grizzly bears, 1973–1985. *Wildlife Society Bulletin* 16: 121–125.

Krebill, R. G., and R. J. Hoff. 1995. Update on *Cronartium ribicola* in *Pinus albicaulis* in the Rocky Mountains, USA. Pages 119–126 in S. Kaneko, ed., *Proceedings of the Fourth IUFRO Rusts of Pines Working Party Confer-*

ence. (2–7 October 1994, University of Tsukuba) Tsukuba, Japan: International Union of Forest Research Organizations.

Lachmund, H. G. 1926. Studies of white pine blister rust in the West. *Journal of Forestry* 24: 874–884.

Lachmund, H. G. 1934. Seasonal development of *Ribes* in relation to spread of *Cronartium ribicola* in the Pacific Northwest. *Journal of Agricultural Research* 49: 93–114.

Lanner, R. M., and B. K. Gilbert. 1992. Nutritive value of whitebark pine seeds and the question of their variable dormancy. Pages 206–211 in W. C. Schmidt, ed., *Proceedings: International Workshop on Subalpine Stone Pines and Their Environment: The Status of Our Knowledge*. Washington, D.C.: USDA, Forest Service.

Lantz, L. 1995. Written Findings of the Washington State Noxious Weed Control Board. Kent, Wash.: Washington State Noxious Weed Control Board. 15 pp.

Liang, X. Z., W. C. Wang, and M. P. Dudek. 1995. Interannual variability of regional climate and its change due to the greenhouse effect. *Global and Planetary Change* 10: 217–238.

Lloyd, M. G., C. A. O'Dell, and H. J. Wells. 1959. A study of spore dispersion by use of silver-iodide particles. *Bulletin of American Meteorological Society* 40: 305–309.

Lonsdale, D., and J. N. Gibbs. 1994. Effects of climate change on fungal diseases of trees. Pages 1–19 in J. C. Frankland, N. Magan, and G. M. Gadd, eds., *Fungi and Environmental Change: Symposium of the British Mycological Society*. Cambridge: Cambridge University Press.

Magan, N., and E. S. Baxter. 1996. Effects of increased CO_2 concentration and temperature on the phyllosphere mycoflora of winter wheat flag leaves during ripening. *Annals of Applied Biology* 129: 189–195.

Magan, N., M. K. Smith, and I. A. Kirkwood. 1994. Effects of atmospheric pollutants on phyllosphere and endophytic fungi. Pages 90–101 in J. C. Frankland, N. Magan, and G. M. Gadd, eds., *Fungi and Environmental Change: Symposium of the British Mycological Society*. Cambridge: Cambridge University Press.

Maloy, O. C. 1997. White pine blister rust control in North America: A case history. *Annual Review of Phytopathology* 35: 87–109.

Manning, W. J., and A. V. Tiedemann. 1995. Climate change: Potential effects of increased atmospheric carbon dioxide (CO_2), ozone (O3), and ultraviolet-B (UV-B) radiation on plant diseases. *Environmental Pollution* 88: 219–245.

Mattes, H. 1992. Coevolution aspects of stone pines and nutcrackers. Pages 31–35 in W. C. Schmidt, ed., *International Workshop on Subalpine Stone Pines and Their Environment: The Status of Our Knowledge*. Washington, D.C.: USDA, Forest Service.

Mattson, D. J. 1990. Human impacts on bear habitat use. Pages 33–56 in Eighth International Conference on Bear Research and Management. British Columbia, Canada. International Association for Bear Research and Management.

———. 1997. Changes in mortality of Yellowstone's grizzly bears. Pages 1–10. International Conference on Bear Research and Management. Grenoble, France. International Association for Bear Research and Management.

Mattson, D. J., and C. Jonkel. 1990. Stone pines and bears. Pages 223–236 in W. C. Schmidt and K. J. McDonald, eds., *Symposium on Whitebark Pine Ecosystems: Ecology and Management of a High Mountain Resource*. Ogden, Utah: USDA Forest Service, Intermountain Research Station.

Mattson, D. J., and D. P. Reinhart. 1989. Whitebark pine on the Mount Washburn Massif, Yellowstone National Park. Pages 106–116 in W. C. Schmidt and K. J. McDonald, eds., *Symposium on Whitebark Pine Ecosystems: Ecology and Management of a High Mountain Resource*. Ogden, Utah: USDA Forest Service, Intermountain Research Station.

———. 1994. Bear use of whitebark pine seeds in North America. In *Proceedings of the International Workshop on Subalpine Stone Pines and Their Environment: The Status of Our Knowledge*. (5–11 September 1992, St. Moritz, Switzerland) Ogden, Utah: Intermountain Research Station.

———. 1997. Excavation of red squirrel middens by grizzly bears in the whitebark pine zone. *Journal of Applied Ecology* 34: 926–940.

Mattson, D. J., D. P. Reinhart, and B. M. Blanchard. 1994. Variation in production and bear use of whitebark pine seeds in the Yellowstone area. Pages 205–220 in D. G. Despain, ed., *Plants and Their Environments: Proceedings of the First Biennial Scientific Conference on the Greater Yellowstone Ecosystem*. (16–17 September 1991, Yellowstone National Park, Wyoming) USDI, National Park Service.

Mattson, D. J., B. M. Blanchard, and R. R. Knight. 1992. Yellowstone grizzly bear mortality, human habituation, and whitebark pine seed crops. *Journal of Wildlife Management* 56: 432–442.

Mattson, D. J., S. Herrero, R. G. Wright, and C. M. Pease. 1995. Science and management of Rocky Mountain grizzly bears. *Conservation Biology* 10: 1013–1025.

Mattson, D. J., K. C. Kendall, and D. P. Reinhart. 2001. Grizzly bears and red squirrels in D. Tomback, S. F. Arno, and R. E. Keane, eds., *Whitebark Pine Communities: Ecology and Restoration*. Washington, D.C.: Island Press.

McCaughey, W. W., and W. C. Schmidt. 1990. Autecology of whitebark pine. Pages 85–96 in W. C. Schmidt and K. J. McDonald, eds., *Symposium on Whitebark Pine Ecosystems: Ecology and Management of a High Mountain Resource*. Ogden, Utah: USDA, Forest Service, Intermountain Research Station, 29–31 March.

McCauley, D. E. 1993. Genetic consequences of extinction and recolonization in fragmented habitats. Pages 217–233 in P. M. Kareiva, J. G. Kingsolver, and R. B. Huey, eds., *Biotic Interactions and Global Change*. Sunderland, Mass.: Sinauer.

McDonald, G. I. 1992. Ecotypes of blister rust and management of sugar pine in California. Pages 137–147 in B. B. Kinloch, M. Moarosy, and M. E. Huddleston, eds., *Sugar Pine, Status, Values, and Roles in Ecosystems*. Davis: University of California, Division of Agriculture and Natural Resources.

Mearns, L. O., R. W. Katz, and S. H. Schneider. 1984. Extreme high-temperature events: Changes in their probabilities with changes in mean temperature. *Journal of Climate and Applied Meteorology* 23: 1601–1613.

Mearns, L., S. Schneider, S. Thompson, and L. McDaniel. 1990. Analysis of climate variability in general circulation models: Comparison with observations and changes in variability in 2 × CO_2. *Journal of Geophysical Research-Atmospheres* 95: 20,469-20,490.

Mearns, L. O., C. Rosenzweig, and R. Goldberg. 1996. The effect of changes in daily and interannual climatic variability on CERES-Wheat: A sensitivity study. *Climatic Change* 32: 275–292.

Mielke, J. L. 1938. Spread of blister rust to sugar pine in Oregon and California. *Journal of Forestry* 36: 695–701.

———. 1943. *White Pine Blister Rust in Western North America*. New Haven: Yale University School of Forestry.

Mielke, J. L., T. W. Childs, and H. G. Lachmund. 1937. Susceptibility to *Cronartium ribicola* of the four principal *Ribes* species found within the commercial range of *Pinus monticola*. *Journal of Agricultural Research* 55: 317–346.

Millspaugh, S. H., and C. Whitlock. 1995. A 750-year fire history based on lake sediment records in Central Yellowstone National Park, USA. *Holocene* 5: 283–292.

Morgan, P., and S. Bunting. 1992. Using cone scars to estimate past cone crops of whitebark pine. *Western Journal of Applied Forestry* 7: 71–73.

Morgan, P., S. C. Bunting, R. E. Keane, and S. F. Arno. 1994. Fire ecology of whitebark pine forest of the northern Rocky Mountains, USA. Pages 136–141 in W. C. Schmidt and F.-K. Holtmeier, eds., *International Workshop on Subalpine Stone Pines and Their Environment: The Status of Our Knowledge*. (5–11 September, St. Moritz, Switzerland) Ogden, Utah: USDA, Forest Service, Intermountain Research Station.

Musante, F. 1998. Parasites Jolt Oystermen from Dreams of Bumper Crops. *New York Times* (25 October 1998): 1, 4.

O'Brien, S. L., and F. G. Lindzey. 1994. *Grizzly Bear Use of Moth Aggregation Sites and Summer Ecology of Army Cutworm Moths in Absaroka Mountains, Wyoming*. Laramie: Wyoming Cooperative Fish and Wildlife Research Unit, University of Wyoming, Dept. of Zoology–Physiology.

Ojima, D. S., T. G. F. Kittlel, T. Rosswall, and B. H. Walker. 1991. Critical issues for understanding global change effects on terrestrial ecosystems. *Ecological Applications* 1: 316–325.

Overpeck, J. T., D. Rind, and R. Goldberg. 1990. Climate-induced changes in forest disturbance and vegetation. *Nature* 343: 51–53.

Patterson, W. A., and A. E. Backman. 1988. Fire and disease history of forests. Pages 603–632 in B. Huntley and T. W., III, eds., *Vegetation History*. Dordrecht: Kluwer.

Pease, C. M., and D. J. Mattson. 1999. Demography of the Yellowstone grizzly bears. *Ecology* 80: 957–975.

Pennington, L. H. 1925. Relation of weather conditions to the spread of white pine blister rust in the Pacific Northwest. *Journal of Agricultural Research* 30: 593–607.

Peters, R. L., and J. D. S. Darling. 1985. The greenhouse effect and nature reserves. *BioScience* 35: 707–717.

Prentice, C. I. 1992. Climate change and long-term vegetation dynamics. Pages 293–339 in D. C. Glenn-Lewin, R. K. Peet, and T. T. Veblen, eds. *Plant Succession: Theory and Prediction*. London: Chapman and Hall.

Price, C., and D. Rind. 1994a. The impact of a $2 \times CO_2$ climate on lightning-caused fires. *Journal of Climate* 7: 1484–1494.

Price, C., and D. Rind. 1994b. Possible implications of global climate change on global lightning distributions and frequencies. *Journal of Geophysical Research* 99: 10,823–10,831.

Renkin, R. A., and D. G. Despain. 1991. Fuel moisture, forest type, and lightning-caused fire in Yellowstone National Park. *Canadian Journal of Forest Research* 22: 37–45.

Richardson, C. W., and D. A. Wright. 1984. *WGEN: A Model for Generating Daily Weather Variables*. Washington, D.C.: USDA, Agricultural Research Service. 83 pp.

Rind, D., R. Goldberg, and R. Ruedy. 1989. Change in climate variability in the twenty-first century. *Climatic Change* 14: 5–37.

Rind, D., R. Goldberg, J. Hansen, C. Rosenzweig, and R. Ruedy. 1990. Potential evapotranspiration and the likelihood of future drought. *Journal of Geophysical Research* 95: 9983–10,004.

Romme, W. H., and D. G. Despain. 1989a. Historical perspective on the Yellowstone fires of 1988. *Bioscience* 39: 695–699.

———. 1989b. The long history of fire in the Greater Yellowstone Ecosystem. *Western Wildlands* Summer 1989: 10–17.

Romme, W. H., and M. G. Turner. 1991. Implications of global climate change for biogeographic patterns in the Greater Yellowstone Ecosystem. *Conservation Biology* 5: 373–386.

Rosenzweig, C. 1989. Global climate change: Predictions and observations. *American Journal of Agricultural Economics* 71: 1265–1271.

Running, S. W., and R. R. Nemani. 1987. Extrapolation of synoptic meteorological data in mountainous terrain and its use for simulating forest evapotranspiration and photosynthesis. *Canadian Journal of Forest Research* 17: 472–483.

Ryan, K. C. 1991. Vegetation and wildland fire: Implications of global climate change. *Environment International* 17: 169–178.

Sapsis, D. B., and R. E. Martin. 1994. Fire, the landscape, and diversity: A theoretical framework for managing wildlands. Pages 270– 277. *Proceedings of the Twelfth Conference on Fire and Forest Meteorology.* (26–28 October 1993, Jekyll Island, Georgia) Bethesda, Md.: Society of American Foresters.

Schmidt, W. C. 1994. Distribution of stone pines. Page 1 in W. C. Schmidt, ed., *International Workshop on Subalpine Stone Pines: The Status of Our Knowledge.* (5–11 September 1992, St. Moritz, Switzerland) Ogden, Utah: USDA, Forest Service.

Schneider, S. H. 1992. Why Build A Model? *Advances in Ecological Research* 22: 1–32.

Shaffer, M. L. 1980. Determining minimum viable population sizes for the grizzly bear. Pages 133–139 in E. C. Meslow, ed., Fifth International Conference on Bear Research and Management. International Association for Bear Research and Management.

Smith, J. P., and J. T. Hoffman. 2000. Status of white pine blister rust in the intermountain west. *Western North American Naturalist* 60: 165–179.

Spaulding, P. 1922. *Investigations of the White Pine Blister Rust.* Washington, D.C.: USDA.

Spaulding, P., and A. Rathburn-Gravatt. 1922. The longevity of the teliospores and accompanying urediospores of *Cronartium ribicola fischer. Journal of Agricultural Research* 33: 397–443.

———. 1926. The Influence of physical factors on the viability of sporidia of *Cronartium ribicola fischer. Journal of Agricultural Research* 33: 397–433.

Tausch, R. J., P. E. Wigand, and J. W. Burkhardt. 1993. Viewpoint: Plant community thresholds, multiple steady states, and multiple successional pathways: Legacy of the Quaternary? *Journal of Range Management* 46: 439–447.

Thornton, P. E., S. W. Running, and M. A. White. 1997. Generating surfaces of daily meteorological variables over large regions of complex terrain. *Journal of Hydrology* 190: 214–251.

Tomback, D. F. 1982. Dispersal of whitebark pine seeds by Clark's nutcracker: A mutualism hypothesis. *Journal of Animal Ecology* 51: 451–467.

———. 1995. Whitebark pine communities in the northern Greater Yellowstone Ecosystem: Patterns of regeneration since the 1988 fires. Nutcracker Notes 6. Available online at http://www.nrmsc.usgs.gov/nutnotes/number6.htm.

Tomback, D., and W. S. F. Schuster. 1994. Genetic population structure and growth form distribution in bird-dispersed pines. Pages 118–126 in W. C. Schmidt and F.-K. Holtmeier, eds., *International Workshop on Subalpine Stone Pines and Their Environment: The Status of Our Knowledge.* (5–11 September, St. Moritz, Switzerland) Ogden, Utah: USDA Forest Service.

Tomback, D. F., S. K. Sund, and L. A. Hoffmann. 1993. Post-fire regeneration of *Pinus albicaulis* height–age relationships, age structure and microsite characteristics. *Canadian Journal of Forest Research* 23: 113–119.

Tomback, D. F., J. K. Clary, J. Koehler, R. J. Hoff, and S. F. Arno. 1995. The effects of blister rust on post-fire regeneration of whitebark pine: The sundance burn of northern Idaho (U.S.A.). *Conservation Biology* 9: 654–664.

Torn, M. S., and J. S. Fried. 1992. Predicting the impacts of global warming on wildland fire. *Climatic Change* 21: 257–274.

Troendle, C. A., and M. R. Kaufmann. 1987. Influence of forests on the hydrology of the subalpine forest. Pages 68–78 in Society of American Foresters, ed., *Management of Subalpine Forests: Building on 50 Years of Research.* Washington, D.C.: USDA, Forest Service.

Turner, M. G., W. H. Romme, R. H. Gardner, and W. Hargrove. 1997. Effects of fire size and pattern on early succession in Yellowstone National Park. *Ecological Monographs* 67: 411–433.

Van Arsdel, E. P. 1961. *Growing White Pine in the Lake States to Avoid Blister Rust.* St. Paul, Minn.: USDA Forest Service, Lake States Forest Experimental Station.

———. 1965. Micrometeorology and plant disease epidemiology. *Phytopathology* 55: 945–950.

———. 1967. The nocturnal diffusion and transport of spores. *Phytopathology* 57: 1221–1229.

Van Arsdel, E. P., and R. G. Krebill. 1995. Climatic distribution of blister rusts on pinyon and white pines in the USA. Pages 127–133 in S. Kaneko, K. Katsuya, M. Kakishima, and Y. Ono, eds., *Proceedings of the Fourth IUFRO Rusts of Pines Working Party Conference.* (2–7 October 1994, Tsukuba, Japan) Washington, D.C.: USDA, Forest Service.

Van Arsdel, E. P., A. J. Riker, and R. F. Patton. 1956a. The effects of temperature and moisture on the spread of white pine blister rust. *Phytopathology* 46: 307–318.

———. 1956b. Microclimatic distribution of white pine blister rust in southern Wisconsin. *Phytopathology* 43: 487–488.

Waddington, J. C. B., and J. H. E. Wright. 1974. Late quaternary vegetational changes on the east side of Yellowstone Park, Wyoming. *Quaternary Research* 4: 175–184.

Webb, T. 1986. Is vegetation in equilibrium with climate? How to interpret late-Quaternary pollen data. *Vegetatio* 67: 75–91.

Whitlock, C. 1993. Postglacial vegetation and climate of Grand Teton and southern Yellowstone National Parks. *Ecological Monographs* 63: 173–198.

Whitlock, C., and P. J. Bartlein. 1991. Spatial variations of Holocene climatic change in the Yellowstone region. *Quaternary Research* 39: 231–238.

————. 1997. Vegetation and climate change in northwest America during the past 125 kyr. *Nature* 388: 57–61.

Whitlock, C., P. J. Bartlein, and K. J. V. Norman. 1994. Stability of Holocene climate regimes in the Yellowstone region. *Quaternary Research* 43: 433–436.

Wigand, P. E., and C. L. Nowak. 1992. Dynamics of northwest Nevada plant communities during the last 30,000 years. Pages 40–62 in C. A. Hall, V. Doyle-Jones, and B. Widawski, eds., *The History of Water: Eastern Sierra Nevada, Owens Valley, White-Inyo Mountains.* Los Angeles: University of California, White Mountain Research Station.

Wilks, D. S. 1992. Adapting stochastic weather generation algorithms for climate change studies. *Climatic Change* 22: 67–84.

Williams, T., and K. Kendall. 1998. Glacier National Park initiates project to restore whitebark pine and limber pine communities. Nutcracker Notes 9. Available online at http://www.nrmsc.usgs.gov/nutnotes/number9.htm.

Zwiers, F. W., and V. K. Viatcheslav. 1998. Changes in the extremes of the climate simulated by CCC GCM2 under CO_2 doubling. *Journal of Climate* 11: 2200–2222.

APPENDIX

Criteria for Evaluation of a
Blister Rust Event
(only the weather between the time period from July 1 to
October 15 was considered for each year)

1. At least 2 consecutive days in which nighttime relative humidity > 90%.
2. At least 2 consecutive days in which daytime relative humidity > 85%.
3. At least 3 consecutive days in which rain > 1 mm and daytime relative humidity is > 85% on any 2 of the 3 days.
4. At least 3 consecutive days in which rain > 1 mm and nighttime relative humidity is > 90% on any 2 of the 3 days.
5. Two consecutive days in which rain is > 1 cm.
6. At least 3 consecutive days in which rain > 1 mm on all days and rain > 1 cm on at least 2 of the days.
7. At least 3 consecutive days in which rain > 1 mm on all days, rain > 1 cm on at least 1 day, and rain > 4 mm on at least 2 of the 3 days, and daytime relative humidity > 85% on at least 1 day.
8. At least 3 consecutive days in which rain > 1 mm on all days, rain > 1 cm on at least 1 day, and rain > 4 mm on at least 2 of the 3 days, and nighttime relative humidity > 90% on at least 1 day.
9. At least 5 consecutive days of rain in which rain is > 1 mm for at least 4 days and rain > 4 mm for at least 1 day.
10. At least 2 consecutive days of rain in which rain > 5 mm on both days, and the sum of the rain over both days > or = 3.5 cm.
11. At least 4 consecutive days in which rain > 4 mm on all days.

Blister Rust Values Evaluated
(as they appear in figures)

Blister Rust Year

1. At least 2 blister rust events must occur during a single season lasting from July 1 to October 15 of a given year.
2. At least 14 days must elapse between the beginning of the first blister rust event and the end of the next blister rust event.
3. No more than 28 days may elapse between the end of 1 blister rust event and the onset of a second blister rust event.
4. If more than 2 blister rust events occur during a given season (lasting from July 1 to October 15), and both criteria 2 and 3 are satisfied for the time period between any of the blister rust events, that year is designated a blister rust year.

Period of Evaluation

Blister Rust values are evaluated over the blister rust season in two five-year periods for each decade considered. The first 5-year period begins in the first year of a decade and extends through the fourth year. The second 5-year period begins in the fifth year of a decade and extends through the ninth year.

Average Blister Rust Years

Average Blister Rust Years equal the average number of blister rust events in a blister rust year divided by the number of blister rust years per 5-year period

Example:

Year 1: 2 BR events

Year 2: 0 BR events

Year 3: 5 BR events

Year 4: 3 BR events

Year 5: 1 BR event ← not a BR Year therefore not included in calculation

Average Blister Rust Years: 2 + 5 + 3 / 3 = 3.33

Average All Years

Average all years equals the average number of blister rust events in a blister rust year divided by 5 (only blister rust years are included).

Example:

Year 1: 2 BR events

Year 2: 0 BR events

Year 3: 5 BR events

Year 4: 3 BR events

Year 5: 1 BR event ← not a BR year therefore not included in calculation

Average All Years: 2+0+5+3+0 / 5 = 2

Blister Rust Index

Blister Rust Index equals the average number of blister rust events that occurred in each blister rust year in a given 5-year period times the number of blister rust years.

Example:

Year 1: 2 BR events

Year 2: 0 BR events

Year 3: 5 BR events

Year 4: 3 BR events

Year 5: 1 BR event ← not a BR Year therefore not included in calculation

Blister Rust Index = (2+5+3 / 3) = 3.33 × 3 = 10

Note: The blister rust index values that appear in the figures are computed through averaging the index values computed for each individual station in an elevational band.

Conclusion: Climate Change and Wildlife—A Look Ahead

PATTY GLICK AND MARK VAN PUTTEN

As we enter the dawn of the twenty-first century, perhaps no environmental problem looms as a greater threat to the world's ecological systems than global climate change.

There is overwhelming agreement—not only among the scientists featured in this volume, but within the scientific community at large—that human activities are sending tremendous quantities of carbon dioxide and other heat-trapping greenhouse gases into the atmosphere (IPCC 2001a). There is also broad consensus that this buildup of gases is causing the average global surface temperature to rise and is altering the Earth's climate—affecting temperature and rainfall patterns, causing a rise in sea level, and possibly increasing the frequency and severity of extreme weather events. Furthermore, scientific evidence of change across a broad range of environmental systems has been building, including melting glaciers, thawing permafrost, earlier spring migration dates for birds and bloom dates for vegetation, range movements up slopes and poleward for butterflies, bleaching of coral reefs, and reduced lake and sea ice cover. These changes are all consistent with the scientific agreement regarding the late-twentieth-century pattern of global warming (IPCC 2001b).

The prospect that climate change will lead to irreversible and costly problems for society has driven a momentous international effort to address the issue. We have enough scientific evidence to warrant immediate action to begin to slow climate change by reducing global greenhouse gas emissions. At the same time, to meet the long-term challenge, we must learn more about how climate change will affect plant and animal life, not to mention its impacts on people. As conservationists, we need a better understanding of the potential regional and localized consequences of climate change, as well as vulnerability of species and ecosystems, to help us build awareness of the problem, and then develop, promote, and implement appropriate solutions.

Through this project, the National Wildlife Federation has sought to build interest and skills among up-and-coming scientists to examine North American wildlife responses to climate change using a range of analytical and observational techniques. We are enormously grateful to Steve Schneider and Terry Root for championing this project from its conception, and to the scientists for their exceptional work and their unwavering dedication to this issue. They truly are scientific and conservation leaders.

From the beginning, our view was that if we could shed scientific light on how climate change may or may not affect the plants and animals we are working so hard to protect—from butterflies to alpine flowers to the endangered grizzly bear—then we would be in a better position to ensure their future well-being. Through observation of historic data, modeling, and controlled experiments, these studies present some compelling findings.

As several of the studies show, recent changes in regional climate variables may already, in fact, be affecting certain species and ecosystems. For example, Lisa Crozier has found that warmer temperatures in the Pacific Northwest have likely contributed to a significant northerly shift in the range of the sachem skipper butterfly over the past 30 years. In a complementary study, Jessica Hellmann has identified a number of ways in which local and regional climate variables affect the Bay checkerspot butterfly and suggests that the sensitivity of this and other butterfly species to climatic and weather variables make them useful indicators of how other wildlife species may respond to climate change.

Both Rafe Sagarin and Eric Sanford show how variations in ocean temperatures and other climate-related variables have had a significant effect on the composition and interactions of intertidal marine species. Sagarin's analyses of long-term trends in the Monterey Bay region of California suggest that observed changes in the ranges of a number of plant and invertebrate species over the past 60 years are likely due, at least in part, to warming sea temperatures in the region. His research shows, for example, that southern invertebrate species have become more abundant and northern species have become less abundant in the region.

Sanford's study demonstrates that, in addition to affecting the geographic distribution of species, changes in sea temperatures may also alter interactions between key species and produce relatively rapid changes in the composition and functioning of natural communities. His experiments show that during periods of cold-water upwelling, the ochre sea star, a keystone predator in the rocky intertidal community of the Pacific Northwest, feeds less intensely on mussels, the dominant competitor for space. This research suggests that changes in the structure and diversity of these rocky shore communities may occur in the future if climate change alters the atmospheric and oceanic conditions that contribute to coastal upwelling.

Francisca Saavedra demonstrated the potential effect of reduced snow cover and associated conditions on the reproductive biology of the *Delphinium nuttallianum*, an early-blooming plant species in the Rocky Mountains of Colorado. Her results suggest that a shift in the timing of annual snowmelt to an average of 10 days earlier, as could occur under a scenario of a doubling of carbon dioxide in the atmosphere, will trigger an earlier flowering time and might affect the fitness of the plants. Should climate change contribute to increased temperatures in the region and lead to earlier snowmelt, it could alter the distribution and growing season of the *Delphinium nuttallianum* and other plants in the region, and could affect interactions with other species that depend on them, including pollinators such as broad-tailed hummingbirds and bumblebees.

On a much broader scale, Elena Shevliakova has modeled potential effects of climate change on the distribution of vegetation in the United States. Her study shows significant changes in the regional composition of plant species under several possible climate change scenarios. For example, the models project a likely shift from forested land to shrub and grassland in the western United States under warmer and drier climate conditions, and an increase in the prevalence of pine species in the Southeast, in a wetter climate.

Erika Zavaleta and Jennifer Royval studied how climate change may affect the susceptibility of U.S. ecosystems to invasions of nonnative species. Using a combination of spatial analysis and climate modeling, they identified potential shifts in the range of two invasive species, the imported red fire ant and *Tamarix* shrubs. Models in each case suggest that changes in regional temperatures and precipitation patterns may contribute to the expansion of both species, with warmer and wetter conditions contributing to the spread of fire ants throughout the southeastern United States and warmer and drier conditions favoring expansion of *Tamarix* in the West.

Finally, Laurie Koteen investigated the potential for environmental change, including climate change, to alter interactions between a range of organisms. In the case of species in the Greater Yellowstone Ecosystem region of the Rocky Mountains, she has found that climate change could promote the spread of white pine blister rust, an exotic fungus that kills whitebark pine trees. Koteen's study suggests that the loss of whitebark pines in the region could alter important symbiotic relationships among species that depend on the trees as an important source of food and shelter, including the Clark's nutcracker and the threatened grizzly bear. She adds that additional factors, such as fire suppression and habitat fragmentation, will likely play a significant role in how populations of these species will respond.

Each of these case studies provides some important information about how biotic systems respond to climatic variables, and how a changing climate may affect them in the future. They also acknowledge the inherent complexities of the problem, and demonstrate the types of scientific questions we need to further explore to improve our understanding of how climate change and other human disturbances affect wildlife and ecosystems.

In their overview chapter, Terry Root and Stephen Schneider explain that it will not be climate change alone, but the synergism between climate change and other environmental problems, such as habitat fragmentation, pollution, and the introduction of exotic species, that will ultimately pose the greatest threat to nature, and the people and wildlife that depend on it. Moreover, they emphasize the need for more interactive, multiscale analyses to provide a more realistic assessment of ecological responses to global change. They also document the convergence of many lines of recent evidence to the effect that the latter half of the twentieth century has seen a discernible impact of climate change on environmental systems like wildlife, plants, and ice systems. The most consistent explanation for the trends is global warming, although uncertainties remain in specific cases.

Accordingly, Shevliakova and Zavaleta and Royval recognize that such factors as land use change, herbicide and pesticide use, and other factors will play a significant role in determining how vegetation and invasive species will ultimately respond to climate change. Hellmann and Koteen point to the importance of piecing together climate impacts with other environmental stresses in order to determine how best to protect endangered species. Saavedra takes the thought one step further and advises that it is worthwhile for ecologists to consider how their research may help decision makers identify appropriate solutions.

This book contributes to unlocking some of the clues to understanding the increasingly complex ways in which human activities, especially global climate change, are altering the world around us. It will provide a meaningful resource for students and scientists studying the effects of climate change on wildlife. It will help focus attention on some of the pivotal questions that researchers and policy makers alike will need to address in the coming years. It will also assist resource managers and other wildlife professionals to better understand factors affecting species they are working so hard to conserve. With a better understanding of how climate affects wildlife, the prospects for successful management are markedly increased. Finally, those who, as stewards of our world's diverse plants and animals, seek to make policy and take actions now to minimize human-induced global climate change and the consequent impacts on our natural world, will find valuable information in these thoughtful and careful studies.

Literature Cited

Intergovernmental Panel on Climate Change (IPCC). 2001a. *Climate Change 2001: The Scientific Basis*, ed. J. T. Houghton, Y. Ding, D. J. Griggs, M. Noguer, P. J. van der Linden, and D. Xiaosu. Contribution of working group I to the third assessment report of the Intergovernmental Panel on Climate Change. Cambridge: Cambridge University Press.

————.2001b. *Climate Change 2001: Impacts, Adaptation, and Vulnerability*, ed. J. J. McCarthy, O. F. Canziani, N. A. Leary, D. J. Dokken, and K. S. White. Contribution of working group II to the third assessment report of the Intergovernmental Panel on Climate Change. Cambridge: Cambridge University Press.

Contributors

LISA CROZIER
Department of Zoology
University of Washington
Seattle, Washington

PATTY GLICK
Climate Change and Wildlife Program
National Wildlife Federation
Washington, D.C.

JESSICA J. HELLMANN
Department of Biological Sciences
Stanford University
Stanford, California

LAURA KOTEEN
Energy and Resources Group
University of California, Berkeley
Berkeley, California

TERRY L. ROOT
School of Natural Resources and Environment
University of Michigan
Ann Arbor, Michigan

JENNIFER L. ROYVAL
Department of Biological Sciences
Stanford University
Stanford, California

FRANCISCA SAAVEDRA
Department of Biology
University of Maryland
College Park, Maryland
and
Rocky Mountain Biological Laboratory
Crested Butte, Colorado

RAPHAEL SAGARIN
Department of Ecology, Evolution, and Marine Biology
University of California, Santa Barbara
Santa Barbara, California
and
Hopkins Marine Station
Stanford University
Pacific Grove, California

ERIC SANFORD
Department of Zoology
Oregon State University
Corvallis, Oregon
and
Hopkins Marine Station
Stanford University
Pacific Grove, California

STEPHEN H. SCHNEIDER
Department of Biological Sciences
Stanford University
Stanford, California

ELENA SHEVLIAKOVA
Department of Ecology and Evolutionary Biology
Princeton University
Princeton, New Jersey

MARK VAN PUTTEN
National Wildlife Federation
Reston, Virginia

ERIKA S. ZAVALETA
Department of Biological Sciences
Stanford University
Stanford, California

Index